AF388899

Frontier Research on the Processing Quality of Cereal and Oil Food

Frontier Research on the Processing Quality of Cereal and Oil Food

Editors

Qiang Wang
Aimin Shi

MDPI • Basel • Beijing • Wuhan • Barcelona • Belgrade • Manchester • Tokyo • Cluj • Tianjin

Editors
Qiang Wang
Insitutute of Food Science
and Technology
Chinese Academy of
Agricultural Sciences
Beijing
China

Aimin Shi
Institute of Food Science and
Technology
Chinese Academy of
Agricultural Sciences
Beijing
China

Editorial Office
MDPI
St. Alban-Anlage 66
4052 Basel, Switzerland

This is a reprint of articles from the Special Issue published online in the open access journal *Foods* (ISSN 2304-8158) (available at: www.mdpi.com/journal/foods/special_issues/processing_cereal).

For citation purposes, cite each article independently as indicated on the article page online and as indicated below:

LastName, A.A.; LastName, B.B.; LastName, C.C. Article Title. *Journal Name* **Year**, *Volume Number*, Page Range.

ISBN 978-3-0365-4492-2 (Hbk)
ISBN 978-3-0365-4491-5 (PDF)

© 2022 by the authors. Articles in this book are Open Access and distributed under the Creative Commons Attribution (CC BY) license, which allows users to download, copy and build upon published articles, as long as the author and publisher are properly credited, which ensures maximum dissemination and a wider impact of our publications.
The book as a whole is distributed by MDPI under the terms and conditions of the Creative Commons license CC BY-NC-ND.

Contents

About the Editors

Qiang Wang

Professor Dr Qiang WANG is a distinguished consultant of the Institute of Food Science and Technology at Chinese Academy of Agricultural Sciences, Director of Processing Laboratory of National Peanut Industrial Technology System, Chief scientist of MOA Cereal and Oil Processing and Comprehensive Utilization Innovation Team, National Agricultural Scientific Research Outstanding Talents, Vice Chairman of Plant-based Food Branch of CIFST, and Chairman of Chinese Peanut Food Association of CCOA.

He has contributed to the field of cereal and oil processing for over 30 years and coordinated over 30 National projects. He has won over ten international and national awards, including Harald Perten Prize, Second prize of National Technological Invention Award, and Outstanding Contribution Award of CIFST, etc. He has obtained 90 patents, 12 software copyrights, and published 6 monographs, 2 book chapters, 6 reports, 9 standards, and over 330 papers, with an H index of 26.

As Chief Scientist, he has planned and designed the National Agricultural Products Processing R&D System and served as the first Director. As a consulting expert, he has written the proposal on high-quality development of the China peanut industry to the State Council and got important approval. He is executive editor-in-chief and editorial board member of 3 international journals. He has chaired over 10 international conferences including International Symposium on Food Components Structural Change and Quality Regulation.

He has actively participated in China's poverty alleviation work, went to poverty-stricken areas many times for training and guidelines, promoting technological upgrading for famers, cooperatives and enterprises and driving local economic and social development. He has trained over 90 technical backbones, Master and Ph.D and post-doctoral, 11 National and provincial talents and was awarded as "National Outstanding Team of Agricultural S&T Innovation".

Aimin Shi

Professor Shi was selected as one of the top talents in the national grain industry, selected as a member of the Young Talent Promotion Project of China Association for Science and Technology, director of the Food and Oil Nutrition Branch of China Cereals and Oils Association, guest editor of the international academic journal "Foods"Special Issue, reviewer of several international journals, peer-review expert of National Natural Science Foundation of China and BeNatural Science Foundation.

Professor Shi has been engaged in food macromolecular colloid science and technology research for a long time, and has made innovative achievements in the fields of plant protein modification, plant protein-based Pickering emulsion construction, controlled release and delivery of active ingredients. He has presided over or participated in more than 10 national and provincial projects such as the National Natural Science Foundation of China, the sub-project of the National Key Research and Development Program of the 13th Five-Year Plan, etc. He has published more than 100 academic papers, including 50 English papers, 27 SCI papers (first/co-first author), 1 published in Angewandte Chemie International Edition (IF>10), and 1 highly cited ESI paper. He has been cited over 1500 times, with 20 in the Google H-index; He has published 2 books as associate editor, participated in the editing of 3 English books; authorized 8 international invention patents, 34 national invention patents; and formulated 4 agricultural industry standards. The research results have passed 2 provincial and ministerial achievements evaluation, earned 2 Shennong China

Agricultural Science and Technology Award, 2 Science and Technology Progress Special Award of China Federation of Commerce, the First Youth Science and Technology Award of China Cereals and Oils Association, and the "Best Poster Award" of the 1st ICC Asia Pacific Food Science and Technology Conference.

Preface to "Frontier Research on the Processing Quality of Cereal and Oil Food"

As everyone knows, cereal and oil are still the main part of our diet and provides essential nutrients and energy every day. With the progress of food processing technology, the quality of cereal and oil food is also improved significantly. Behind this, major nutrients of grain and oil, including protein, carbohydrate, lipid, and functional components, have experienced a variety of physical, chemical, and biological reactions during food processing. Moreover, research in this field also covers the multi-scale structural changes of characteristic components, such as component interaction and formation of key domains, which is essential for the quality enhancement of cereal and oil food.

Based on the increasing consumer demand for nourishing, healthy, delicious cereal and oil food, it might be interesting to report the latest research on the application of novel technology in food processing, multi-scale structural changes of characteristic components in food processing, structure-activity mechanism of food functional components. This reprint aimed to provide useful reference and guidance for the processing and utilization of cereal and oil food so as to provide technical support for the healthy development of cereal and oil food processing industry wordwide.

The members involved in the preparation, collation and experiment processes include Shi Aimin, Liu Hongzhi, Liu Li, Hu Hui, Ma Xiaojie, Jiao Bo, Liu Yue, Liu Zhe, Li Jiaxiao, and other members, who have made their contributions. Additionally, during the writing, the works and papers of experts and scholars domestic and foreign have been referenced simultaneously. We want to show our wholehearted thanks.

This book is supported by the following funding "National Key Research & Development Program (2021YFD2100400), National Natural Science Foundation of China (32172149, U21A20270), and Agricultural Science and Technology Innovation Project (CAAS-ASTIP-201X-IAPPST)".

Qiang Wang and Aimin Shi
Editors

 *foods*

Article

Study on Key Aroma Compounds and Its Precursors of Peanut Oil Prepared with Normal- and High-Oleic Peanuts

Hui Hu [1,2], Aimin Shi [1], Hongzhi Liu [1], Li Liu [1], Marie Laure Fauconnier [2,*] and Qiang Wang [1,*]

[1] Institute of Food Science and Technology, Chinese Academy of Agricultural Sciences/Key Laboratory of Agro-Products Processing, Ministry of Agriculture, Beijing 100193, China; huhui@caas.cn (H.H.); sam_0912@163.com (A.S.); lhz0416@126.com (H.L.); liulicaas@126.com (L.L.)

[2] Laboratory of Chemistry of Natural Molecules, Gembloux Agro-Bio Tech, Liege University, Passage des Déportés 2, 5030 Gembloux, Belgium

* Correspondence: marie-laure.fauconnier@ulg.ac.be (M.L.F.); wangqiang06@caas.cn (Q.W.); Tel.: +32-81622289 (M.L.F.); +86-(10)-62815837 (Q.W.)

Abstract: High-oleic acid peanut oil has developed rapidly in China in recent years due to its high oxidative stability and nutritional properties. However, consumer feedback showed that the aroma of high-oleic peanut oil was not as good as the oil obtained from normal-oleic peanut variety. The aim of this study was to investigate the key volatile compounds and precursors of peanut oil prepared with normal- and high-oleic peanuts. The peanut raw materials and oil processing samples used in the present study were collected from a company in China. Sensory evaluation results indicated that normal-oleic peanut oil showed stronger characteristic flavor than high-oleic peanut oil. The compounds methylpyrazine, 2,5-dimethylpyrazine, 2-ethyl-5-methylpyrazine and benzaldehyde were considered as key volatiles which contribute to dark roast, roast peanutty and sweet aroma of peanut oil. The initial concentration of volatile precursors (arginine, tyrosine, lysine and glucose) in normal-oleic peanut was higher than in high-oleic peanut, which led to more characteristic volatiles forming during process and provided a stronger oil aroma of. The present research will provide data support for raw material screening and sensory quality improvement during high-oleic acid peanut oil industrial production.

Keywords: peanut; high-oleic; peanut oil; volatiles; precursors

Citation: Hu, H.; Shi, A.; Liu, H.; Liu, L.; Fauconnier, M.L.; Wang, Q. Study on Key Aroma Compounds and Its Precursors of Peanut Oil Prepared with Normal- and High-Oleic Peanuts. *Foods* **2021**, *10*, 3036. https://doi.org/10.3390/foods10123036

Academic Editor: Victor Rodov

Received: 19 November 2021
Accepted: 4 December 2021
Published: 7 December 2021

Publisher's Note: MDPI stays neutral with regard to jurisdictional claims in published maps and institutional affiliations.

Copyright: © 2021 by the authors. Licensee MDPI, Basel, Switzerland. This article is an open access article distributed under the terms and conditions of the Creative Commons Attribution (CC BY) license (https://creativecommons.org/licenses/by/4.0/).

1. Introduction

Peanut is one of the most important oil crops in the world. Worldwide, the production of peanuts reached 49.62 million tons in 2020/21, and the production of peanut oil was 6.43 million tons, among which approximately 50% was produced in China [1]. The total amount of unsaturated fatty acid is over 85% in peanut oil. The fatty acid profile of peanut oil resembles that of olive oil, which could reduce the risk of cardiovascular disease [2].

The flavor, nutritional quality, and shelf-life of peanut and its products are related to the relative proportion of various fatty acids [3]. With more than 72% oleic acids, high-oleic peanut is well recognized by processors for its low oxidative and ability to extend the shelf life of products [4]. Wang Qiang research group reported that high-oleic peanut oil could attenuate diet-induced Metabolic Syndrome, associated with modulating gut microbiota [5]. The breeding of high-oleic acid peanut in China has developed rapidly in recent years. Since the first high-oleic natural mutant discovered in 1987, over 190 high oleic peanut cultivars have been developed in China [6]. More and more peanut processing companies are trying to use high-oleic acid peanut in oil processing. More than five brands of high-oleic peanut oil have entered the market in China in the last three years. All these products use high-oleic runner peanut raw materials from the USA. However, the consumer feedback showed that the aroma of high-oleic peanut oil was not as good as that of normal-oleic peanut oil.

Compared with other edible vegetable oils, aromatic roasted peanut oil obtained by thermal processing is more popular for consumers because of its strong nutty and roasty flavor [7]. The unique flavors of thermally processed foods are commonly generated through the Strecker degradation during the Maillard reaction, which is responsible for generating various heterocyclic compounds, including pyrazines, pyrroles, pyridines, etc. [8]. Correlation of volatile compounds to peanut sensory evaluation has attracted researcher's attention. A previous study reported that aspartic acid, glutamic acid, glutamine, asparagine, histidine, and phenylalanine contributed to the characteristic peanut flavor formation, and monosaccharides are highly related to pyrazine component [9]. Pyrazine compounds are responsible for the roasted flavor and aroma during peanut roasting [10]. Over 100 volatile components were identified in hot-pressed peanut oil, including pyrazines, aldehydes, furans, alcohols and pyrroles. Pyrazines are considered to be the major volatile compounds responsible for the typical roasted/nutty flavor of hot-pressed peanut oil [11]. The compounds 2/3-methyl-1H-pyrrole, 5-methyl-2-furancarboxaldehyde, benzeneacetaldehyde, 2,3 dimethyl-1H-pyrrole, 2,5 dimethyl pyrazine, 5-methyl-2-furanmethanol, and maltol were considered the most important volatile components which positively correlated with the peanutty and roasted aroma [12].

The major precursors for volatiles in peanut are proteins, sugars, and lipids [13]. Different kinds of sugars and proteins mixtures react differently, which lead to different volatiles formation. Compared with glycine and diglycine, triglycine has the highest capability to formed pyrazines in Maillard model systems. Major pyrazines were identified as 2,5-dimethylpyrazine and trimethylpyrazine [14]. Glutamine and asparagine have shown high reactivities to produce high content of pyrazines [15]. The rapeseed peptides subsequently reacted with D-xylose to largely produce methylpyrazine and ethyl-2,5-dimethylpyrazine [16]. Methylpyrazine and 2,5-dimethylpyrazine were identified in the D-glucose and L-theanine Maillard model systems but were not detectable in thermal reactions with single D-glucose or L-theanine [17]. The compounds 2,6-dimethyl-3-ethyl pyrazine, 2,5-diethylpyrazine and 2-methyl-3,5-diethylpyrazine were formed in the reaction between 1,4-13C-labeled L-ascorbic acid and L-glutamic acid. The α-amino carbonyl or α-amino hydroxy compounds were found to be the precursors of pyrazines [18].

The sensory quality difference between normal- and high-oleic peanut has also been studied. There were small differences in the roasting, astringency, over-roasting, and nuttiness attributes between these two kinds of peanuts. High-oleic lines exhibiting slightly greater intensities of those attributes [19]. Variation among individual lines for several sensory attributes (dark roasted, raw/beany, roasted peanutty, sweet aromatic, sweet, bitter, wood-hulls-skins, and "off flavors" stale/cardboard, fruity/fermented and plastic/chemical) suggest the flavor of high-oleic cultivars is at least as good as the profiles of normal-oleic cultivars [20].

Studies of characteristic volatile compounds and precursors of normal- and high-oleic peanut oil are still lacking. The object of this study was to compare the sensory quality and the key aroma components of normal- and high-oleic peanut oil produced industrially. For a possible precursor study, the amino acids and reducing sugar profile of peanut have also been monitored during oil processing. The results of this study will provide data support for raw material screening and sensory quality improvement during high-oleic acid peanut oil industrial production.

2. Materials and Methods

2.1. Materials

Normal- and high-oleic peanut raw materials and oil processing samples (roasted peanut, peanut oil, and peanut meal) were collected from the industrialized production line in factory (Jinsheng Cereals & Oils Group, Shandong Province, China). The varieties of normal- and high-oleic peanut are Baisha 1016 (China) and high-oleic runner (USA), respectively. The peanut raw material was roasted at 150 °C for 45 min. After this, roasted peanuts were pressed at 120 °C to obtain peanut oil. All reagents used in this research were obtained from Sigma-Aldrich (St. Louis, MO, USA), including methyl-pyrazine, 2,5-dimethylpyrazine, 2-ethyl-5-methylpyrazine, 1-methyl-1H-pyrrole, furfural, benzaldehyde, 2-furanmethanol, hexanal, pentanal, 1,2,3-trichloropropane, etc.

2.2. Sensory Evaluation

Sensory evaluation was performed at room temperature. Twelve panelists (7:5 male: female) participated to sensory evaluation. All of the panelists are well-trained researchers with a minimum of 300 h experience in sensory evaluation. Details on the methods, lexicon and attribute definitions have been previously published [21–23]. The sensory attributes used were roast peanutty aroma, dark roast aroma, sweet aroma, raw/beany aroma, woody/hulls/skins aroma, and a 9-point scale was used (1 = very weak, 9 = very strong). The lexicon of flavor sensory attributes is shown in Table 1.

Table 1. Flavor sensory attributes as obtained from the expert panel.

Sensory Attribute	Description [a]
Roast Peanutty	The aromatic associated with medium-roast peanuts having a fragrant character such as methyl pyrazine
Dark Roast	The aromatic associated with dark roasted peanuts having a very browned or toasted character
Sweet Aromatic	The aromatics associated with sweet material such as caramel, vanilla or molasses
Raw/Beany	The aromatics associated with light roast peanuts having a legume like character
Woody/Hulls/Skins	The aromatics associated with base peanut character (absence of fragrant top notes) related to dry wood, peanut hulls and skins

[a] Lexicon and method defined in the literature [21–23].

2.3. Volatile Compounds Analysis

Volatile compounds in normal- and high-oleic peanut and oil processing samples were analyzed by headspace-solid phase micro-extraction (HS-SPME). SPME fiber (50/30 μm divinylbenzene/Carboxen/polydimethylsiloxane, Stableflex, Supelco Co., Bellefonte, PA, USA) was utilized for flavor extraction. The fiber was previously conditioned at 270 °C for 30 min before the first measurement. The sample (5 g) were weighed into a 20 mL glass vial which was sealed with an aluminum cover and a Teflon septum. A 25 μL aliquot of 1,2,3-trichloropropane (0.25 mg/mL in methanol) as internal standard was added. It was pre-equilibrated for 10 min at 55 °C in shaken incubator. After the equilibration time, an auto SPME holder containing fiber was inserted into the vial, and the fiber was exposed to the headspace for 40 min. The volatiles absorbed by the fiber were thermally desorbed in the hot injection port of the GC for 150 s at 260 °C. GC-MS analysis was performed using GC system (Agilent 7890B, Agilent Technologies, Santa Clara, CA, USA) and mass selective detector (Agilent 5977B) equipped with a VF-WAX column (30 m × 0.25 mm i.d., 0.25 μm film thickness; Agilent CP9205, Agilent Technologies, Santa Clara, CA, USA). The analysis was carried out in the splitless mode, using helium as the carrier gas (1 mL/min flow rate). The detector temperature was 250 °C. The oven temperature program was initially set at 40 °C for 5 min, and programmed at 5 °C/min to 250 °C which was held for 5 min. Mass spectra were recorded in electron impact ionization mode (70 eV) scanning a mass range (*m/z*) from 35 to 500 amu. The ion source temperature was maintained at 230 °C. For the identification of volatiles, the peanut oils were analyzed by GC-MS under the experimental conditions mentioned above. Volatiles were primarily identified by comparison of the mass spectra with data from the commercially available mass spectra NIST databases. In addition, the volatiles were identified by matching the retention indices (RI) with data found in the literature [24] and comparing them with commercial standards. Based on the series of n-alkanes (C7-C30), RI were calculated according to the following formula:

$$RIx = 100n + 100 \, (tRx - tRn)/(tRn + 1 - tRn) \tag{1}$$

where retention time (tR) of tRn < tRx < tRn + 1; n = number of atom carbon.

2.4. GC-MS-O Analysis of Volatile Compounds

GC-MS-O analysis was performed using GC system (Agilent 7890B, Agilent Technologies, Santa Clara, CA, USA) and mass selective detector (Agilent 5973B) equipped with Olfactory detection port (ODP3, Gerstel, Germany). The GC-MS system parameters were the same as in 2.3. The connector temperature of the olfactometer was 150 °C. The end effluent of capillary, respectively, flows into the MS and olfactometer at a split ratio of 1:1. The odor strength was set up to a 5-point scale (1 = very weak, 5 = very strong).

2.5. Amino Acid Profile Analysis

Amino acids determination followed the method described in Reference [25]. The amino acid profile was measured using ion exchange chromatography. The sample (100 mg) was hydrolyzed with 10 mL 6 N HCl containing 0.1% phenol, followed by nitrogen flushing for 1 min and closing the hydrolysis bottle. Bottles were heated at 110 °C for 24 h in an oven and cooled with ice. After this, 30 mL of citrate buffer at pH 2.2 was poured (with continuous stirring) into bottles while they were still on ice. Then pH was adjusted between 0.5 and 1 using 7.5 N NaOH and then readjusted to 2.2 using 1 N NaOH. This solution was added in a 100 mL volumetric flask already containing 1 mL solution of 500 μM norleucine in citrate buffer at 2.2 pH. The volume of this flask was made 100 mL by adding citrate buffer at 2.2 pH. This solution was stirred and filtered through a 0.2 μm filter. The filtered solution was used to measure amino acids separately using Biochrom 20 plus amino acid analyzer (Biochrom Limited, Cambridge, UK).

2.6. Soluble Reducing Sugar Profile Analysis

Soluble reducing sugars determination followed the method described in Reference [26]. Defatted sample (0.5 g) was weighted in centrifuge tube. Soluble reducing sugars were extracted with 10 mL 70% ethanol under ultrasonic condition for 20 min. The supernatant was collected after 2000 r/min centrifugation for 10 min. The ethanol extraction and centrifuge procedure were repeated with the residue. Two parts of supernatant were filtered and vacuum rotary evaporated under 50 °C. The volume was made constant at 1 mL with 70% ethanol for analysis. The detection was performed on HPLC (Agilent 1260 Infinity, Agilent Technologies, Santa Clara, CA, USA) with diode array detector (G4212B). Spherisorb column (4.6 mm × 250 mm, 5 μm, Waters, Milford, MA, USA) was used. The mobile phase was 70% acetonitrile at a flow rate of 1 mL/min. The results were expressed as gram sugar per kilogram samples.

2.7. Statistical Analysis

The experiments were performed in triplicate. The least significant difference (LSD) method was used to determine the significant difference between mean values. A confidence level was set at $p < 0.05$, and the software SPSS (IBM SPSS 22.0, Chicago, IL, USA) was used for statistical analysis.

3. Results and Discussion

3.1. Sensory Evaluation of Oil Processing Samples

Flavor is the most important quality of peanut products. The sensory evaluation results of peanut raw materials and thermal processed samples are shown in Figure 1. There are significant differences in sensory attributes between the raw material and thermal processed sample. Raw/beany and woody are the main flavors of peanut raw material. Normal-oleic peanut has a slightly stronger sweet aromatic (3.15) than high oleic peanut (2.40). Flavor attributes of high-oleic raw peanuts have been reported to be very similar to the normal oleic cultivars [19]. After roasting, raw/beany and woody flavor attributes significantly reduced. Dark roast, roast peanutty and sweet aromas make a great contribution to the roasted peanut flavor. Under the same processing conditions, roasted normal-oleic peanut has stronger roast (4.28), peanutty (4.80) and sweet (4.65) flavors, which were 16.33%, 20.75% and 29.17% higher than those of roasted high-oleic peanut, respectively. Roasted high-oleic peanuts have a stronger raw/beany (3.6) and woody (3.45) aroma than roasted normal-oleic peanuts (3 for raw/beany, 2.4 for woody). After high temperature press, the dark roast, roast peanutty and sweet aroma of samples continuously increased. Normal-oleic peanut oil has stronger roast (6.00), peanutty (7.2) and sweet (5.85) flavors, which were 21.21%, 29.73% and 18.18% higher than those of roasted high oleic peanut, respectively. The raw/beany and woody aromas of normal- and high oleic peanut oil were all around 2 with no difference. Statistically significant variation among 59 roasted peanuts was reported [20]. High oleic peanut cultivars showed a wide range of several sensory attributes (dark roasted, raw/beany, roasted peanutty, sweet aromatic, wood-hulls-skins, and "off flavors" stale/cardboard). The upper limit of positive sensory attributes for the high-oleic peanuts was greater than the normal cultivars. The differences in sensory quality between normal- and high oleic peanut products maybe caused by the composition and relative concentration of characteristic key volatile components.

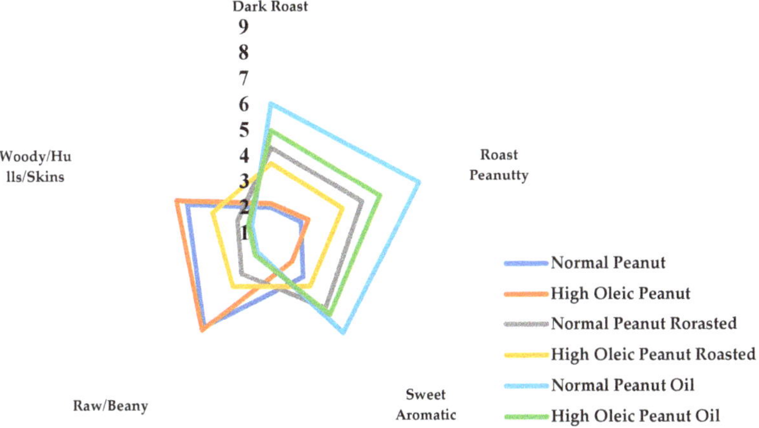

Figure 1. Sensory evaluation result of normal- and high-oleic peanut and oil processing samples.

3.2. Comparison between Volatile Components of Normal- and High Oleic Peanut Oils

As shown in Table 2, a total of 93 volatile components were identified in normal- and high-oleic peanut oil (NPO and HOPO), including 20 aldehydes, 17 alcohols, 10 alkanes, 8 acids, 5 ketones, 5 alkenes, 2 esters, 7 Pyrazines, 3 Pyridines, 3 Pyrroles, 11 furans and 2 pyrans. Most of the identified volatile components have been reported [11,24]. Several pyrazines, pyridines, pyrroles, furans and pyrans were firstly reported in the present study, including 2-methoxy-3-(1-methylethyl)-pyrazine, 1-(2-pyridinyl)-ethanone, 1-methyl-1H-pyrrole, 2-methyl-furan, 5-methyldihydro-2(3H)-furanone, 4-methyldihydro-2(3H)-furanone, 5-pentyldihydro-2(3H)-furanone, 3-hydroxydihydro-4,4-dimethyl-2(3H)-furanone, tetrahydro-2H-pyran-2-one and 3-hydroxy-2-methyl-4H-pyran-4-one. The composition and relative content of N-heterocyclic, O-heterocyclic and nonheterocyclic between normal- and high-oleic peanut oil were significantly different. The HOPO contains 39.40% N-heterocyclic, which is twice that of NPO. Among them, 35.14% 1-methyl-1H-Pyrrole in HOPO is three times that of NPO. Pyrroles were formed in the Maillard reaction and highly correlated to roast flavor and aroma [27]. Pyrazines are diverse heterocyclic nitrogen-containing compounds derived from nonenzymatic protein–sugar interactions. These volatile compounds contribute to the roasted/nutty flavor [13]. The pyrazine content of NPO and HOPO were 3.16% and 4.17%, respectively. O-heterocyclic accounting for 7.44% and 4.24% volatile components of NPO and HOPO, respectively. Ten furans in HOPO account for 6.78% of volatile components, which is twice the amount in HOPO. Furan derivatives have been identified as the second largest volatiles in roasted peanut oil [28]. They were considered to contribute to the thermally processed food flavor, including caramel-like, sweet, fruity, and nutty. Nonheterocyclic compounds were derived from lipid decomposition [29]. Aldehyde compounds were the most important nonheterocyclic compounds which appear as green, painty, metallic, beany and rancid and are also responsible for the undesirable flavors of oils [30]. The NPO contains 39.36% aldehydes which is 1.61 times that of HOPO. Hexanal accounts for 17.29% and 5.74% of total volatiles in NPO and HOPO, respectively. High oleic acid improves the oxidative stability of peanut products and reduces the formation of aldehydes.

Table 2. Composition and relative content of volatile components in normal- and high-oleic peanut oil.

Order	Retention Index	Volatile Compound	Normal-Oleic Peanut Oil Volatile Compound (%)	High-Oleic Peanut Oil Volatile Compound (%)
		N-heterocyclic		
		Pyrazines		
1	1265	Pyrazine, methyl-	0.57	0.79
2	1319	Pyrazine, 2,5-dimethyl-	1.27	1.73
3	1334	Pyrazine, ethyl-	0.24	0.22
4	1344	Pyrazine, 2,3-dimethyl-	0.08	0.03
5	1389	Pyrazine, 2-ethyl-5-methyl	0.75	1.20
6	1434	Pyrazine, 2-methoxy-3-(1-methylethyl)	0.14	
7	1440	Pyrazine, 3-ethyl-2,5-diemethyl	0.09	0.19
		Pyridines		
8	1178	Pyridine	0.11	
9	1577	Pyridine, 3-methoxy-	0.49	
10	1599	Ethanone, 1-(2-pyridinyl)	0.10	0.09
		Pyrroles		
11	1140	1H-Pyrrole, 1-methyl-	12.94	35.14
12	1976	Ethanone, 1-(1H-pyrrol-2-yl)	0.13	
13	2032	1H-Pyrrole-2-caboxaldehyde	0.05	
		Total	16.98	39.40
		O-heterocyclic		
		Furans		
14	1233	Furan, 2-methyl-	0.17	
15	1235	Furan, 2-pentyl-	2.39	1.12
16	1608	2(3H)-Furanone, dihydro-5-methyl		0.18
17	1614	2(3H)-Furanone, dihydro-4-methyl	1.04	1.05
18	1629	2(3H)-Furanone, dihydro-	0.53	0.74
19	1665	2-Furanmethanol	0.91	0.22
20	1697	Furan, 2-pentyl-	0.14	
21	1730	2,5-Furandione, 3,4-dimethyl	0.09	
22	2027	2(3H)-Furanone, dihydro-5-pentyl	0.07	
23	2042	2(3H)-Furanone, dihydro-3-hydroxy-4,4-dimethyl	0.41	
24	2407	2,3-dihydro-benzofuran	1.03	
		Pyrans		
25	1804	2H-Pyran-2-one, tetrahydro-		0.06
26	1968	4H-Pyran-4-one, 3-hydroxy-2-methyl-	0.67	0.87
		Total	7.44	4.24

Table 2. *Cont.*

Order	Retention Index	Volatile Compound	Normal-Oleic Peanut Oil Volatile Compound (%)	High-Oleic Peanut Oil Volatile Compound (%)
		Nonheterocyclic		
		Aldehydes		
27	<1000	Butanal, 2-methyl-	1.19	1.12
28	<1000	Butanal, 3-methyl-	1.16	0.88
29	<1000	Pentanal	2.87	0.48
30	1078	Hexanal	17.29	5.74
31	1185	Heptanal	0.76	1.89
32	1218	2-Hexenal, (E)	0.48	0.13
33	1290	Octanal	1.08	3.75
34	1324	2-Heptenal, (Z)	4.94	1.06
35	1395	Nonanal	1.33	4.95
36	1429	2-Octenal, (E)	0.63	
37	1468	Furfural	1.27	0.76
38	1518	Benzaldehyde	2.69	2.77
39	1531	2-Nonenal, (E)	0.47	
40	1643	benzeneacetaldehyde	1.15	0.10
41	1704	Benzaldehyde, 4-ethyl-	0.12	0.09
42	1762	2,4-Decadienal	0.56	
43	1783	3-Phenylbutanal	0.13	
44	1806	2,4-Decadienal, (E,E)-	1.21	
45	1829	2-Propenal, 3-phenyl	0.05	
46	2405	Benzaldehyde, 4-methyl		0.70
		Alcohols		
47	<1000	2-Propanol		0.33
48	<1000	Ethanol	0.44	
49	<1000	2-Butanol		0.05
50	1092	1-Propanol, 2-methyl-	0.06	0.31
51	1207	1-Butanol, 3-methyl		2.16
52	1256	1-Pentanol	4.08	2.16
53	1359	1-Hexanol	5.79	2.29
54	1453	1-Octen-3-ol	2.33	0.36
55	1558	1-Octanol	0.46	2.45
56	1582	2,3-Butanediol	0.08	
57	1618	Ethanol, 2-(2-ethoxyethoxy)	0.18	0.26
58	1659	1-Nonanol		0.62
59	1791	Ethanol, 2-(2-butoxyethoxy)-	0.15	
60	1912	Phenylethyl Alcohol	1.33	2.89
61	2018	Phenol	0.08	0.07
62	2184	Phenol, 2-(1-methylpropyl)-	0.19	
63	2309	5-Thiazoleethanol, 4-methyl	0.35	
		Alkanes		
64	<1000	Pentane	6.34	
65	<1000	Heptane	1.01	1.59
66	<1000	Octane	1.92	2.00
67	<1000	Heptane, 2,4-dimethyl		0.65
68	<1000	2-Propanone		1.35
69	<1000	Octane, 4-methyl		0.43
70	<1000	Heptane, 2,2,4,6,6-pentamethyl-	0.51	4.52
71	<1000	Decane	0.03	2.29
72	1197	Dodecane		1.07
73	1401	Tetradecane	0.80	
		Acids		
74	1496	Acetic acid	1.58	0.73
75	1769	Pentanoic acid	0.35	0.11
76	1875	Hexanoic acid	1.87	0.96

Table 2. *Cont.*

Order	Retention Index	Volatile Compound	Normal-Oleic Peanut Oil Volatile Compound (%)	High-Oleic Peanut Oil Volatile Compound (%)
77	1981	Heptanoic acid	0.44	0.49
78	2087	Octanoic acid	0.46	0.28
79	2164	Benzoic acid	0.55	
80	2192	Nonanoic acid	0.99	0.50
81	2298	Decanoic acid	0.07	
		Ketones		
82	1129	3-Penten-2-one, 4-methyl-	0.50	
83	1182	2-Heptanone	0.49	
84	1286	2-Octanone	0.14	0.20
85	1340	6-Methyl-5-hepten-2-one	0.06	0.06
86	1407	3-Octen-2-one	0.33	
		Alkenes		
87	<1000	2-Octene, (Z)	0.49	
88	<1000	2-Octene, (E)	0.23	
89	<1000	Alpha-Pinene	0.27	
90	1096	Undecane		0.42
91	1195	Limonene	0.70	0.24
		Esters		
92	1071	Acetic acid, butyl ester		0.12
93	1635	Decanoic acid, ethyl ester	0.55	
		Total	75.58	56.37

3.3. Comparison between Key Volatile Components of Normal- and High Oleic Peanut Oils

The contribution of volatile components to the whole flavor of peanut oil was based on their relative concentration, odor classification and odor strength. Quantitative determination and odor strength evaluation are used for characteristic volatile components study. The correlation between characteristics volatile compounds and sensory characteristics has been studied. The compounds 2,5-dimethylpyrazine (correlated with nutty and roasted odors) and 1-methyl-1H-pyrrol (correlated with sweet and woody odor) are two of the most reported volatile components in roasted peanut products. The compounds 2/3-methyl-1H-pyrrole, 5-methyl-2-furancarboxaldehyde, benzeneacetaldehyde, 2,3-dimethyl-1H-pyrrole, 2,5-dimethylpyrazine, 5-methyl-2-furanmethanol and maltol are positively correlated to peanutty and roast aroma [31]. Having the second highest relative concentration of furan derivatives, 2-furaldehyde contributes to the sweet and caramel-like aromas of heated foods [32]. As a Strecker degradation product of phenylalanine amino acid, benzaldehyde provided an almond-like aroma [33].

As shown in Table 3, several pyrazines, pyrroles, furans and aldehydes were screened out as possible key volatiles, which contribute to the nutty, roasty and sweet flavors of peanut oil. Among all pyrazines, 2,5-dimethylpyrazine was most highly correlated to roasted peanut flavor and aroma. Comparing with 1-methy-1H-pyrrole, 3 pyrazine showed stronger nutty odor with GC-MS-O evaluation in the present study. 2,5-dimethylpyrazine has the strongest nutty odor (3.67). The nutty odor strength of methylpyrazine and 2-ethyl-5-methylpyrazine are 3.00 and 2.67. Although the 1-methyl-1H-pyrrole has the highest relative concentration (6.29 mg/kg in NPO, 7.28 mg/kg in HOPO), its nutty odor strength is only 1.33. A similar result was reported in which the correlation coefficient of 1-methyl-1H-pyrrole to nutty flavor was relatively low [31]. It can be determined that methylpyrazine, 2,5-dimethylpyrazine and 2-ethyl-5-methylpyrazine are key volatiles that contribute to the nutty and roast flavor of peanut oil. The NPO contains 0.28 ± 0.02 mg/kg methylpyrazine, 0.62 ± 0.05 mg/kg 2,5-dimethylpyrazine and 0.37 ± 0.03 mg/kg 2-ethyl-5-methylpyrazine, which are 75%, 72% and 48% higher than those of HOPO, respectively. The sensory comparison between normal- and high-oleic peanut has been reported. Compared with normal-oleic peanut, four high-oleic breeding lines (derived by the Florida high-oleic gene) showed a stronger roasted peanut sensory attribute [34]. High-oleic peanuts contribute higher roast (1.83 vs. 1.57, $p < 0.05$) and nutty (2.69 vs. 2.53, $p < 0.05$) aromas than normal-oleic peanuts among 14 peanut genetic entries (5 high-oleic, 9 normal-oleic). Roasted high-oleic peanuts have a wider roasted peanutty (3.92–5.15) odor range than roasted normal-oleic peanut (4.26–4.89) [19].

Table 3. Odor strength and relative concentration of key volatile components in normal- and high-oleic peanut oil.

Retention Time	Key Volatile Compounds	Odor Description	Odor Strength	Normal-Oleic Peanut Oil Concentration (mg/kg)	High-Oleic Peanut Oil Concentration (mg/kg)
7.4–7.5	Pentanal	Nutty	1.33 ± 0.58	1.39 ± 0.08	0.10 ± 0.01
10.3–10.6	Hexanal	Green, Beany	1.00 ± 0.00	8.40 ± 0.74	1.19 ± 0.06
12.1–12.7	1H-Pyrrole, 1-methyl-	Nutty, Sweet	1.33 ± 0.58	6.29 ± 0.69	7.28 ± 0.42
14.9–15.1	Furan, 2-pentyl-	Green, Earthy, Beany	1.00 ± 0.00	1.16 ± 0.10	0.23 ± 0.01
16.0–16.2	Pyrazine, methyl-	Nutty, Roasted, Cocoa	3.00 ± 0.00	0.28 ± 0.02	0.16 ± 0.02
17.6–17.8	Pyrazine, 2,5-dimethyl-	Nutty, Roasted, Cocoa	3.67 ± 0.58	0.62 ± 0.05	0.36 ± 0.01
19.8–20.0	Pyrazine, 2-ethyl-5-methyl	Nutty, Roasted, Grassy	2.67 ± 0.58	0.37 ± 0.03	0.25 ± 0.00
21.3–21.5	Furfural	Sweet	1.67 ± 0.58	0.62 ± 0.05	
22.8–23.0	Benzaldehyde	Sweet	2.33 ± 0.58	1.30 ± 0.09	0.57 ± 0.01
25.8–26.0	2-Furanmethanol	Sweet	1.67 ± 0.58	0.44 ± 0.05	0.05 ± 0.00
32.4–32.6	4H-Pyran-4-one, 3-hydroxy-2-methyl-	Sweet	2.00 ± 0.00	0.32 ± 0.01	0.18 ± 0.01

The compounds 1-methyl-1H-pyrrole, furfural, benzaldehyde, 2-furanmethanol and 3-hydroxyl-2-methyl-4H-pyran-one contribute to the sweet aroma of peanut oil (Table 3). Among of them, benzaldehyde and 3-hydroxyl-2-methyl-4H-pyran-one are considered to be the key volatiles with 2.33 and 2.00 sweet aroma strength, respectively. Compared with normal-oleic peanut oil, high-oleic peanut oil has a higher relative concentration of furfural (0.62 vs. 0.00), benzaldehyde (1.30 vs. 0.57), 2-furanmethanol (0.44 vs. 0.05) and 3-hydroxyl-2-methyl-4H-pyran-one (0.32 vs. 0.18), which lead to a stronger sweet aroma. This is consistent with the previous sensory evaluation result. There is no significant difference on sweet aroma between high-oleic roasted (2.44) and normal-oleic roasted peanuts (2.39) [19]. Sweet aromatic strength ranged from 2.41 to 3.24 fiu for high-oleics peanuts and 2.71 to 3.24 fiu for normal-oleic [20]. Hexanal is one of the primary oxidation products of linoleic acid, which contributes the green and grassy flavors to the oil. The relative content of hexanal in normal-oleic peanut oil was much higher (8.40 vs. 1.19) than in high-oleic peanut oil. This is attributed to the oxidative stability of high-oleic oil. However, the green odor strength of hexanal is very weak (1.00).

The variety and origin of collected peanut greatly influenced the sensory comparison results between normal- and high-oleic peanut samples. In the present study, with the same processing condition, normal-oleic peanut oil has a higher relative concentration of key volatile components, which contribute to stronger roasted, nutty and sweet aromas. The differences in peanut oil flavor maybe caused by the composition of volatile precursors in raw material.

3.4. Comparison between Amino Acids and Reducing Sugars Profile of Normal- and High Oleic Peanut Oil Processing Samples

Proteins and sugars are considered the major precursors for volatiles in peanuts. Reactivities of amino acids in Maillard model systems have drawn much attention. Dimethylpyrazine and 3-ethyl-2,5-dimethylpyrazine were largely synthesized in an aspartic acid–ascorbic acid model system [35]. Similarly, nine pyrazines were identified in the L-glutamic acid and 1,4-13C-labeled-ascorbic acid Maillard model system, and the total content of pyrazines was 63.52 mg/mol. 2,5-dimethylpyrazine (34.42 mg/mol) and ethyl-5-methylpyrazine (21.17 mg/mol) were the major pyrazines formed in the model system [18]. The structure of the N-terminal amino acid determined the overall formation of pyrazines, and the C-terminal amino acid showed less influence. The production of 2,5(6)-dimethylpyrazine and trimethylpyrazine was very high in the case of glycine, alanine or serine, whereas it was low for proline, valine or leucine [36].

The Maillard reaction between characteristic amino acids and sugars has also been studied. A quantity of 17,280 µg pyrazines was formed in a leucine (0.5 mol/L)-rhamnose (2.0 mol/L) model system, and 2-isoamyl-6-methylpyrazine (780 µg) was highly branched [37]. Eight pyrazines (0.805 mg/g of ribose) were synthesized in cysteine-ribose Maillard model system. 5H-5-methyl-6,7-dihydrocyclopentapyrazine (0.042 mg/g of ribose) was identified as a distinctive volatile component among all the pyrazines [38]. Volatile compounds formed by the reaction of 2-deoxyglucose with glutamine, glutamic acid, asparagine and aspartic acid were studied [39]. Compared with other amino acids-involved model systems, 2-deoxyglucose and asparagine generated the highest content of methylpyrazine. Results also indicated the importance of the 2-hydroxy group on glucose molecules for the effective generation of flavor compounds. A reactive Maillard reaction intermediate derived from xylose and phenylalanine was synthesized by using a stepwise increase of heating temperature. The Maillard Reaction intermediate reacted with cysteine to form various pyrazines [40].

The amino acids and reducing sugars profile of peanut samples during the oil processing were investigated in the present study. As shown in Table 4, there is no significant difference in amino acids between high-oleic peanuts and normal-oleic peanuts. Arginine, tyrosine and lysine were continuously decreased during the thermal processing. Among of them, arginine has the highest relative concentration in peanut raw materials. During the roasting procedure, the relative concentration of arginine in normal-oleic peanut decreased from 2.63 g/100 g to 1.13 g/100 g, which is also the highest loss of all the amino acids. The relative concentration of arginine in high-oleic peanuts decreased from 2.51 g/100 g to 1.08 g/100 g. There was no tyrosine detected in roasted peanut and oil samples, which indicates that all the tyrosine was reacted in the roasting procedure. As shown in Table 5, glucose was the only sugar which was continuously consumed during the thermal processing. The relative concentration of glucose in normal-oleic peanuts decreased from 0.18 mg/g to 0.12 mg/g during the thermal procedure. The content of glucose in high-oleic peanuts decreased from 0.07 mg/g to 0.03 mg/g during peanut oil processing. The relative concentration of arginine, tyrosine, lysine and glucose in peanut samples had a significant negative correlation with characteristic pyrazines, which indicated these compositions could be precursors of key volatile components. The initial relative concentration and process consumption of characteristic precursors (arginine, tyrosine, lysine and glucose) in normal-oleic peanuts was higher than in high-oleic peanuts, which led to the formation of more specific volatile components. This is consistent with sensory evaluation results for normal- and high-oleic peanut oil. Similarly, a quantity of 2229.66 mg/mol pyrazines were formed with the Maillard model system between 0.5 mol/L tyrosine and 0.5 mol/L glucose under 130 °C for 2.5 h. 2,5-Dimethylpyrazine and 2-ethyl-3-methylpyrazine were the majority of 15 formed pyrazines [35]. The effects of high-intensity ultrasound on Maillard reaction in a model system of D-xylose and L-lysine were studied [41]. 2-Methylpyrazine, 2,5-Dimethylpyrazine, 2,3-Dimethylpyrazine and 2,3,5-Trimethylpyrazine were formed in the thermal model. The ultrasonic-assisted Maillard model system could produce 3-ethyl-2,5-dimethylpyrazine, butyl amine and maltol, which were absent from thermal model. The capacity of glucose for pyrazine formation during the Maillard reaction was reported [42]. The glucose produced by Maillard reaction generated 56.7 ng/g 2-methylpyrazine, which is 18.62–32.17% higher than the fructose, ribose and xylose produced by Maillard reaction.

Table 4. Amino acids profile of normal- and high-oleic peanut oil processing samples (g/100 g).

	Normal-Oleic Peanut	Roasted Normal-oleic Peanut	Normal-Oleic Peanut Meal	High Oleic Peanut	Roasted High-Oleic Peanut	High-Oleic Peanut Meal
Aspartic acid	2.83 ± 0.11 [a]	2.80 ± 0.14 [a]	3.37 ± 0.24 [a]	2.65 ± 0.13 [a]	2.75 ± 0.08 [a]	3.10 ± 0.03 [a]
Threonine	0.66 ± 0.07 [a]	0.67 ± 0.04 [a]	0.75 ± 0.04 [a]	0.64 ± 0.05 [a]	0.65 ± 0.07 [a]	0.71 ± 0.05 [a]
Serine	1.37 ± 0.06 [a]	1.38 ± 0.07 [a]	1.53 ± 0.03 [a]	1.25 ± 0.09 [a]	1.29 ± 0.12 [a]	1.44 ± 0.09 [a]
Glutamic acid	4.63 ± 0.12 [ab]	4.65 ± 0.02 [ab]	5.45 ± 0.31 [a]	4.37 ± 0.16 [b]	4.52 ± 0.15 [ab]	5.08 ± 0.16 [ab]
Proline	1.06 ± 0.05 [a]	1.07 ± 0.06 [a]	1.16 ± 0.14 [a]	0.97 ± 0.04 [a]	1.02 ± 0.04 [a]	1.09 ± 0.05 [a]
Glycine	1.22 ± 0.06 [a]	1.22 ± 0.09 [a]	1.42 ± 0.09 [a]	1.43 ± 0.07 [a]	1.45 ± 0.05 [a]	1.55 ± 0.08 [a]
Alanine	0.92 ± 0.08 [a]	0.93 ± 0.05 [a]	1.06 ± 0.07 [a]	0.86 ± 0.00 [a]	0.91 ± 0.07 [a]	0.99 ± 0.03 [a]
Cystine	0.35 ± 0.02 [a]	0.35 ± 0.01 [a]	0.39 ± 0.02 [a]	0.35 ± 0.04 [a]	0.35 ± 0.02 [a]	0.40 ± 0.01 [a]
Valine	1.05 ± 0.04 [a]	1.04 ± 0.07 [a]	1.23 ± 0.14 [a]	1.05 ± 0.07 [a]	1.09 ± 0.05 [a]	1.21 ± 0.07 [a]
Isoleucine	0.75 ± 0.05 [a]	0.74 ± 0.06 [a]	0.95 ± 0.08 [a]	0.74 ± 0.05 [a]	0.79 ± 0.11 [a]	0.86 ± 0.05 [a]
Leucine	1.55 ± 0.11 [a]	1.55 ± 0.09 [a]	1.87 ± 0.09 [a]	1.57 ± 0.08 [a]	1.56 ± 0.13 [a]	1.72 ± 0.12 [a]
Tyrosine	0.96 ± 0.07 [a]	0.00 ± 0.00 [b]	0.00 ± 0.00 [b]	0.90 ± 0.00 [a]	0.00 ± 0.00 [b]	0.00 ± 0.00 [b]
Phenylalanine	1.23 ± 0.14 [a]	0.95 ± 0.04 [a]	1.14 ± 0.04 [a]	1.15 ± 0.07 [a]	1.01 ± 0.07 [a]	1.35 ± 0.08 [a]
Histidine	0.71 ± 0.05 [b]	1.25 ± 0.13 [a]	1.43 ± 0.05 [a]	0.70 ± 0.03 [b]	1.25 ± 0.09 [a]	0.80 ± 0.00 [b]
Lysine	1.02 ± 0.07 [a]	0.70 ± 0.03 [b]	0.69 ± 0.03 [b]	1.00 ± 0.04 [a]	0.74 ± 0.03 [b]	0.73 ± 0.04 [b]
Arginine	2.63 ± 0.16 [a]	1.13 ± 0.05 [b]	1.09 ± 0.08 [b]	2.51 ± 0.07 [a]	1.08 ± 0.07 [b]	1.07 ± 0.01 [b]

Volume in a row with different superscripts were significantly different ($p < 0.5$).

Table 5. Sugars profile of normal- and high-oleic peanut oil processing samples (g/kg).

	Normal-Oleic Peanut	Roasted Normal-Oleic Peanut	Normal-Oleic Peanut Meal	High-Oleic Peanut	Roasted High-Oleic Peanut	High-Oleic Peanut Meal
Fructose	0.26 ± 0.06 [b]	0.81 ± 0.09 [a]	0.94 ± 0.05 [a]	0.24 ± 0.10 [b]	0.62 ± 0.04 [ab]	0.57 ± 0.05 [ab]
Glucose	0.18 ± 0.06 [a]	0.14 ± 0.03 [a]	0.12 ± 0.04 [a]	0.07 ± 0.01 [a]	0.04 ± 0.01 [a]	0.03 ± 0.02 [a]
Sucrose	50.99 ± 1.37 [b]	58.36 ± 3.18 [ab]	68.57 ± 0.81 [a]	56.73 ± 3.51 [ab]	64.15 ± 3.29 [ab]	60.19 ± 1.27 [ab]
Maltose	3.06 ± 0.16 [bc]	4.31 ± 0.20 [ab]	4.83 ± 0.34 [a]	1.87 ± 0.23 [d]	3.27 ± 0.34 [bc]	2.64 ± 0.12 [cd]
Starchyose	0.69 ± 0.06 [b]	2.36 ± 0.37 [ab]	3.27 ± 0.45 [ab]	2.57 ± 0.23 [ab]	4.53 ± 0.63 [a]	3.05 ± 0.99 [ab]
Raffinose	2.30 ± 0.02 [b]	3.59 ± 0.12 [a]	3.66 ± 0.05 [a]	2.42 ± 0.15 [b]	2.64 ± 0.17 [b]	2.89 ± 0.07 [b]

Volumes in a row with different superscripts were significantly different ($p < 0.5$).

4. Conclusions

Significant differences in sensory attributes were found between peanut raw materials and thermal processed samples. Sensory evaluation results showed that normal-oleic peanut oil has a stronger dark roast, roast peanutty and sweet aroma than high-oleic peanut oil under the same processing conditions. Methylpyrazine, 2,5-dimethylpyrazine and 2-ethyl-5-methylpyrazine are considered to be the key volatiles contributing to the nutty and roasty flavor of peanut oil. Benzaldehyde and 3-hydroxyl-2-methyl-4H-pyran-one play important roles in the sweet aroma of peanut oil. The initial concentration of characteristic precursors (arginine, tyrosine, lysine and glucose) in normal-oleic peanuts was higher than in high-oleic peanuts, which led to the formation of more specific volatile components and contributed to the stronger, specific aroma of the oil. The formation mechanism of key volatiles in peanut oil needs to be further investigated. The results of this study could provide data to support the screening of suitable high-oleic peanut varieties for industrial oil processing and improve the characteristic flavor of peanut oil.

Author Contributions: Conceptualization, H.H. and M.L.F.; methodology, H.H.; software, H.L.; formal analysis, H.H.; investigation, H.H.; resources, L.L.; data curation, H.H.; writing—original draft preparation, H.H.; writing—review and editing, M.L.F. and A.S.; supervision, Q.W. and M.L.F.; project administration, Q.W.; funding acquisition, Q.W. All authors have read and agreed to the published version of the manuscript.

Funding: This research was funded by Technology Innovation Program of the Chinese Academy of Agricultural Sciences [CAAS-ASTIP-2020-IFST]; TaiShan Industrial Leading Talent Program of Shandong Province, China (LJNY201711).

Data Availability Statement: The data presented in this study are available on request from the corresponding author.

Conflicts of Interest: The authors declare no conflict of interest.

References

1. USDA. Available online: https://www.fas.usda.gov/data/oilseeds-world-markets-and-trade/ (accessed on 9 November 2021).
2. Wang, Q.; Liu, L.; Wang, L.; Guo, Y.; Wang, J. Introduction. In *Peanuts: Processing Technology and Product Development*, 1st ed.; Wang, Q., Ed.; Academic Press: Cambridge, MA, USA, 2016; pp. 1–22. [CrossRef]
3. Derbyshire, E.J. A review of the nutritional composition, organoleptic characteristics and biological effects of the high oleic peanut. Int. *J. Food Sci. Nutr.* **2014**, *65*, 781–790. [CrossRef]
4. Yu, H.; Liu, H.; Wang, Q.; Van Ruth, S. Evaluation of portable and benchtop NIR for classification of high oleic acid peanuts and fatty acid quantitation. *LWT-Food Sci. Technol.* **2020**, *128*, 109398. [CrossRef]
5. Zhao, Z.; Shi, A.; Wang, Q.; Zhou, J. High Oleic Acid Peanut Oil and Extra Virgin Olive Oil Supplementation Attenuate Metabolic Syndrome in Rats by Modulating the Gut Microbiota. *Nutrients* **2019**, *11*, 3005. [CrossRef]
6. Norden, A.J.; Borget, D.W.; Knauft, D.A.; Young, C.T. Variability in oil quality among peanut genotypes in the Florida breeding program. *Peanut Sci.* **1987**, *4*, 7–11. [CrossRef]
7. Hu, H.; Liu, H.; Shi, A.; Liu, L.; Fauconnier, M.L.; Wang, Q. The Effect of Microwave Pretreatment on Micronutrient Contents, Oxidative Stability and Flavor Quality of Peanut Oil. *Molecules* **2018**, *24*, 62. [CrossRef]
8. Salehi, F. Physico-chemical properties of fruit and vegetable juices as affected by pulsed electric field: A review. *Int. J. Food Prop.* **2020**, *23*, 1036–1050. [CrossRef]
9. Newell, J.A.; Mason, M.E.; Matlock, R.S. Precursors of typical atypical roasted peanut flavor. *J. Agric. Food Chem.* **1967**, *15*, 767–772. [CrossRef]
10. Baker, G.L.; Cornell, J.A.; Gorbet, D.W.; O'Keefe, S.F.; Sims, C.A.; Talcott, S.T. Determination of pyrazine and flavor variations in peanut genotypes during roasting. *J. Food Sci.* **2003**, *68*, 394–400. [CrossRef]
11. Qian, D.; Yao, L.; Deng, Z.; Li, H.; Li, J.; Fan, Y.; Zhang, B. Effects of hot and cold-pressed processes on volatile compounds of peanut oil and corresponding analysis of characteristic flavor components. *LWT-Food Sci. Technol.* **2019**, *112*, 107648. [CrossRef]
12. Dimitrios, L.; Vincenzo, F.; Edoardo, C. Flavor of roasted peanuts (Arachis hypogaea)—Part I: Effect of raw material and processing technology on flavor, color and fatty acid composition of peanuts. *Food Res. Int.* **2016**, *89*, 860–869. [CrossRef]
13. Davis, J.P.; Dean, L.L. Peanut composition, flavor and nutrition. In *Peanuts Genetics, Processing, and Utilization*, 1st ed.; Stalker, H.T., Wilson, R.F., Eds.; Academic Press: Cambridge, MA, USA; AOCS Press: Urbana, IL, USA, 2016; pp. 289–345.
14. Lu, C.Y.; Hao, Z.; Payne, R.; Ho, C.T. Effects of water content on volatile generation and peptide degradation in the Maillard reaction of glycine, diglycine, and triglycine. *J. Agric. Food Chem.* **2005**, *53*, 6443–6447. [CrossRef]
15. Ho, C.T.; Zhang, J.; Hwang, H.I.; Riha, W.E. Release of ammonia from peptides and proteins and their effects on Maillard flavor generation. In *Maillard Reactions in Chemistry, Food and Health*, 1st ed.; Labuza, T.P., Reineccius, G.A., Monnier, V.M., O'Brien, J., Baynes, J.W., Eds.; Woodhead Publishing: Cambridge, UK, 2005; pp. 126–130.
16. He, S.; Zhang, Z.; Sun, H.; Zhu, Y.; Zhao, J.; Tang, M.; Wu, X.; Cao, Y. Contributions of temperature and l-cysteine on the physicochemical properties and sensory characteristics of rapeseed flavor enhancer obtained from the rapeseed peptide and d-xylose Maillard reaction system. *Ind. Crops Prod.* **2019**, *128*, 455–463. [CrossRef]
17. Guo, X.; Song, C.; Ho, C.T.; Wan, X. Contribution of L-theanine to the formation of 2,5-dimethylpyrazine, a key roasted peanutty flavor in Oolong tea during manufacturing processes. *Food Chem.* **2018**, *263*, 18–28. [CrossRef]
18. Yu, A.N.; Tan, Z.W.; Wang, F.S. Mechanistic studies on the formation of pyrazines by Maillard reaction between l-ascorbic acid and l-glutamic acid. *LWT-Food Sci. Technol.* **2013**, *50*, 64–71. [CrossRef]
19. Isleib, T.G.; Pattee, H.E.; Sanders, T.H.; Hendrix, K.W.; Dean, L.O. Compositional and sensory comparisons between normal- and high-oleic peanuts. *J. Agric. Food Chem.* **2006**, *54*, 1759. [CrossRef] [PubMed]
20. Isleib, T.G.; Pattee, H.E.; Tubbs, R.S.; Sanders, T.H.; Dean, L.O.; Hendrix, K.W. Intensities of Sensory Attributes in High- and Normal-Oleic Cultivars in the Uniform Peanut Performance Test. *Peanut Sci.* **2015**, *42*, 83–91. [CrossRef]
21. Johnsen, P.; Civille, G.; Vercellotti, J.; Sanders, T.; Dus, C. Development of a lexicon for the description of peanut flavor. *J. Sens. Stud.* **1988**, *3*, 9–17. [CrossRef]
22. Sanders, T.H.; Vercellotti, J.R.; Crippen, K.L.; Civille, G.V. Effect of maturity on roast color and descriptive flavor of peanuts. *J. Food Sci.* **1989**, *54*, 475–477. [CrossRef]
23. Schirack, A.; Drake, M.A.; Sanders, T.H.; Sandeep, K.P. Characterization of aroma-active compounds in microwave blanched peanuts. *J. Food Sci.* **2006**, *71*, C513. [CrossRef]
24. Liu, X.; Jin, Q.; Liu, Y.; Huang, J.; Wang, X.; Mao, W.; Wang, S. Changes in volatile compounds of peanut oil during the roasting process for production of aromatic roasted peanut oil. *J. Food Sci.* **2011**, *76*, C404–C412. [CrossRef] [PubMed]

25. Paul, A.; Frederich, M.; Uyttenbroeck, R.; Malik, P.; Filocco, S.; Richel, A.; Heuskin, S.; Alabi, T.; Megido, R.C.; Franck, T.; et al. Nutritional composition and rearing potential of the meadow grasshopper (Chorthippus parallelus Zetterstedt). *J. Asia-Pac. Entomol.* **2016**, *19*, 1111–1116. [CrossRef]
26. Tahir, M.; Vandenberg, A.; Chibbar, R.N. Influence of environment on seed soluble carbohydrates in selected lentil cultivars. *J. Food Compos. Anal.* **2011**, *24*, 596–602. [CrossRef]
27. Nursten, H. Flavour and off-flavour formation in nonenzymic browning. In *The Maillard Reaction Chemistry, Biochemistry and Implications*, 1st ed.; Nursten, H., Ed.; Atheneum Press Ltd.: London, UK, 2005; pp. 62–89.
28. Vranov, J.; Ciesarov, Z. Furan in food—A review. *Czech J. Food Sci.* **2009**, *27*, 1–10. [CrossRef]
29. Ho, C.T.; Shahidi, F. Flavor components of fats and oil. In *Bailey's Industrial Oil and Fat Products*, 6th ed.; Shahidi, F., Ed.; John Wiley & Sons, Inc.: Hoboken, NJ, USA, 2005; pp. 387–411. [CrossRef]
30. Kalua, C.M.; Allen, M.S.; Bedgood, D.R., Jr.; Bishop, A.G.; Prenzler, P.D.; Robards, K. Olive oil volatile compounds, flavour development and quality: A critical review. *Food Chem.* **2007**, *100*, 273–286. [CrossRef]
31. Dimitrios, L. Flavor of roasted peanuts (Arachis hypogaea)—Part II: Correlation of volatile compounds to sensory characteristics. *Food Res. Int.* **2016**, *89*, 870–881. [CrossRef]
32. Flament, I. The individual constituents: Structure, nomenclature, origin, chemical and organoleptic properties. In *Coffee Flavor Chemistry*, 1st ed.; Flament, I., Ed.; John Wiley & Sons: Chichester, UK, 2002; pp. 81–346.
33. Ho, C.W.; Wan Aida, W.M.; Maskat, M.Y.; Osman, H. Changes in volatile compounds of palm sap (Arenga pinnata) during the heating process for production of palm sugar. *Food Chem.* **2007**, *102*, 1156–1162. [CrossRef]
34. Pattee, H.E.; Knauft, D.A. Comparison of selected high oleic acid breeding lines, Florunner and NC 7 for roasted peanut, sweet and other sensory attribute intensities. *Peanut Sci.* **1995**, *22*, 26–29. [CrossRef]
35. Yu, A.N.; Tan, Z.W.; Shi, B.A. Influence of the pH on the formation of pyrazine compounds by the Maillard reaction of L-ascorbic acid with acidic, basic and neutral amino acids. *Asia-Pac. J. Chem. Eng.* **2012**, *7*, 455–462. [CrossRef]
36. Van Lancker, F.; Adams, A.; De Kimpe, N. Impact of the N-terminal amino acid on the formation of pyrazines from peptides in Maillard model systems. *J. Agric. Food Chem.* **2012**, *60*, 4697–4708. [CrossRef]
37. Ara, K.M.; Taylor, L.T.; Ashraf-Khorassani, M.; Coleman, W.M. Alkyl pyrazine synthesis via an open heated bath with variable sugars, ammonia, and various amino acids. *J. Sci. Food Agric.* **2017**, *97*, 2263–2270. [CrossRef]
38. Chen, Y.; Xing, J.; Chin, C.-K.; Ho, C.-T. Effect of urea on volatile generation from Maillard reaction of cysteine and ribose. *J. Agric. Food Chem.* **2000**, *48*, 3512–3516. [CrossRef]
39. Lu, G.; Yu, T.H.; Ho, C.T. Generation of flavor compounds by the reaction of 2-deoxyglucose with selected amino acids. *J. Agric. Food Chem.* **1997**, *45*, 233–236. [CrossRef]
40. Cui, H.; Jia, C.; Hayat, K.; Yu, J.; Deng, S.; Karangwa, E.; Duhoranimana, E.; Zhang, X. Controlled formation of flavor compounds by preparation and application of Maillard reaction intermediate (MRI) derived from xylose and phenylalanine. *RSC Adv.* **2017**, *7*, 45442–45451. [CrossRef]
41. Yu, H.; Seow, Y.X.; Ong, P.K.; Zhou, W. Effects of high-intensity ultrasound on Maillard reaction in a model system of d-xylose and l-lysine. *Ultrason. Sonochem.* **2017**, *34*, 154–163. [CrossRef] [PubMed]
42. Ni, Z.J.; Liu, X.; Xia, B.; Hu, L.T.; Thakur, K.; Wei, Z.J. Effects of sugars on the flavor and antioxidant properties of the Maillard reaction products of camellia seed meals. *Food Chem. X* **2021**, *11*, 100127. [CrossRef]

Article

Effect of Hydrothermal Cooking Combined with High-Pressure Homogenization and Enzymatic Hydrolysis on the Solubility and Stability of Peanut Protein at Low pH

Jiaxiao Li [1,2], Aimin Shi [1,*], Hongzhi Liu [1], Hui Hu [1], Qiang Wang [1,*], Benu Adhikari [3], Bo Jiao [1] and Marc Pignitter [4]

[1] Institute of Food Science and Technology, Chinese Academy of Agricultural Sciences, Key Laboratory of Agro-Products Processing, Ministry of Agriculture and Rural Affairs, P.O. Box 5109, Beijing 100193, China; lijiaxiao721@outlook.com (J.L.); liuhongzhi@caas.cn (H.L.); huhui@caas.cn (H.H.); jiaobo@caas.cn (B.J.)

[2] Shaanxi Provincial Land Engineering Construction Group, Xi'an 710075, China

[3] School of Applied Sciences, City Campus, RMIT University, Melbourne, VIC 3001, Australia; benu.adhikari@rmit.edu.au

[4] Department of Physiological Chemistry, Faculty of Chemistry, University of Vienna, Althanstrasse 14, 1090 Vienna, Austria; marc.pignitter@univie.ac.at

* Correspondence: shiaimin@caas.cn (A.S.); wangqiang06@caas.cn (Q.W.); Tel.: +86-10-6281-5837

Abstract: A novel method combining high-pressure homogenization with enzymatic hydrolysis and hydrothermal cooking (HTC) was applied in this study to modify the structure of peanut protein, thus improving its physicochemical properties. Results showed that after combined modification, the solubility of peanut protein at a pH range of 2–10 was significantly improved. Moreover, the Turbiscan stability index of modified protein in the acidic solution was significantly decreased, indicating its excellent stability in low pH. From SDS-PAGE (Sodium Dodecyl Sulfate Poly Acrylamide Gel Electrophoresis), the high molecular weight fractions in modified protein were dissociated and the low molecular weight fractions increased. The combined modification decreased the particle size of peanut protein from 74.82 to 21.74 μm and shifted the isoelectric point to a lower pH. The improvement of solubility was also confirmed from the decrease in surface hydrophobicity and changes in secondary structure. This study provides some references on the modification of plant protein as well as addresses the possibility of applying peanut protein to acidic beverages.

Keywords: peanut protein; hydrothermal cooking; combined modification; low pH; physicochemical properties; protein structure

Citation: Li, J.; Shi, A.; Liu, H.; Hu, H.; Wang, Q.; Adhikari, B.; Jiao, B.; Pignitter, M. Effect of Hydrothermal Cooking Combined with High-Pressure Homogenization and Enzymatic Hydrolysis on the Solubility and Stability of Peanut Protein at Low pH. *Foods* **2022**, *11*, 1289. https://doi.org/10.3390/foods11091289

Academic Editor: Harjinder Singh

Received: 4 April 2022
Accepted: 20 April 2022
Published: 29 April 2022

Publisher's Note: MDPI stays neutral with regard to jurisdictional claims in published maps and institutional affiliations.

Copyright: © 2022 by the authors. Licensee MDPI, Basel, Switzerland. This article is an open access article distributed under the terms and conditions of the Creative Commons Attribution (CC BY) license (https://creativecommons.org/licenses/by/4.0/).

1. Introduction

Peanut (*Arachis hypogaea* Linn.) is one of the most important oil crops in the world. China's peanut production accounts for more than 40% of the world's total production [1], and 50–65% of the total China's peanut production is used for oil production.

The peanut meal contained more than 50% protein (dry basis) after pressing or removing the oil [2]. It is estimated that at least 3.5 million tons of defatted peanut meal is utilized in China each year, which is equivalent to approximately 2 million tons of peanut protein. Peanut protein is currently the third largest source of plant protein in China after wheat and soybean [3,4]. Serious protein denaturation happens after the traditional oil extraction process, which results in a huge waste of protein resources. Wang Qiang's group identified that high-quality peanut protein with a low degree of modification (the nitrogen solubility index is above 70%) could be obtained via improving defatting process [1,5]. Hence, peanut protein extracted from the wasted oil meal by this process is a high-quality and cheap protein source for food industries.

In the recent decade, peanut protein has received increasing attention because of its large output and high nutritional value [6]. At the present, peanut protein with high commercial value can be used as powered ingredient for direct use in high-protein foods and as food additives in foods such as pasta, meat products, and milk powder to improve appearance, water retention, texture characteristics and flavor, and nutritional value of foods [7,8]. Compared with soy protein, the utilization of peanut protein is still relatively low, primarily due to the relatively high cost of production, low solubility, and other characteristics [3]. As stated above, better use of peanut meal can take full advantage of valuable resource and increase the profits of the vegetable protein industry. Therefore, it is an important topic for peanut processing enterprises to further enhance the utilization of peanut protein.

Acidic beverages account for 60–70% of soft drink products and are popular with consumers because of their good taste [9]. However, due to their simple composition and lack of protein, they cannot meet the needs of consumers today for nutrition balance [10–12]. The addition of vegetable protein to traditional acidic beverages not only enhances the flavor of the product, but also increases the protein content. Thus, increasing protein content in acidic beverages and making them more stable in acidic beverages are research topics of high interest [13].

The isoelectric points of plant proteins from peanut, soy, pea, and walnut proteins fall under acidic pH range, which is also the pH range of most acidic beverages (pH 3.0 to 4.5). Plant proteins either flocculate or precipitate at their isoelectric point, limiting their use in acidic beverages [14]. The existing vegetable protein-based beverages on the market are mostly neutral or alkaline and have low protein content. Thus, the question of how to expand the application of vegetable protein in acidic beverages is currently a challenge and is an important issue regarding application of plant proteins. At present, methods such as adding polysaccharide stabilizers such as pectin, seaweed, and xanthan gums are used in beverages [15]. These additives increase the viscosity of the beverages and stabilize them by inhibiting the precipitation of protein particles. However, when beverages stabilized in this way are stored for a long time, they are susceptible to phase separation and/or precipitation, which affect their quality [16]. Therefore, the modification of natural plant proteins to achieve their stability in acidic conditions is of great practical significance. It is important to develop a green and efficient plant protein modification method.

Due to the limitations of a single physical, chemical, and enzymatic modification method, this study used a combination of high-pressure homogenization, enzymatic hydrolysis, and hydrothermal cooking to modify peanut protein. High pressure homogenization is commonly used in the food industry. The high-pressure homogenization process simultaneously involves the processes of cavitations, shear, turbulence, and temperature rise [17]. Before the enzymatic hydrolysis of the protein, the high-pressure homogenization pretreatment can loosen the protein structure, break the disulfide bond of the protein, expose more enzyme cleavage sites, and enable the protease to better act on the peptide bond, thereby breaking peptide bonds and accelerating protein breakdown [18].

Hydrothermal cooking is a steam injection process, commonly known as jet cooking, which combines high temperature and shearing force. The HTC equipment is mainly composed of a water heater, a material maintenance pipe, and a cooling system. High-speed turbulent materials mix with compressed air and high-temperature steam passing through the nozzle. Due to the sudden release of pressure in the flash chamber, there is a fast transfer of heat energy, and the temperature of the material rises sharply; the material enters the cooling system after the temperature is maintained for a certain period. This processing method has been used to improve protein extraction and re-functionalize heat-denatured soybean protein in extruded meals [19,20]. HTC can increase the solubility of protein by improving the swelling of protein particles and dissolving the protein aggregates of protein particles, so that the protein has better functional properties. The study of the Xia et al. (2012) [21] proved that HTC improved the extraction rate and purity of the protein in rice bran, as well as improved the solubility of rice protein, providing theoretical support for the industrial application of plant-insoluble protein extraction by HTC technology. The functional properties of protein have been improved after HTC treatment, which may be attributed to the expansion and exposure of protein structure and the increased of protein hydrophilicity.

In this study, the physicochemical properties of peanut protein through the treatments of HPH, HPH-E, and HPH-E-HTC were characterized, including solubility, zeta potential, secondary structure and average droplet size, and the stability and aggregation status. The systematic analysis of the properties of peanut protein can provide a scientific basis for its comprehensive development and utilization in the future.

2. Materials and Methods

2.1. Materials

Low-temperature defatted peanut protein powder was obtained from Qingdao Changshou Food Co. Ltd. (Qingdao, China). We purchased Neutrase, standard protein marker, and bovine serum albumin (BSA, Standard Grade) from Solarbio Company. (Beijing, China). Neutrase was used in its original form without further purification, and according to the manufacturer its activity was 1000 U/g. The 1, 8-anilinonaphthalenesulfonate (ANS) was obtained from Cayman Chemical Company (Ann Arbor, MI, USA).

2.2. Preparation of Modified Peanut Protein

The combined modification of peanut protein was performed in an order of high-pressure homogenization, enzymatic hydrolysis, and hydrothermal cooking. First, 7% (w/v) peanut protein aqueous solution was homogenized using a nano homogenize machine (AH-100D, ATS Engineering Limited Company, Shanghai, China) under the pressure of 800 bar for 20 min. Half of this homogenized sample (HPH) was freeze-dried and stored at 18 °C before analysis. The remaining HPH protein sample was treated by enzymatic modification at pH 7.0. According to the manufacturer instructions, Neutrase was added to start the hydrolysis at 50 °C for 50 min. Once the hydrolysis was complete, the hydrolysates were heated to 85 °C for 10 min to inactivate the enzyme. Half of this HPH-E sample was freeze-dried and stored at 18 °C before analysis. The final HPH-E-HTC sample was prepared by subjecting the remaining HPH-E sample into a hydrothermal cooking treatment using the HTC system (ESCS-M104, Xiaoledongchao Food Machine Company (Co. Ltd., Shanghai, China) at 130 °C for 120 s and then cooled to 25 °C. The obtained HPH-E-HTC sample was freeze-dried and preserved at 4 °C before analysis. The natural peanut protein was used as comparison.

2.3. Determination of Protein Solubility

A 1% (w/v) protein solution was prepared with distilled water and the pH was adjusted to 2.0–10.0 with 0.5 M HCl or 0.5 M NaOH. After stirring for 40 min, each sample was centrifuged at 10,000× g for 10 min at 20 °C. The protein content of supernatant was measured by Lowry's method [22] (Lowry, Rosebrough, Farr, and Randall 1951) and estimated from the abovine serum albumin (BSA) calibration curve (r^2 = 0.9994). The results were expressed as grams of soluble protein/100 g of protein used.

2.4. Determination of Main Protein Fractions

SDS-PAGE experiments were performed with a precast 5~13% gradient polyacrylamide gel according to the previously reported method by Guo et al. [23]. Briefly, 4 mg protein sample was dissolved in 1.0 mL of loading buffer and then heated in boiling water for 10 min. After centrifugation (4500× g, 10 min), 8 μL of supernatant was loaded into the gel. Electrophoresis was performed under 80 V followed by 110 V. Subsequently, the gel was stained with 0.1% (w/v) Coomassie brilliant blue (R-250) for 2 h and destained with 5% methanol/10% acetic acid/85% deionized water for 12 h. The images were finally taken using a Bio-Rad gel imaging system.

2.5. Determining the Stability of Protein Solution

The stability of protein solution was determined by a dispersion stability analyzer (multiple light scattering instrument) (Turbiscan Lab, Formulaction, Toulouse, France). The light source detector scans from the bottom of the sample bottle to the top of the bottle, and the sample scanning range is 2 to 45 mm. The final application software analyzes the graph to obtain the stability index (Turbiscan stability index, TSI). A lower TSI value indicates better stability of the system [24,25].

2.6. Measurement of Particle Size

The protein sample was dispersed in distilled water with a solution concentration of 1% (w/v) and stirred continuously for 2 h until testing. The particle size distribution of protein samples was measured with Mastersizer 3000 (Malvern instruments, Worcestershire, UK). The refractive index of the dispersed phase is 1.570 and that of the continuous phase (water) is 1.333.

2.7. Determination of Zeta-Potential

The titer zeta potential of the protein was measured using Nano-ZS and MPT-2 zeta potential and nanometer particle size distribution instrument (Malvern Instrument Ltd., Worcestershire, UK). Protein solution of 0.5% (w/v) was stirred at 100 rpm under ambient temperature for 1.5 h and the pH was adjusted with 0.25 M NaOH and 0.25 M HCl. Using the test zeta potential function, we set the measurement interval to pH 2.0–10.0. All determinations were made with at least two freshly prepared undiluted samples.

2.8. Determination of Surface Hydrophobicity

Surface hydrophobicity (H_0) was measured using 8-aniline-1-naphthalenesulfonic acid ammonium salt (ANS) as a fluorescent probe. The protein sample to be tested was dissolved in a phosphate buffer solution (0.01 M, pH 7.0) to prepare 1 mg/mL protein dispersion. The supernatant was obtained after centrifugation at $10,000 \times g$ for 20 min.

The protein content in the supernatant was determined by the Folin phenol protein quantification method. The above protein solutions were then diluted to different concentrations (0.019, 0.038, 0.075, 0.150 mg/mL) using the same buffer. Next, 4 mL of the diluted solution and 20 μL of ANS solution (8.0 mM) were added together and mixed well. The fluorescence intensity was quickly measured with a Hitachi-F-2500 fluorescence spectrophotometer (F-2500, Hitachi Co., Tokyo, Japan). The emission and excitation wavelengths were set to 470 nm and 390 nm, and H_0 is the slope of the fluorescence intensity versus protein concentration plot.

2.9. Determination of Secondary Structure of Protein

FTIR spectroscopic tests were conducted using TRUORS 27 spectrophotometer (TRUORS 27, Buruker, Germany). The spectra were obtained at 60 cm scan at 4 cm^{-1} resolution. The dried samples were tableted onto the ATR accessory of Spectro technology with KBr at 4 cm^{-1} resolution. Each sample was scanned 64 times and each sample was measured 3 times and averaged. The wavelength range in these tests was 500–4000 cm^{-1} scanned at 4 cm^{-1} resolution.

2.10. Observing Microstructure

The morphological properties of the protein were measured with a scanning electron microscope (SU8010, Hitachi Science Systems, Tokyo, Japan).

2.11. Statistical Analysis

The tests were carried out three times. We used v17.0 SPSS Statistics software (IBM Corp., Armonk, NY, USA) to perform one-way ANOVA on the dataset, and the results were expressed as mean $\pm$ standard deviation. The significant difference between the two mean values at 95% confidence level was determined by multiple-range test of Duncan ($p < 0.05$).

3. Results and Discussion

3.1. Protein Solubility

The solubility of untreated peanut protein (PP), high pressure homogenized peanut protein (HPH), HPH and enzyme treated peanut protein (HPH-E), and subjected to all these treatments peanut protein (HPH-E-HTC) in the range pH 2.0–10.0 are presented in Figure 1. The solubility of HPH, HPH-E, and HPH-E-HTC in the range pH 2.0–10.0 were significantly higher ($p < 0.05$) than PP (%). The protein solubility of all samples reached a minimum around pH 4.5; the protein solubility of HPH-E-HTC at pH 4.5 was still 37.3% compared to 4.0% of PP. As pH > 7.0, the solubility of all hydrolysates exceeded 50%, significantly higher than that of PP ($p < 0.05$). The HPH and HPH-E samples exhibited different protein dissolution with changes in pH. The solubility of the PP suspension increased significantly after high-pressure homogenization treatment. The improvement in solubility can be attributed to the fact that turbulence forces and cavitation involved in the high-pressure homogenization had broken down the molecular chain because of order-to disorder transitions, and opened the molecular structure. The mechanical energy supplied by the high-pressure homogenizer breaks down the structure of protein and expose more charged moieties so that the protein can bind better with water, which improves protein solubility [17]. Enzymatic hydrolysis significantly increased the solubility of peanut protein over the entire tested pH range. Solubility of protein can also be improved by enzymatic hydrolysis due to the breaking down of protein to more soluble peptides and increased exposure of ionizable amino and carboxyl groups. The unfolded protein can more readily interact with water through the exposed hydrophilic groups, and the looser tertiary conformation can promote hydration [12]. Therefore, combination of HPH, enzymatic hydrolysis, and HTC can be used as an effective modification technique for improving the solubility of peanut protein.

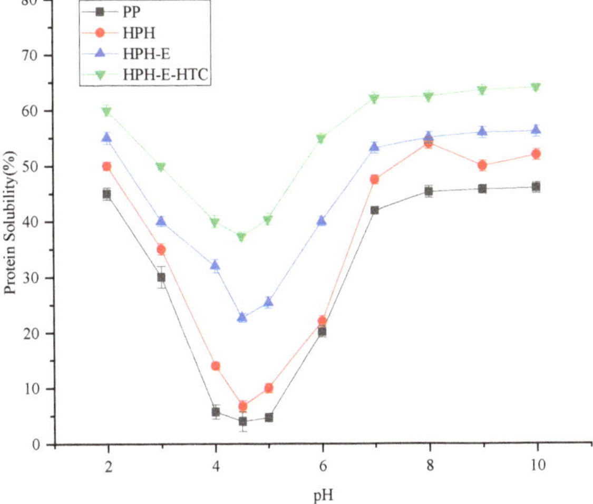

Figure 1. Solubility of peanut protein (PP), high pressure homogenized protein (HPH), high pressure homogenized and enzyme treated protein (HPH-E), and HPH-E-hydrothermally cooked (HPH-E-HTC) protein as a function of pH.

3.2. Morphology of Modified Peanut Proteins

Figure 2 shows SEM micrographs of peanut proteins obtained through different modification methods. PP appeared as a sheet-like surface structure with ridge-like protrusions. No large pore-like structure was observed. HPH-PP appeared dense and less porous, whereas the HPH-E-PP became loose and disordered with pores becoming smaller and more ordered after high-pressure homogenization and moderate enzymatic hydrolysis.

The surface area of these treated protein appeared large due to very high unevenness. Protein with this structure may disperse in water quite rapidly. The protein morphology was consistent with those of protein after jet cooking shearing and heat treatment in the study of Gong et al. (2016) [3].

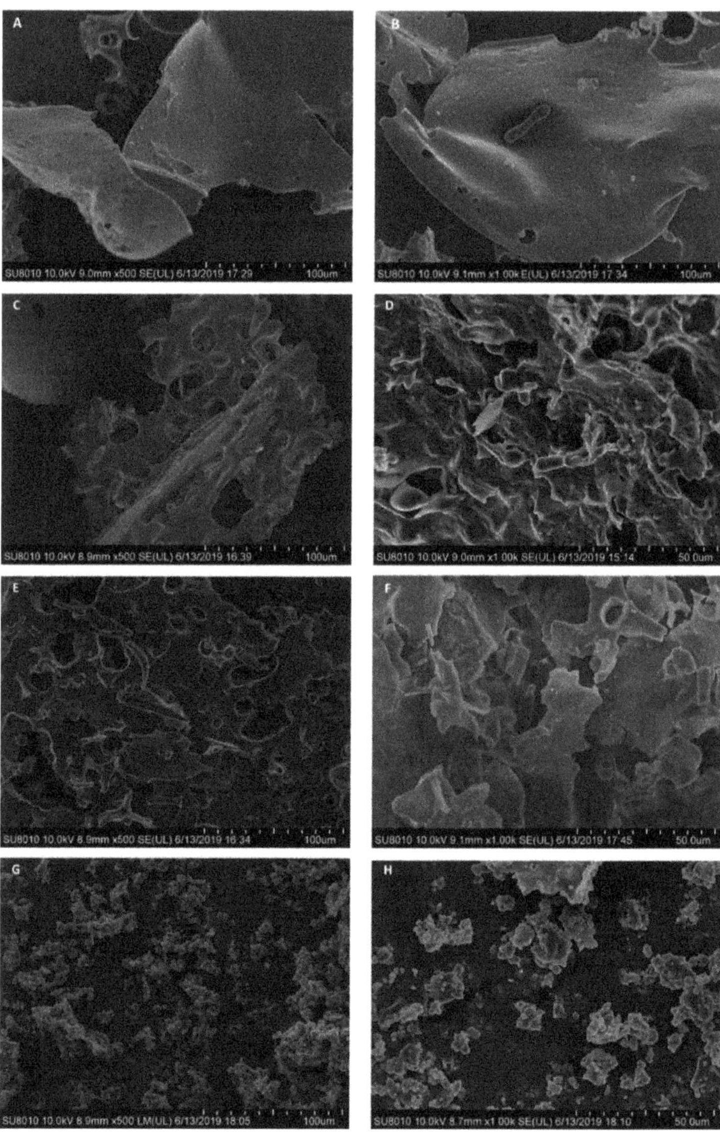

Figure 2. SEM micrographs of peanut proteins obtained through different modification methods: (**A**): peanut protein (PP) ×500; (**B**): peanut protein (PP) ×1 k; (**C**): high pressure homogenized protein (HPH) ×500; (**D**): high pressure homogenized protein (HPH) ×1 k; (**E**): high pressure homogenized and enzyme treated protein (HPH-E) ×500; (**F**): high pressure homogenized and enzyme treated protein (HPH-E) ×1 k; (**G**): HPH-E-hydrothermally cooked (HPH-E-HTC) protein ×500; (**H**): HPH-E-hydrothermally cooked (HPH-E-HTC) protein ×1 k.

3.3. SDS–PAGE Analysis

Figure 3 shows the SDS-PAGE profiles of PP, HPH, HPH-E, and HPH-E-HTC. A typical SDS-PAGE profile of peanut protein consists of 7 s and 11 s fractions [23] and their molecular weight ranges 48.0–245.0 kDa, 20.0–48.0 kDa and 20.0 kDa, representing bands I, II and III, respectively. The SDS-PAGE profile (reducing) observed in this study (Figure 2) is consistent with those previous reported. There are four bands of peanut globulin with molecular weights of 40.5 kDa, 37.5 kDa, 35.5 kDa, and 23.5 kDa. Conarachin I have three bands with molecular weight 18 kDa, 17 kDa, or 15.5 kDa. Similarly, conarachin II has one band with a molecular weight of 61 kDa [26]. All the proteins were electrophoresed with equal amounts of soluble proteins (15 ug), and the result showed that there was no alteration in molecular weight distribution of protein fractions treated with high-pressure homogenization (HPH) showing that HPH treatment did not affect the protein composition. The HPH-E and HPH-E-HTC treated peanut proteins showed significant alteration in all the bands; band I1 with molecular weight 64 kDa and band II with molecular weight 48 kDa completely disappeared, reflecting the complete degradation of these protein fractions. Band II2, having molecular weight about 24 kDa, was significantly altered in HPH-E and HPH-E-HTC treated peanut protein. In contrast, band III3, having molecular weight 15 kDa, 17 kDa, and 19 kDa, increased after HPH-E treatment and then slightly decreased after HPH-E-HTC treatment. SDS-PAGE images also show a light band with molecular weight above 100 kDa in the HPH treated sample, which indicates HPH induced denaturation and aggregation, which is consistent with previous study [27].

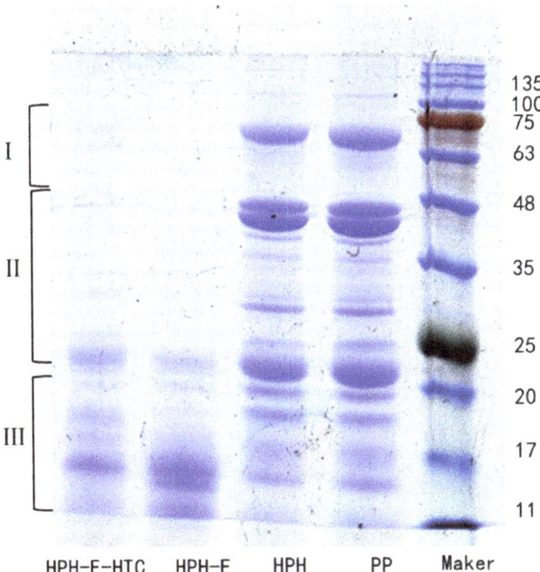

Figure 3. Sodium dodecyl sulphate–polyacrylamide gel electrophoresis profiles of peanut protein (PP), high pressure homogenized protein (HPH), high pressure homogenized and enzyme treated protein (HPH-E), and HPH-E-hydrothermally cooked (HPH-E-HTC) protein.

Further treatment of HPH treated protein with enzyme did not show aggregated protein bands, and bands II and II also disappeared due to enzymatic digestion. Interestingly, enzyme treatment of HPH treated peanut increased the intensity of band III, suggesting digestion (breakdown) of high molecular weight fractions into smaller ones followed by their aggregation [21]. Further treatment of HPH-E with heat treatment, i.e., HPH-E-HTC, further reduced the intensity of low molecular weight protein fractions slightly.

3.4. Physical Stability in Aqueous Solution Analyzed by Turbiscan

Figure 4 shows the backscattering (BS) profiles of peanut proteins subjected to HPH, HPH-E, and HPH-E-HTC as a function of tube height with storage time. As can be observed, these treatments had a significant effect on the stability of the peanut protein.

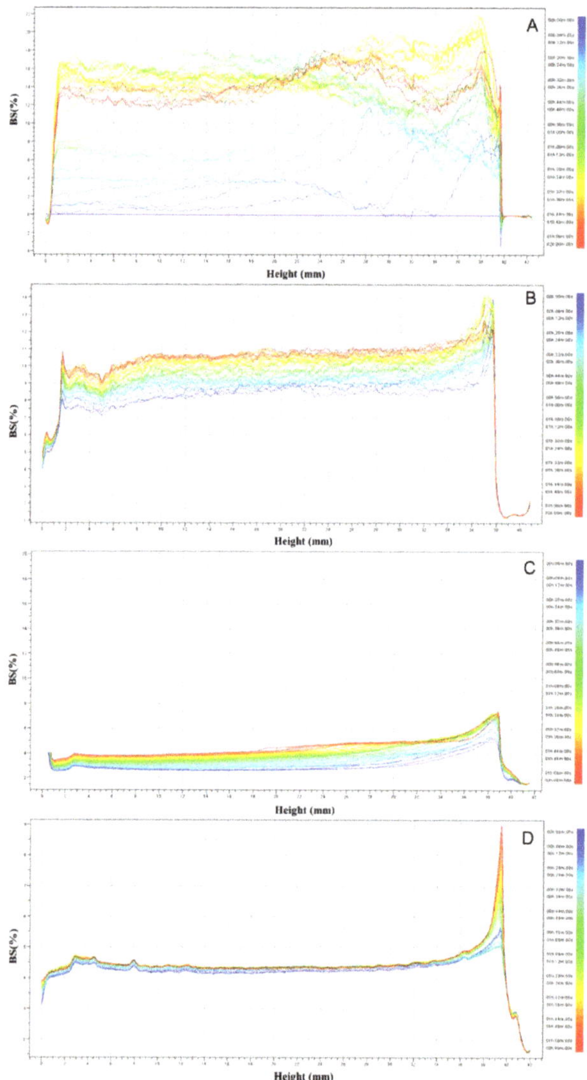

Figure 4. Backscattering (BS) profiles of peanut proteins obtained through different modification methods as a function of the tube height with storage time: (**A**): peanut protein (PP); (**B**): high pressure homogenized protein (HPH); (**C**): high pressure homogenized and enzyme treated protein (HPH-E); (**D**): HPH-E-hydrothermally cooked (HPH-E-HTC) protein.

The change of back scattering (ΔBS) with time, as shown in Figure 4A, indicated that there was sedimentation of protein at the bottom indicated by the ΔBS value 4–16% in the 4–34 mm height range. The PP sample had relatively large flocculation and coalescence in the middle and creaming on the top. Figure 4B represents the data for the sample subjected to high pressure homogenization. The ΔBS data fluctuated from 10% to 15% in the 4.00–34.00 mm height range of the HPH treated sample. The ΔBS value of 19% on the top part (36.00–40.00 mm) was the upper peak. The smaller ΔBS value indicates better stability of protein solution, so we can know that the stability of the HPH sample is not good, but the HPH produced more stable solution than the untreated PP. As seen from Figure 4C, at the top portion of the sample (height = 36.00–42.00 mm), ΔBS gradually increased to 7.50%, indicating a slight creaming of protein. There was a convex peak at the top of 36.00–41.00 mm. There was some sedimentation at the bottom and flocculation at top of the HPH-E sample. However, overall, HPH-E was more uniform and more stable than HPH. In Figure 4D, the HPH-E-HTC sample solution was homogeneous, and the fluctuation of ΔBS bottom and middle portion of sample was small, indicating that the precipitation and flocculation of protein particles were not significant. The peak of ΔBS gradually increased to 4.7% in the height range of 0.0 to 0.5 mm. There is a small raised peak at 38~40 mm with ΔBS value of 9.0%, and there was a very small fraction of protein that floated. It is apparent from the figure that the stability of the HPH-E-HTC treated sample was significantly better than that of the untreated sample PP.

Figure 5 shows the Turbiscan stability index (TSI of PP, HPH, HPH-E, and HPH-E-HTC) protein as a function of time. Larger deviation from the starting line (TSI = 0) and higher slope indicate greater instability and vice versa.

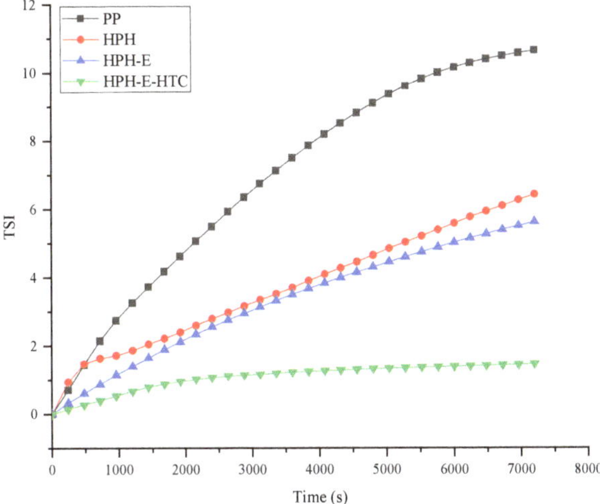

Figure 5. The instability index of peanut protein (PP), high pressure homogenized protein (HPH), high pressure homogenized and enzyme treated protein (HPH-E), and HPH-E-hydrothermally cooked (HPH-E-HTC) protein as a function of time.

The slope of HPH-E-HTC was the smallest and the stability of the solution system was the best, whereas the slope of the PP was the steepest indicating that the stability of its solution was the poorest.

These TSI data shows that the order of stability of these samples was: HPH-E-HTC > HPH > HPH-E > PP, which corroborates their solubility data (Section 3.1) backscattering (BS) profiles. These data indicate that high pressure homogenization plays an important role in stabilizing protein solutions. This is because high pressure homogenization can provide a strong mechanical force that improves the solubility of protein in aqueous medium and make solutions homogeneous. Enzymatic hydrolysis and HTC unfold the protein structure and expose a higher number of hydrophilic groups and thus improves interaction with water. This result is consistent with the surface hydrophobicity data discussed in Section 3.7.

3.5. Zeta-Potential

The relationship between the zeta (ζ) potential of various protein samples as a function of pH is shown in Figure 6. The zeta-potential of PP changed from +21.0 mV to −18.8 mV in the range from pH 2.0 to 10.0. The isoelectric point was observed at pH 3.8. The relationship between zeta-potential and pH are significantly affected by different treatments. There was no significant change in zeta-potential after HPH treatment. The HPH-E and HPH-E-HTC treated proteins were similar with PP at pH > 5, but at pH < 5, the zeta-potential of HPH-E and HPH-E-HTC protein decreased significantly and was negatively charged, and the isoelectric point showed a tendency to migrate to a lower pH. PP has no charge under low pH conditions (near pH = 3.8), so aggregation and precipitation occur between proteins, and water solubility and stability are poor.

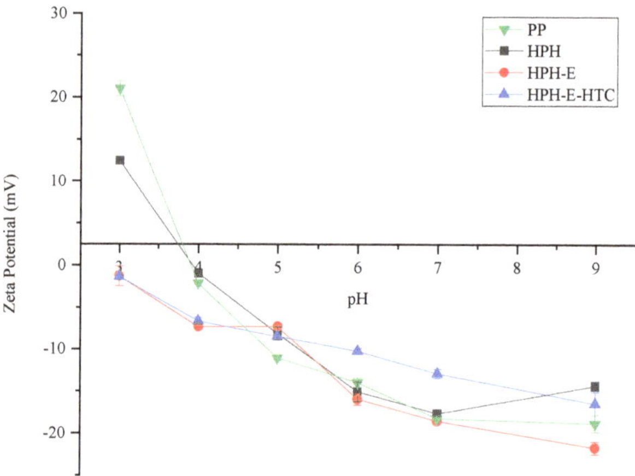

Figure 6. Zeta potential (ζ) profiles of peanut protein (PP), high pressure homogenized protein (HPH), high pressure homogenized and enzyme treated protein (HPH-E), and HPH-E-hydrothermally cooked (HPH-E-HTC) protein as a function of pH at 25 °C.

The HPH-E and HPH-E-HTC treated proteins are negatively charged under these conditions. The most noteworthy feature of these profiles is that isoelectric point was observed only in native PP and high pressure homogenized PP (HPH), and the HPH-E-HTC and HPH-E were still negatively charged at pH 3.0. This can be attributed to the increased exposure of hydrophilic groups of protein due to high temperature and high pressure [18]. The presence or dominance of charged or polar amino acids on the surface of the aggregate prevents their precipitation, and thus the solubility and stability of protein solubility is improved. This is consistent with previous research on soy protein [25].

3.6. Particle Size Distribution

Figure 7 shows the particle size distributions of PP, HPH, HPH-E, and HPH-E-HTC, and their volume-weighted mean diameters (D(4,3)) are listed in Table 1.

The D(4,3) value of the untreated peanut protein powder was 74.82 µm. The protein powder prepared by HTC showed significantly ($p < 0.05$) lower D (4,3)value as compared to the control samples.

The extent of decrease in particle size in HTC depended on the temperature and protease pretreatment. Proteolytic pretreatment prior to HTC significantly ($p < 0.05$) reduced the average droplet diameter of the dispersion. The combined modification of peanut protein led to a decrease in the average particle size. Smaller protein particles are easier to disperse uniformly in aqueous solution and less precipitation will happen, resulting in improved solubility.

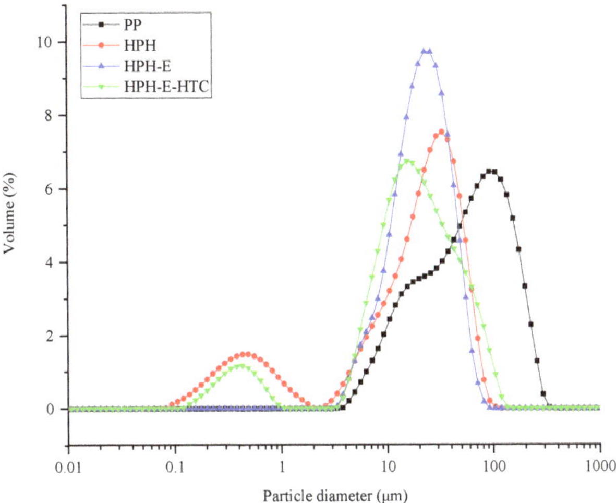

Figure 7. Particle size (D (4,3)) distributions of peanut protein (PP), high pressure homogenized protein (HPH), high pressure homogenized and enzyme treated protein (HPH-E), and HPH-E-hydrothermally cooked (HPH-E-HTC) protein.

Table 1. Particle-size distributions of peanut protein (PP), high pressure homogenized protein (HPH), high pressure homogenized and enzyme (E) treated protein (HPH-E), and HPH-E-hydrothermally cooked (HPH-E-HTC) protein.

Samples	D (4,3) (μm)
PP	74.82 ± 0.57 c
HPH	24.18 ± 0.32 a
HPH-E	23.65 ± 0.39 ab
HPH-E-HTC	21.74 ± 0.17 b

Data are expressed as means ± SD. Superscript characters (a–c) indicate significant ($p < 0.05$) differences in the same column.

This may be because the treated peanut proteins show more disordered conformation, leading to increased exposure of hydrophobic domains evidenced by surface hydrophobicity data (Section 3.7). Unfolded proteins are more easily dispersed in water than naturally folded molecules and form a stable dispersion.

When subjected to high-pressure homogenization and moderate enzymatic hydrolysis, the particle size of peanut protein aggregates decreased to a similar extent (24.2 μm and 23.6 μm, respectively). This indicated that high pressure homogenization and moderate enzymatic hydrolysis had similar effects on protein particle size. After HTC treatment, the particle size was reduced to 21.7 μm. This may be caused by high shearing and impact by the HTC process, which destroys the ordered structure and reduces the particle size. After high-pressure homogenization-enzymatic followed by hydrolysis-high temperature cooking, the particles had narrow (more uniform) size distribution.

3.7. Surface Hydrophobicity

The change in hydrophobicity (H_0) of the protein surface is presented in Table 2. The H_0 of natural untreated peanut protein powder was 644.4. After high-pressure homogenization and combined enzymatic hydrolysis and HTC, it decreased to 297.5. This indicates that high pressure, enzymatic hydrolysis, and HTC can hinder the hydrophobic groups of proteins from being exposed to the surface and reduce surface hydrophobicity, increase hydrophilic groups, and better bind proteins to water.

Table 2. Surface hydrophobicity (H_0) of peanut protein (PP), high pressure homogenized protein (HPH), high pressure homogenized and enzyme (E) treated protein (HPH-E), and HPH-E-hydrothermally cooked (HPH-E-HTC) protein.

Samples	PP	HPH	HPH-E	HPH-E-HTC
H_0	644.41 ± 4.68 a	582.67 ± 2.11 b	346.20 ± 1.15 c	297.45 ± 0.75 d

Data are expressed as means ± SD. Superscript characters (a–d) indicate significant ($p < 0.05$) differences in the same row.

Similar findings were reported by Xia et al. [21] and Feng et al. [28] in the case of heat-stabilized rice bran prepared using amylase pretreatment combined with hydrothermal cooking amylase pretreatment. The peanut protein is expected to undergo unfolding during these treatments then expose hydrophilic groups and increase interaction with water.

3.8. Change in Peanut Proteins' Secondary Structure

As can be seen from Table 3, the HP-H-E-HTC had the highest β-sheet content (45.9%) among all the modified samples. However, the α-helics and random coil structures showed gradual decrease, and the former decreased the most. The β-turn and β-sheet structures of native PP increased from 14.36% and 25.18% to 25.34% and 45.93% after HP-H-E-HTC treatment, indicating that the α-helical and random coil structures changed to the β-sheet and β-turn structure when subjected to high pressure, high temperature treatment and enzymatic hydrolysis. HPH and HPH-E treated samples also had lower α-helix and random structures, but there was no definite trend in in β-sheet and β-turn structures. The fact that these treatments decreased the α-helical and random coil structures indicated that the protein structure was first unfolded and then reaggregated to improve protein solubility. These results are supported by the research of Blanc et al. [29] and Rao et al. [30], both of which reported a significant increase of β-sheet and decrease of α-helix in the secondary structure of heat-treated peanut protein after boiling or roasting.

Table 3. The secondary structure of peanut protein (PP), high pressure homogenized protein (HPH), high pressure homogenized and enzyme (E) treated protein (HPH-E), and HPH-E-hydrothermally cooked (HPH-E-HTC) protein.

Samples	PP	HPH	HPH-E	HPH-E-HTC
Helix	43.41 ± 0.86 a	36.02 ± 1.06 b	21.08 ± 0.85 c	19.72 ± 0.73 d
Beta-turn	14.36 ± 0.78 a	10.92 ± 0.97 b	19.96 ± 0.80 c	25.34 ± 0.53 d
Beta-sheet	25.18 ± 0.46 a	34.82 ± 0.52 b	30.05 ± 0.62 c	45.93 ± 0.47 d
Random	17.05 ± 0.67 a	18.24 ± 0.69 b	28.90 ± 0.76 c	9.00 ± 0.89 d

Data are expressed as means ± SD. Superscript characters (a–d) indicate significant ($p < 0.05$) differences in the same row.

4. Conclusions

This study shows that the novel combined high-pressure homogenization, enzymatic hydrolysis, high temperature cooking (HPH-E-HTC) treatment significantly improved the solubility and stability of peanut protein solution. Smaller pieces of peanut protein prepared by HPH-E-HTC were observed due to high pressure, enzymatic, and shearing force treatment. The sample prepared by HPH-E-HTC showed lower surface hydrophobicity and particle size and the zeta-potential with negative charge than PP and HPH-E in low pH conditions (near pH = 3.8), indicating improved solubility and stability. These results indicated that HPH-E-HTC can be used to improve the quality functional properties of peanut protein obtained from low-temperature defatted peanut meal. Furthermore, this kind of modified peanut protein has a great potential as a food additive for acidic beverages.

Author Contributions: J.L. experimental operation and experimental design, data analysis, writing—original draft preparation; A.S. and Q.W. conceptualization, project administration, funding acquisition; H.L. and H.H. investigation, software; B.A., B.J. and M.P. writing—review and editing. All authors have read and agreed to the published version of the manuscript.

Funding: This study was supported by the National Key Research and Development Program (2021YFD2100402), Central Public-interest Scientific Institution Basal Research Fund (Y2022QC11), National Natural Science Foundation of China (32172149, U21A20270), Agricultural Science and Technology Innovation Project (CAAS-ASTIP-201X-IAPPST).

Institutional Review Board Statement: Not applicable.

Informed Consent Statement: Not applicable.

Data Availability Statement: The datasets generated for this study are available on request to the corresponding author.

Acknowledgments: Authors present greetings and thanks to the people who provided their comments and suggestions for the work: Zelong Liu (COFCO nutrition and Health Research Institute), Yumeng Du (COFCO nutrition and Health Research Institute), and Faisal Shah (Institute of Food Science and Technology, Chinese Academy of Agriculture Sciences).

Conflicts of Interest: The authors declare no conflict of interest.

References

1. Jiao, B.; Shi, A.; Wang, Q.; Binks, B.P. High-internal-phase pickering emulsions stabilized solely by peanut-protein-isolate microgel particles with multiple potential applications. *Angew. Chem. Int. Ed.* **2018**, *130*, 9418–9422. [CrossRef]
2. Faisal, S.; Zhang, J.; Meng, S.; Shi, A.; Liu, L.; Wang, Q.; Maleki, S.; Adhikari, B. Effect of high-moisture extrusion and addition of transglutaminase on major peanut allergens content extracted by three step sequential method. *Food Chem.* **2022**, *385*, 132569. [CrossRef] [PubMed]
3. Gong, K.J.; Shi, A.M.; Liu, H.Z.; Liu, L.; Hu, H.; Adhikari, B.; Wang, Q. Emulsifying properties and structure changes of spray and freeze-dried peanut protein isolate. *J. Food Eng.* **2016**, *170*, 33–40. [CrossRef]
4. Zhang, J.; Liu, L.; Zhu, S.; Wang, Q. Texturisation behaviour of peanut–soy bean/wheat protein mixtures during high moisture extrusion cooking. *Int. J. Food Sci. Technol.* **2018**, *53*, 2535–2541. [CrossRef]
5. Hu, Y.; Sun-Waterhouse, D.; Liu, L.; He, W.; Zhao, M.; Su, G. Modification of peanut protein isolate in glucose-containing solutions during simulated industrial thermal processes and gastric-duodenal sequential digestion. *Food Chem.* **2019**, *295*, 120–128. [CrossRef] [PubMed]
6. Hertzler, S.R.; Lieblein-Boff, J.C.; Weiler, M.; Allgeier, C. Plant proteins: Assessing their nutritional quality and effects on health and physical function. *Nutrients* **2020**, *12*, 3704. [CrossRef] [PubMed]
7. Etemadian, Y.; Ghaemi, V.; Shaviklo, A.R.; Pourashouri, P.; Mahoonak, A.R.S.; Rafipour, F. Development of animal/plant-based protein hydrolysate and its application in food, feed and nutraceutical industries: State of the art. *J. Clean. Prod.* **2021**, *278*, 123219. [CrossRef]
8. Riveros, C.G.; Martin, M.P.; Aguirre, A.; Grosso, N.R. Film preparation with high protein defatted peanut flour: Characterization and potential use as food packaging. *Int. J. Food Sci. Technol.* **2018**, *53*, 969–975. [CrossRef]
9. Miglietta, N.; Battisti, E.; Campanella, F. Value maximization and open innovation in food and beverage industry: Evidence from US market. *Br. Food J.* **2017**, *119*, 2477–2492. [CrossRef]
10. Oltman, A.E.; Lopetcharat, K.; Bastian, E.; Drake, M.A. Identifying key attributes for protein beverages. *J. Food Sci.* **2015**, *80*, S1383–S1390. [CrossRef]
11. Lin, Y.; Mouratidou, T.; Vereecken, C.; Kersting, M.; Bolca, S.; de Moraes, A.C.F.; Cuenca-García, M.; Moreno, L.A.; González-Gross, M.; Valtueña, J.; et al. Dietary animal and plant protein intakes and their associations with obesity and cardio-metabolic indicators in European adolescents: The HELENA cross-sectional study. *Nutr. J.* **2015**, *14*, 10. [CrossRef] [PubMed]
12. Shori, A.B. Influence of food matrix on the viability of probiotic bacteria: A review based on dairy and non-dairy beverages. *Food Biosci.* **2016**, *13*, 1–8. [CrossRef]
13. Zhao, G.; Liu, Y.; Zhao, M.; Ren, J.; Yang, B. Enzymatic hydrolysis and their effects on conformational and functional properties of peanut protein isolate. *Food Chem.* **2011**, *127*, 1438–1443. [CrossRef]
14. Tan, M.; Nawaz, M.A.; Buckow, R. Functional and food application of plant proteins—A review. *Food Rev. Int.* **2021**, *38*, 1–29. [CrossRef]
15. Li, J.M.; Nie, S.P. The functional and nutritional aspects of hydrocolloids in foods. *Food Hydrocoll.* **2016**, *53*, 46–61. [CrossRef]
16. Qin, Y. Bioactive seaweeds for food applications. In *Seaweed Hydrocolloids as Thickening, Gelling, and Emulsifying Agents in Functional Food Products*; Qin, Y., Ed.; Academic Press Inc.: New York, NY, USA, 2018; pp. 135–152.

17. Jiao, B.; Shi, A.M.; Liu, H.Z.; Sheng, X.J.; Liu, L.; Adhikari, B.; Wang, Q. Effect of electrostatically charged and neutral polysaccharides on the rheological characteristics of peanut protein isolate after high-pressure homogenization. *Food Hydrocoll.* **2018**, *77*, 329–335. [CrossRef]
18. Girgih, A.T.; Chao, D.; Lin, L.; He, R.; Jung, S.; Aluko, R.E. Enzymatic protein hydrolysates from high pressure-pretreated isolated pea proteins have better antioxidant properties than similar hydrolysates produced from heat pretreatment. *Food Chem.* **2015**, *188*, 510–516. [CrossRef] [PubMed]
19. Wang, H.; Wang, T.; Johnson, L.A. Effect of alkali on the refunctionalization of soy protein by hydrothermal cooking. *J. Am. Oil Chem. Soc.* **2005**, *82*, 451–456. [CrossRef]
20. Wang, H.; Wang, T.; Johnson, L.A. Mechanism for refunctionalizing heat-denatured soy protein by alkaline hydrothermal cooking. *J. Am. Oil Chem. Soc.* **2006**, *83*, 39–45. [CrossRef]
21. Xia, N.; Wang, J.; Yang, X.; Yin, S.; Qi, J.; Hu, L. Preparation and characterization of protein from heat-stabilized rice bran using hydrothermal cooking combined with amylase pretreatment. *J. Food Eng.* **2012**, *110*, 95–101. [CrossRef]
22. Lowry, O.H.; Rosebrough, N.J.; Farr, A.L.; Randall, R.J. Protein measurement with the folin phenol reagent. *J. Biol. Chem.* **1951**, *193*, 265–275. [CrossRef]
23. Guo, Y.; Hu, H.; Wang, Q.; Liu, H. A novel process for peanut tofu gel: Its texture, microstructure and protein behavioral changes affected by processing conditions. *LWT* **2018**, *96*, 140–146. [CrossRef]
24. Wiśniewska, M.; Urban, T.; Nosal, A. Comparison of stability properties of poly (acrylic acid) adsorbed on the surface of silica, alumina and mixed silica-alumina nanoparticles—Application of turbidimetry method. *Open Chem.* **2014**, *12*, 476–479. [CrossRef]
25. Lam, M.; Shen, R.; Paulsen, P.; Corredig, M. Pectin stabilization of soy protein isolates at low pH. *Food Res. Int.* **2007**, *40*, 105–110. [CrossRef]
26. Bianchi-Hall, C.M.; Keys, R.D.; Stalker, H.T.; Murphy, J.P. Diversity of seed storage protein patterns in wild peanut (*Arachis, Fabaceae*) species. *Plant Syst. Evol.* **1993**, *186*, 1–15. [CrossRef]
27. He, X.H.; Liu, H.Z.; Liu, L.; Zhao, G.L.; Wang, Q.; Chen, Q.L. Effects of high pressure on the physicochemical and functional properties of peanut protein isolates. *Food Hydrocoll.* **2014**, *36*, 123–129. [CrossRef]
28. Feng, X.L.; Liu, H.Z.; Shi, A.M.; Liu, L.; Wang, Q.; Adhikari, B. Effects of transglutaminase catalyzed crosslinking on physicochemical characteristics of arachin and conarachin-rich peanut protein fractions. *Food Res. Int.* **2014**, *62*, 84–90. [CrossRef]
29. Blanc, F.; Vissers, Y.M.; Adel-Patient, K.; Rigby, N.M.; Mackie, A.R.; Gunning, A.P.; Mills, E.C. Boiling peanut Ara h 1 results in the formation of aggregates with reduced allergenicity. *Mol. Nutr. Food Res.* **2011**, *55*, 1887–1894. [CrossRef]
30. Rao, H.; Tian, Y.; Tao, S.; Tang, J.; Li, X.; Xue, W.T. Key factors affecting the immunoreactivity of roasted and boiled peanuts: Temperature and water. *LWT-Food Sci. Technol.* **2016**, *72*, 492–500. [CrossRef]

 *foods*

Article

Quality Formation of Adzuki Bean Baked: From Acrylamide to Volatiles under Microwave Heating and Drum Roasting

Xinmiao Yao [1,2], Xianzhe Zheng [3], Rui Zhao [1], Zhebin Li [1], Huifang Shen [1], Tie Li [4], Zhiyong Gu [5], Ye Zhou [1], Na Xu [1], Aimin Shi [2], Qiang Wang [2] and Shuwen Lu [1,*]

[1] Food Processing Research Institute, Heilongjiang Academy of Agricultural Sciences, Harbin 150086, China; cocoyococo@163.com (X.Y.); lilyamongthorns@163.com (R.Z.); lizhebin2010@163.com (Z.L.); shenhuifang_1987@126.com (H.S.); zhouye614@163.com (Y.Z.); doctorserena@163.com (N.X.)

[2] Institute of Food Science and Technology, Chinese Academy of Agricultural Sciences, Beijing 100193, China; shiaimin@caas.cn (A.S.); wangqiang06@caas.cn (Q.W.)

[3] China School of Engineering, Northeast Agricultural University, Harbin 150030, China; zhengxz@163.com

[4] Crop Resources Institute, Heilongjiang Academy of Agricultural Sciences, Harbin 150086, China; haas2005@163.com

[5] Gansu United Testing Standards Technical Service Co., Ltd., Lanzhou 730030, China; my_road@163.com

* Correspondence: sunny@haas.cn

Abstract: Baked adzuki beans are rich in tantalizing odor and nutritional components, such as protein, dietary fiber, vitamin B, and minerals. To analyze the final quality of baked beans, the acrylamide and volatile formation of adzuki beans were investigated under the conditions of microwave baking and drum roasting. The results indicate that the acrylamide formation in baked adzuki beans obeys the exponential growth function during the baking process, where a rapid increase in acrylamide content occurs at a critical temperature and low moisture content. The critical temperature that leads to a sudden increase in acrylamide content is 116.5 °C for the moisture content of 5.6% (*w.b.*) in microwave baking and 91.6 °C for the moisture content of 6.1% (*w.b.*) in drum roasting. The microwave-baked adzuki beans had a higher formation of the kinetics of acrylamide than that of drum-roasted beans due to the microwave volumetric heating mode. The acrylamide content in baked adzuki beans had a significant correlation with their color due to the Maillard reaction. A color difference of 11.1 and 3.6 may be introduced to evaluate the starting point of the increase in acrylamide content under microwave baking and drum roasting, respectively. Heating processes, including microwave baking and drum roasting, for adzuki beans generate characteristic volatile compounds such as furan, pyrazine, ketone, alcohols, aldehydes, esters, pyrroles, sulfocompound, phenols, and pyridine. Regarding flavor formation, beans baked via drum roasting showed better flavor quality than microwave-baked beans.

Keywords: adzuki bean; acrylamide; volatile; microwave baking; drum roasting

Citation: Yao, X.; Zheng, X.; Zhao, R.; Li, Z.; Shen, H.; Li, T.; Gu, Z.; Zhou, Y.; Xu, N.; Shi, A.; et al. Quality Formation of Adzuki Bean Baked: From Acrylamide to Volatiles under Microwave Heating and Drum Roasting. *Foods* **2021**, *10*, 2762. https://doi.org/10.3390/foods10112762

Academic Editor: Alain Le-Bail

Received: 15 September 2021
Accepted: 4 November 2021
Published: 10 November 2021

Publisher's Note: MDPI stays neutral with regard to jurisdictional claims in published maps and institutional affiliations.

Copyright: © 2021 by the authors. Licensee MDPI, Basel, Switzerland. This article is an open access article distributed under the terms and conditions of the Creative Commons Attribution (CC BY) license (https:// creativecommons.org/licenses/by/ 4.0/).

1. Introduction

The adzuki bean, a well-known agricultural product, has a high protein content, low fat, and diverse ingredients. The baking processes for adzuki beans may produce products with a unique flavor–texture combination. The baked powder of adzuki beans is very popular in Asian countries due to its pleasant odor, rich nutritional qualities, and ease of serving. The baked powder from whole beans contains functional ingredients in the bean's coating, in accordance with the health concept of whole-cereal nutrition. Rotated drum roasting is a conventional method of baking treatment of adzuki beans, with a high processing capacity and the use of technology. As a new baking technology, microwave heating improves the functional properties and flavor qualities of the processed material [1] and may be used to bake adzuki beans. The principal mechanisms of microwave heating are inherent ionic conduction and dipolar relaxation of polar molecules at the rapid frequency of 2.45 GHz to generate volumetric heat inside the material within a very short time [2]. In the microwave-baking process, microwave heating causes a rapid increase in temperature and rapidly dehydrates the adzuki beans to achieve a brown color and tantalizing aroma in the final product. Compared to convection heating mechanisms, such as drum roasting, for adzuki beans, the radiant heating mechanism in microwave baking exhibits significant advantages in terms of high efficiency, easy control, and easy-to-attain equipment. However, there have been few comparisons of the two baking methods to evaluate the quality of the baked adzuki beans in terms of safety and flavor generation.

In addition, the safety of baked food has been a public concern. Heating processing may induce the production of harmful substances. Acrylamide formation in food material led to particularly high starch content when sugars and an amino acid (asparagine) reacted in starchy foods undergoing a high-temperature treatment of over 120 °C [3]. Acrylamide was detected in plant foods, such as potato products, grain products, cocoa, or coffee beans during high-temperature cooking processes, such as frying, roasting, and baking [4,5]. In laboratory studies, acrylamide, a potential carcinogen, caused cancer in animals, but at much higher acrylamide levels than those in foods [6]. The formation of acrylamide in baked food may weaken its antioxidant capacity to reduce the functional value [7]. The FDA (Food and Drug Administration) initiated acrylamide-related research on toxicology, the development of analytical methodologies, food surveys, exposure assessments, and formation migration. The formation and content of acrylamide in plant-origin material containing starch that will be heated are given more attention due to their importance for the safety and quality of food [8]. Due to the relevance of food safety, the prediction and evaluation of acrylamide content in baked food play a key role in parameter optimization and technology selection during baking processes. From the perspective of food engineering, Özge and Gökmen developed a viable approach with which to evaluate the risk factors related to acrylamide formation in cookies based on the function of acrylamide content in baking cookies with regard to temperature for a risk threshold value of 200 ppb of acrylamide [9]. Radio-frequency (RF) heating following convection heating contributes to the occurrence of excessive browning in the internal portion of baked thin biscuits with less acrylamide formation [10]. Besides temperature and components [11], the heating method, such as microwave baking or drum roasting, also dominates the formation of acrylamide in the baked material [12]. The acrylamide content of cookies processed by vacuum baking was 30% lower than in those baked the conventional way due to the low temperature in the vacuum conditions [13]. However, not enough research has evaluated acrylamide content in adzuki beans baked using microwave and roasting methods. Color may be introduced as an intuitive index to evaluate the quality of the baked product's appearance. When baking cookies, the acrylamide content has a high correlation with the browning index, which follows the first-order reaction kinetics equation [14] and elucidates the effect of baking temperature on surface color [15]. In the Maillard reaction pathways of French fries, the glucose and fructose content directly contribute to the acrylamide formation via sugar–asparagine glycoconjugates [14]. Both the color parameter and the moisture content of the French fries have a significant correlation with the acrylamide content [5]. Therefore, the acrylamide content in baked products may be evaluated based on changes in its color.

Color, as an exterior quality index, may be used to evaluate the appearance quality of baked beans, and flavor, as an interior quality index, may characterize the smell quality of baked beans, relating to the volatile component contents. Volatile components from baked materials, for example, furan, lead to a pleasant baked aroma [16], which plays a key role in quality evaluations of bean food products, related to the optimization of baking technology parameters. The formation of and changes in the volatile components of beans are complex in baking processes [17]. Scant research has analyzed the changes in the volatile components of baked beans.

To our knowledge, little information has been published on the kinetics of acrylamide and volatile components of the formation of adzuki beans under drum-roasting and microwave-baking modes. This defect limits the quality of baked adzuki beans, and the acrylamide content cannot be controlled. Therefore, the research objectives were developed as follows.

(1) To elucidate the acrylamide formation kinetics of adzuki beans baked using microwave baking and drum roasting.
(2) To investigate the changes in the volatiles and color of adzuki beans after microwave baking and drum roasting.
(3) To analyze the relevance of acrylamide formation and color changes in adzuki beans during roasting.

2. Materials and Methods

2.1. Raw Material

Adzuki beans with a variety of Longyin 09-05 were collected from the planting base at Heilongjiang Academy of Agricultural Sciences (Harbin, China).

2.2. Microwave-Baking Processing

In the microwave-baking experiment for adzuki beans, the research focused on the formation of the kinetics of acrylamide as a function of temperature and moisture, other than the effects and optimization of microwave-baking technology. Therefore, we selected fixed-input power and beans' mass with only one level. One hundred grams of adzuki beans were selected as experimental material and immersed in purified water for 4 h at a room temperature of 20–22 °C to achieve a moisture content of 55.04 $\pm$ 0.01% (*w.b.*). The microwave-baking experiments were conducted in a microwave workstation (MWS, FISO Technologies Inc., Quebec City, QC, Canada) and connected to a temperature sensor, as shown in Figure 1a. The immersed adzuki beans of 100 g with an initial moisture content of 55.04 $\pm$ 0.01% (*w.b.*) were placed on a glass tray inside the microwave cavity under 800 W input power. In the microwave-baking experiments, at every 1 min interval, the temperature of the experimental material was recorded, and the material was taken out to measure moisture and acrylamide content, until the baked beans achieved a mass corresponding to a moisture content below 3% (*w.b.*).

To measure the beans' temperature during microwave heating, a fiber-optic sensor (FOT-L-SD, FISO Co., Quebec City, QC, Canada) was plugged into a single bean kernel placed in the center of the glass tray to measure the temperature. There may exist a deviation in the temperature in a single bean located in the center of a microwave workstation and the average temperature of the total beans baked. In further research, a feasible method will be developed to record the online temperature of material processed by microwave.

2.3. Drum-Roasting Experiments

Comparing the kinetics of acrylamide formation in adzuki beans to that under microwave baking, the input power of the roasting drum and beans' mass was fixed at only one level. One kilogram of adzuki beans (moisture content of 12.03 $\pm$ 0.01% by weight) was roasted using an apparatus (BD-CR-D1001BB, Bideli, Guangzhou, China), as shown in Figure 1b. Experiments were carried out at atmospheric pressure under an input power of 1500 W. The experimental material's temperature was recorded at every 1 min interval using an embodied infrared radiation thermometer in the apparatus, then taken out to measure moisture, as in Figure 1b. The roasting drum was used for the baking experiment until the baked beans achieved a mass corresponding to a moisture content below 3% (*w.b.*).

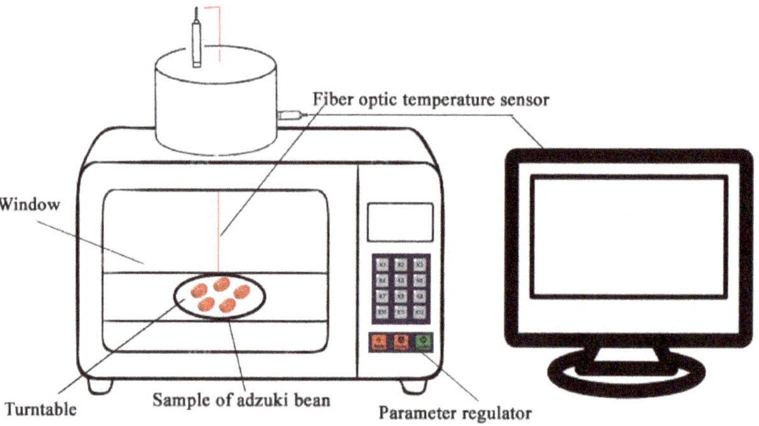

(a)

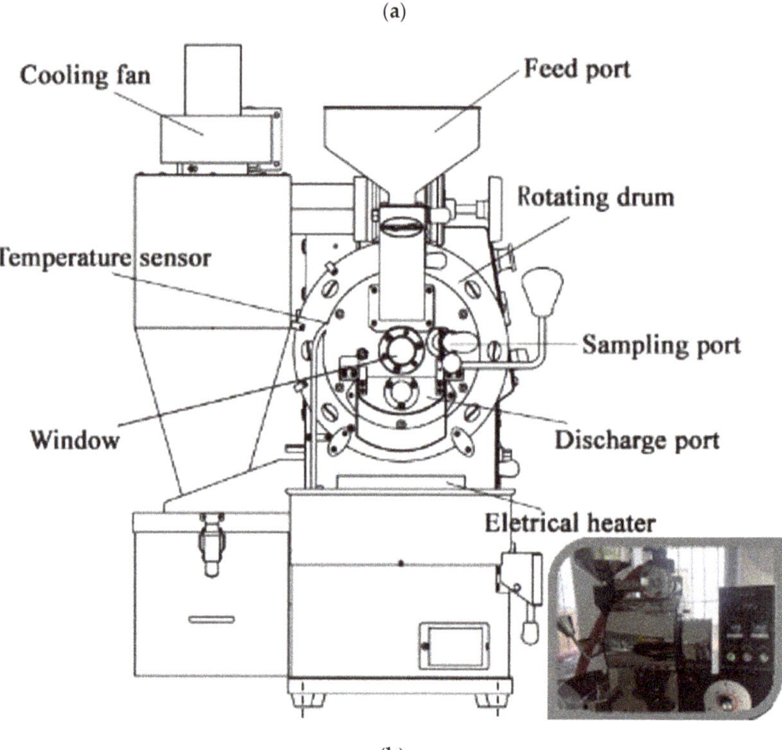

(b)

Figure 1. *Cont.*

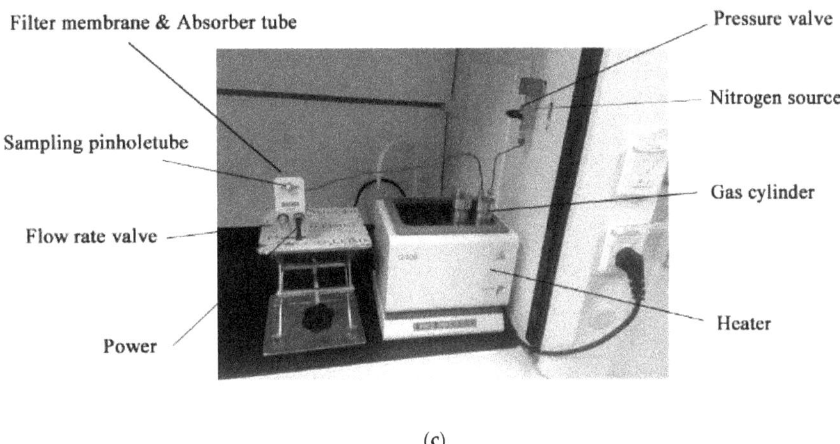

(c)

Figure 1. (**a**) Diagram of microwave workstation for baking treatment of adzuki beans. (**b**) Roasting drum device for baking treatment of adzuki beans. (**c**) Purge and trap device for detection of volatile compounds in baked beans.

2.4. Color Analysis

The color was measured for whole-kernel samples using a pulsed xenon lamp tristimulus colorimeter (CR-400, Minolta, Osaka, Japan) with an 8 mm diameter test area. The instrument was standardized against a white tile, a standard fitting for the instrument, before measurements. The color was expressed in L^*, a^* and b^* scale parameters. To evaluate the overall color value in L^*, a^* and b^*, the combined expression of the color difference value ∇E was calculated using Equation (1).

$$\nabla E = \sqrt{\left(L^* - L_0^*\right)^2 + \left(a^* - a_0^*\right) + \left(b^* - b_0^*\right)} \tag{1}$$

where L^*, a^*, and b^* are the brightness value, red–green value, and yellow–blue value of the baked beans, respectively. L_0^*, a_0^*, and b_0^* are the brightness value, red–green value, and yellow–blue value of the raw beans, respectively.

2.5. Measurement of Moisture Content

A total of 3000 g of each sample was accurately weighed and then dried to a constant mass in an oven at 105 °C for 24 h, according to the Association of Official Analytical Chemists (AOAC, Gaithersburg, MD, USA) method (1995), to determine the moisture content of each sample.

2.6. Acrylamide Content Analysis

Acrylamide content in adzuki beans at different stages of the roasting process was determined according to the method of Bortolomeazzi, Munari, Anese, and Verardo [18], with some modifications. Fifty grams of samples was crushed by a grinder and frozen at −20 °C. Each sample was accurately weighed to 1 g (accurate to 0.001 g), and 10 μL of $_{13}C_3$-acrylamide (10 mg/L) was added as an internal standard working solution, followed by the addition of 10 mL ultra-pure water. Following shaking operation for 30 min, centrifugation was performed at 4000 r/m for 10 min, and the supernatant was taken for further purification.

Five milliliters of n-hexane was added to the supernatant extracted from the sample and then shaken for 10 min. The organic phase was removed following centrifugal treatment at 10,000 r/m for 5 min. The extraction was repeated with 5 mL of n-hexane, and 6 mL of aqueous phase was rapidly filtered using a 0.45 μm membrane. An HLB

SPE (Solid Phase Extraction) column (5 µm 2.1 mm I.D. × 150 mm Atlantis C18 column, Waters, Milford, MA, USA) was activated using 3 mL of methanol and 3 mL of water for further experiments. A total of 5 mL of filtrate was extracted using the HLB SPE column, and the extract liquor was collected. This was followed by elution with 4 mL of methanol aqueous solution (80%). All the eluent was collected and combined with the extract liquor for purification. The Bond Elut-Accucat SPE column (3 mL, 200 mg, Agilent, Santa Clara, CA, USA) was activated using 3 mL of methanol and 3 mL of water. All the extract liquor discharged under gravity was collected, concentrated to dryness under nitrogen flow, and diluted to 1.0 mL using 0.1% formic acid solution for the test.

The information of the liquid chromatogram and analysis parameters were as follows.
Chromatograph: Xevo TQ-S Triple Quad Mass Spectrometer, Waters, Milford, MA, USA.
Chromatographic column: Atlantis C18 (5 µm, 2.1 mm I.D. × 150 mm).
Precolumn: C18 protection column (5 µm, 2.1 mm I.D. × 30 mm).
Mobile phase: methanol/0.1% formic acid (10:90, volume fraction).
Flow rate: 0.2 mL/min.
Injection volume: 25 µL.
Column temperature: 26 °C.
Mass spectrometry conditions:
Capillary voltage: 3500 V.
Cone voltage: 40 V.
RF lens 1 voltage: 30.8 V.
Ion source temperature: 80 °C.
Desolvation temperature: 300 °C.
Ion collision energy: 6 eV.
Acrylamide: parent ion m/z 72, daughter ion m/z 55, daughter ion m/z 44.
$_{13}C_3$ acrylamide: parent ion m/z 75, daughter ion m/z 58, daughter ion m/z 45.
Quantification ion: m/z 55 for acrylamide, m/z 58 for $_{13}C_3$ acrylamide.

The method of signal-to-noise ratio (SNR) was used to determine the quantification limit (LOQ) and the detection limit (LOD). The concentration corresponding to 10 times the SNR was used as the LOQ, and the concentration corresponding to 3 times the SNR was used as the LOD, represented by Equations (2) and (3).

$$LOQ = 10 \text{ N/S} \tag{2}$$

$$LOD = 3 \text{ N/S} \tag{3}$$

where N is noise, and S is detector sensitivity.

The LOQ of the method is 10 µg/kg, and the absolute difference between the two independent measurements obtained under repeatability conditions should not exceed 20% of the arithmetic mean.

2.7. Volatile Components Measurement

A purge and trap sampling device (homemade, tailored to experiments) was used to detect volatile components of the baked beans, as shown in Figure 1c. A 20 g baked bean sample was placed into the sampling jar with a volume of 250 mL. Considering the boiling point of volatile components, the sample was heated using an air flow of 40 mL/min under a temperature of 40–120 °C with 20 °C intervals. The purge and trap sampling duration was set as 60 min to collect volatiles at a volume of 2.4 L.

The analysis of volatile substances was carried out by concentrating volatile compounds using the purging and trapping method, then introducing volatile compounds into GC-MS (Gas Chromatography-Mass Spectrometer) using a thermal desorption supplementary collector for qualitative and quantitative analysis. This method focuses on the relative strength of these volatiles, rather than the absolute composition of flavor-producing substances. Therefore, two quantitative methods were used: one was the normalization method, and the second quantified all peak substances using the relative correction factor of toluene, which represents the relative content of the volatiles collected using this test method. The response factor of toluene (F) was obtained by the standard curve method. The concentration of volatiles (C) in baked beans was determined by Equations (4) and (5).

$$F = \frac{\text{Peak area of toluene}}{\text{Toluene content in standard curve}} \tag{4}$$

$$C(\text{Volatile}) = \frac{\text{Peak area of volatile} \times F}{\text{Mass of sample}} \times \text{Unit conversion factor} \tag{5}$$

where the mass of sample was 20 g in Equation (5).

The information of gas chromatogram and analysis parameters were as follows.

Chromatograph: Unity-xr/7890B-5977B Thermal Desorption Gas Chromatography-Mass Spectrometry, Agilent, Santa Clara, CA, USA.

Chromatographic column: DB-5ms 60 m $\times$ 0.25 mm $\times$ 0.25 μm, Agilent, Santa Clara, CA, USA.

Initial temperature: 40 °C, maintain 5 min; 3 °C/min to 160 °C, maintain 2 min; 8 °C/min to 260 °C, maintain 2 min.

Carrier gas: helium.

Flow rate: 1 mL/min.

Mass spectrometry:

Electron energy: 70 eV.

Ion source temperature: 230 °C.

Quadrupole temperature: 180 °C.

Transmission line temperature: 230 °C.

Scan range: 31 m/z to 500 m/z.

Thermal desorption temperature: 280 °C for 5 min.

Filling temperature of cold trap: -15 °C.

Analytical temperature of cold trap: 300 °C for 5 min.

2.8. Model Development

According to the acrylamide changes in beans processed by microwave baking and drum roasting, the SigmaPlot software (Version 12.5, Systat Software, Inc., San Jose, CA, USA) method was employed to fit the power function. In the preliminary regression treatment of experimental data, it was observed that the change trends of acrylamide in adzuki bean baked had high consistency with the curves of the power function. According to the data of acrylamide content in adzuki beans with baking duration (t), temperature (T), moisture content (M), and ratio of temperature to moisture content (T/M) under microwave baking and drum roasting, the power model types from the SigmaPlot software (Version 12.5, Systat Software, Inc.) were selected to be fitted by using the iterative regression method to constant and kinetic coefficients. The determined coefficient (R^2) and relative square error (RSE) were introduced to evaluate the fitting models. The closer to 1 the R^2 value, the less the RSE indicates the more reasonable fitting model. The developed power models characterize the formation kinetics of acrylamide. The function characterizes the formation kinetics of acrylamide. The determined coefficient and relative error were introduced to evaluate the fitting models.

2.9. Statistical Analysis

Three repeated experiments were conducted for every operation, and average data were expressed by mean $\pm$ standard deviation (SD). Three repetitions were performed in individual determination such as analysis of dry matter, acrylamide content, and composition of volatile compounds. The data were statistically processed by using analysis of variance (ANOVA) and Tukey's honestly significant difference (HSD) test, which were performed using the SPSS statistics software (Version 19.0, SPSS Inc., Chicago, IL, USA). Statistically significant differences were tested at a 5% probability level ($p < 0.05$).

3. Results and Discussion

3.1. Formation Kinetic of Acrylamide in Adzuki Beans under Microwave Baking and Drum Roasting

Adzuki beans are abundant in starch, which easily forms acrylamide during heating treatments as a result of the Maillard reaction, as the main chain from asparaginase in amino acids forms acrylamide molecules [19]. The reducing sugar reacts with free amino groups of amino acids or protein to produce the Maillard reaction in baking processes with a high temperature [20].

3.1.1. Changes in the Acrylamide Content of Adzuki Beans during Microwave Baking

The change in the acrylamide content of adzuki beans during microwave heating is presented in Figure 2a.

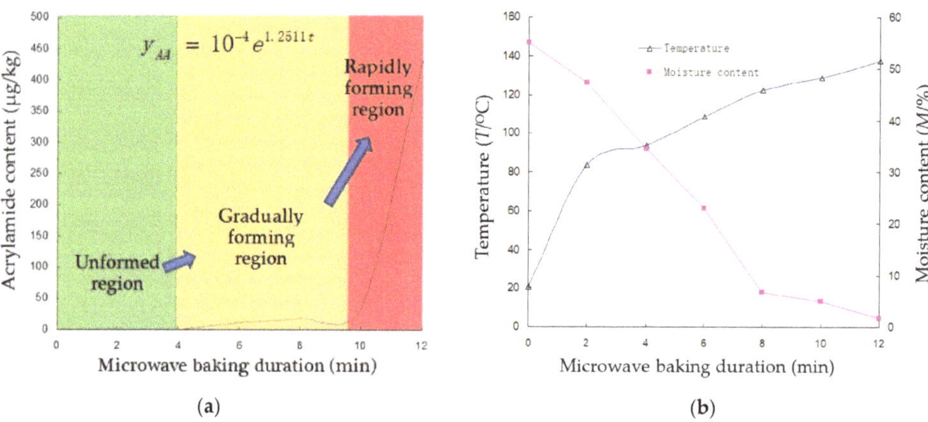

Figure 2. (**a**) Acrylamide content of adzuki beans under microwave baking. (**b**) Temperature and moisture content of adzuki beans after microwave baking.

As shown in Figure 2a, no acrylamide was detected inside adzuki beans until after 4 min of microwave baking; then, acrylamide gradually formed, reaching 32.8 μg/kg at 9 min, followed by a rapid increase to 432.9 μg/kg. These changes are attributed to the three stages of the Maillard reaction, as follows [21]. In Stage 1, with the absorption of the thermal energy into baked adzuki beans, the amino acids react by reducing sugar to form Amadori molecular product rearrangement (APR). In this stage, the carbonyl group of reducing sugars and the amino of amino acids form Schiff with a C=N bond. As a high-activity medium substance, the rearrangement of Schiff generates the APR. In Stage 2, with the further accumulation of thermal energy inside baked adzuki beans, the degradation of APR generates hundreds of volatile substances. In this stage, the rising temperature causes the degradation of amino to form carbonyl substances during the Maillard reaction, including aldehydes, ammonia, and ribose, under the Strecker degradation mechanism. In Stage 3, when the thermal accumulation achieves the active energy of the Maillard reaction at the critical temperature of acrylamide formation, asparagine and carbonyl compounds, as the precursor substances, rapidly synthesize the acrylamide inside the baked material.

The change in the acrylamide content of adzuki beans under microwave baking conditions followed the exponential growth function with a kinetic constant of 1.25 (1/s), embodied in Figure 2a. Acrylamide content is at low levels until a critical point, which provides guidance for controlling the acrylamide content of adzuki beans under microwave baking.

In the baking processes of adzuki beans, the formation of acrylamide is an endothermic reaction, where the reaction rate constant, as a function of T and M, is characterized by a modified Arrhenius equation (Equation (6)).

$$k = k_{ref} \times \exp\left[\frac{-E_a}{R} \times \left(\frac{1}{T} - \frac{1}{T_0}\right)\right] \times \exp\left[\frac{-H_a}{R \times T} \times (M - M_0)\right] \tag{6}$$

where k is the reaction rate (1/s), min^{-1}; k_{ref} is the constant of the reaction rate (1/s); E_a is the reaction active energy, kJ/mol; R is the gas constant, 8.314 J/(mol·K); T is the temperature, °C; T_0 is the initial temperature of bean processed at room temperature, °C (20 °C); H_a is the reaction heating, kJ/kg; M is the moisture content (%); and M_0 is the initial moisture content (w.b., %).

From Equation (1), the kinetics of acrylamide formation in adzuki beans are shown to depend on the changes in temperature and moisture content. As shown in Figure 2b, microwave volumetric heating causes the temperature to rise with the reduction in adzuki beans' moisture content. Both temperature and moisture content influence the chemical reaction of acrylamide formation, which was presented using the temperature to moisture ratio (T/M) as an exponential growth function, as shown in Equation (7).

$$y_{MR} = 6.94e^{0.05 \times \frac{T}{M}}, R_2 = 0.9985, SEE = 6.78 \tag{7}$$

where y_{MR} is the acrylamide content inside the baked beans considering the ratios of temperature (T, °C) and moisture content (M, %, w.b.). R^2 is the determinate coefficient, and SEE is the sum of squared residuals. When adzuki beans were baked in a microwave, the acrylamide resultant was formed with increasing temperature and decreasing moisture content. The effect of moisture on the acrylamide formation of adzuki beans obeys the exponential rise to maximum function shown in Equation (7), and the effect of temperature on the acrylamide formation obeys the exponential growth function, as shown in Equation (8).

$$y_{MM} = 1602.68 - 1598.74\left(1 - e^{-0.79M}\right), R_2 = 0.9990, SEE = 6.31 \tag{8}$$

$$y_{MT} = e^{0.29(T - 116.51)}, R_2 = 0.9979, SEE = 8.01 \tag{9}$$

where y_{MM} is the acrylamide content in beans baked by microwave heating considering the moisture content (M, w.b., %), and y_{MT} is the acrylamide content in beans baked by microwave heating considering temperature (T, °C). The rate constant k of acrylamide formation in Equation (1) has a positive correlation with the temperature of adzuki beans due to the high temperatures needed for acrylamide formation.

According to the results of Equation (9), the critical temperature needed for acrylamide content was at 116.51 °C, corresponding to a moisture content of 5.6% (w.b.). Therefore, microwave baking may produce baked beans with no detectable acrylamide content when the heating temperature is below 116 °C, with a moisture content higher than 5.6%.

3.1.2. Changes in the Acrylamide Content of Adzuki Beans during Drum Roasting

The changes in the acrylamide content, temperature, and moisture content of adzuki beans are presented in Figure 3a,b.

In drum roasting, no acrylamide content is detectable in adzuki beans for 7.5 min; then, there is a clear rising trend up to 87.4 µg/kg for 23 min, as shown in Figure 3a. The change in acrylamide content in adzuki beans under drum-roasting conditions obeys the exponential growth function with a kinetic constant of 0.1225 (1/s), as shown in Figure 3a.

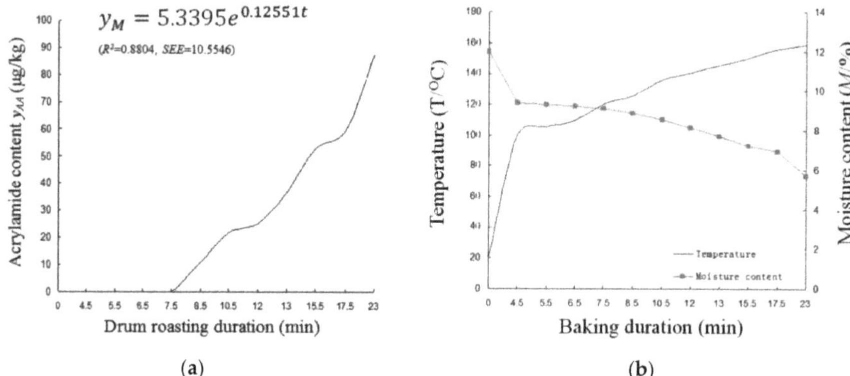

Figure 3. (**a**) Acrylamide content of adzuki beans during drum roasting. (**b**) Temperature and moisture content of adzuki beans with drum roasting.

In the drum-roasting processes, the acrylamide content of adzuki beans as a function of the temperature to moisture content ratio (shown in Figure 3b) obeys the exponential growth function, as shown in Equation (10).

$$y_D = 2.44e^{0.13 \times \frac{T}{M}}, R_2 = 0.8834, SEE = 10.42 \tag{10}$$

where y_D is the kinetic constant of the acrylamide content of adzuki beans, considering the temperature to moisture ratio; T is the temperature of the baked beans (°C); and M is the moisture content (%, *w.b.*).

The temperature to moisture content ratio in drum roasting (0.13) in Equation (10) is higher than that in microwave baking (0.05) in Equation (7). However, the kinetic coefficients of moisture in Equation (11) and temperature in Equation (12) for drum-roasted adzuki beans are lower than those of microwave-baked beans, as shown in Equations (8) and (9).

$$y_{DM} = 926.84 - 951.42\left(1 - e^{-0.36M}\right), R_2 = 0.9305, SEE = 8.48 \tag{11}$$

$$y_{DT} = e^{0.07(T-91.62)}, R_2 = 0.9796, SEE = 4.36 \tag{12}$$

where y_{DM} is the acrylamide content in drum-roasted beans, considering moisture content (M, %), and y_{DT} is the acrylamide content in drum-roasted beans, considering temperature (T, °C). According to Equation (8), the critical temperature causing acrylamide content to rise in adzuki beans during drum roasting is 91.62 °C at a corresponding moisture content of 6.1% (*w.b.*).

In summary, the kinetics of acrylamide formation in adzuki beans during microwave baking are higher than those found during drum roasting. The difference in the kinetic behavior of the acrylamide formation in adzuki beans is attributed to the greater heating intensity of microwave volumetric heating compared to drum roasting. The greater heating intensity results in higher temperatures and a quicker reduction in moisture content in adzuki beans baked using microwave volumetric heating. When baking adzuki beans using a microwave, the microwave field causes the rapid rotation of polar molecules, at 2.45×10^9 times per second. The rapid movements of polar molecules cause fierce friction and collision, producing volumetric heating, and increasing the entropy value. The increment of entropy inside microwave-baked beans may reduce the activation energy. Therefore, the low kinetic coefficient of acrylamide formation in baked beans is due to the reduction in the activation energy of precursor substance molecules, including asparagine and carbonyl compounds. In drum-roasted adzuki beans, material is heated from the surface, and inside via the convection transfer route. Compared to microwave baking, the low heating intensity in drum roasting results in a slowly increasing temperature and decreasing moisture content in the baked materials, as the precursor substance molecules need more activation energy to form acrylamide. Therefore, the acrylamide content in adzuki beans remains at low levels until high temperatures (116 °C) are reached during microwave baking, which improves the baking efficiency. However, the initial temperature of acrylamide formation under drum roasting is 91.6 °C, which is lower than the consensus temperature of 120 °C [22]. If baking processes have a long-enough duration, high temperature, and low moisture content, they contain the conditions for acrylamide formation.

3.2. Changes in the Color of Adzuki Beans during Microwave Baking and Drum Roasting

As shown in Figure 4a,b, the L^* value of adzuki beans tends to decrease, but the a^* and b^* values show fluctuating changes due to the formation of Amadori intermediates and Maillard reaction products during microwave baking and drum roasting.

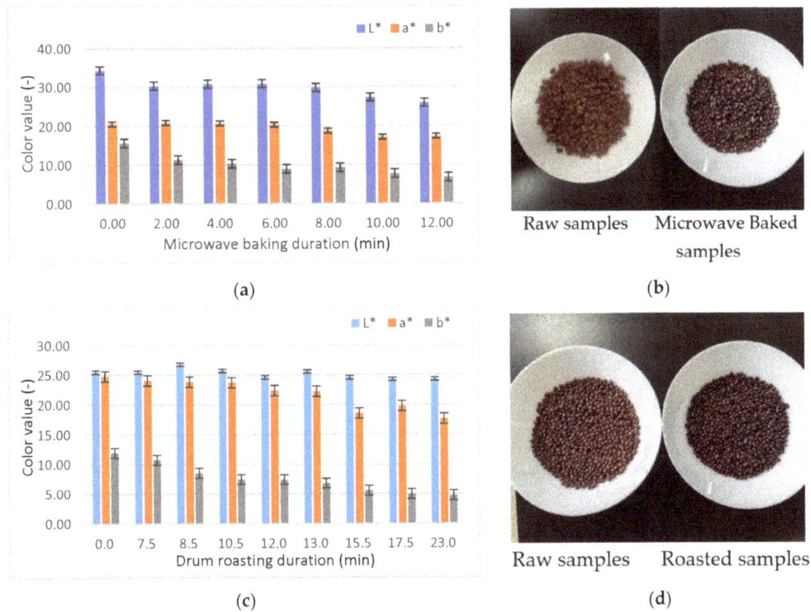

Figure 4. (**a**,**b**) Color changes in adzuki beans during microwave baking. (**c**,**d**) Color changes in adzuki beans during drum roasting.

In the microwave-baking process, the color changes obey the exponential decay with two stages, as shown in Equation (13).

$$\nabla E = 16.47 e^{(-2.95E-11)^t} - 15.38 e^{-0.11t} \tag{13}$$

From Equation (13), the overall color of adzuki beans tended to slow, then showed a sharp decline due to the thermal accumulation inside baked beans resulting from microwave volumetric heating ($p < 0.05$), which caused an obvious Maillard reaction in adzuki beans in a short time period.

In the drum-roasting processes, the color changes obey the same rule as those in microwave baking, as shown in Equation (14).

$$\nabla E = 0.73 e^{0.08t} + 1.24 e^{0.08t} \tag{14}$$

From Equation (14), the overall color of adzuki beans shows a gentle decline at two stages due to the heat diffusion of convection and conduction inside baked adzuki beans.

The acrylamide formation of baked adzuki beans has a high correlation with its color level due to the Maillard reaction. Although no statistically significant correlation was established for acrylamide content with a single color value of L^*, a^*, and b^*, the second term in Equations (13) and (14) presents the changes in the color of adzuki beans as acrylamide content rapidly rose under microwave-baking and drum-roasting technologies, respectively. According to the critical duration of 9.5 min and 7.5 min in Figures 2a and 3a, presenting a rapidly rising acrylamide content ($p < 0.05$), ∇E was determined from Equations (13) and (14) as 11.1 and 3.6 under microwave baking and drum roasting, respectively. There is reason to infer that the initial point of the formation of acrylamide content in baked beans may be evaluated based on its overall color ∇E level under different baking methods, although further quantitative validation is needed.

3.3. Changes in Volatile Compounds in Baked Bean in Baking Processes

The chromatograms of baked beans, measured by the purge and trap sampling method, are shown in Figure 5 under airflow temperatures of 40 °C, 60 °C, 80 °C, 100 °C, and 120 °C. More volatile components were collected under higher airflow temperatures.

The selected measurement temperature was 100 °C, referring to the brewing temperature of adzuki bean powder with boiling water. Based on the species and concentration of volatiles from baked adzuki beans, shown in Figure 5 at 100 °C, we classified and added the test results to find the changes in the characteristic curves of volatile compounds in baked beans during the microwave-baking (Figure 6a–e) and drum-roasting processes (Figure 6a′–e′). The data in detailed in Figure 6 were presented in Table S1 at the Supplementary Materials.

According to the GC-MS results for the baked beans, as shown in Figure 6, an average of 48 volatile components are detected in microwave baking, and an average of 56 volatile components are detected after drum roasting. Characteristic volatile compounds in baked beans are determined as furan, pyrazine, ketone, alcohols, aldehydes, esters, pyrroles, sulfocompound, phenols, and pyridine. The precursor of these volatile substances is produced during the thermal treatment of baked beans with the formation of organic acid, which results from the thermal degradation of polysaccharide to produce the acidity. The thermal decomposition of oligosaccharide in beans generates the reducing sugar. The Maillard reaction between the reducing sugar and free amino acid produces volatile substances, including pyrazines, furans, and aldehydes [23]. Further high-temperature baking, and the stark degradation reaction between α-amino acid and dicarbonyl compounds, form aldehydes and ketone. These characteristic compounds present odor as a cream from aldehyde, delicate fragrance from alcohols, fruit odor from esters, and sour from acids. In baking processes, including microwave and convective heating, high temperatures result in the production of characteristic volatile compounds in baked beans. In a general trend, the maximum contents of volatile compounds were higher in drum roasting than that in microwave baking. As shown in Figure 6a,a′, in microwave baking and drum roasting, the formation of furan led to the pleasant smell of baked beans. With different bean-baking processes, a high temperature resulted in a clear reduction in furan. Excessive baking with high temperatures and low moisture contents caused the fine volatiles to almost disappear. The aldehyde content of baked beans has similar trends as the furnace changes in microwave and drum roasting baking, as shown in Figure 6b,b′. The alcohol content increases then decreases with the microwave- and drum-roasting processes shown in Figure 6c,c′. Figure 6d,d′ shows that the ester content of baked beans increases in microwave baking and first increases then decreases in drum roasting. From Figure 6e,e′, drum roasting leads to the formation of pyrazine, an attractive volatile with a cocoa aroma, and microwave baking is not good for the formation of pyrazine. Although little research has been conducted on the formation of acrylamide content and volatile compounds in baked beans, in the existing research, cocoa bean roasting increased the content of acrylamide 2–3-fold, which indicates that roasting cocoa beans increases the content of furan by 25.1–34.8 ng g^{-1} [4]. Milling baked adzuki beans may be used as the raw material for instant food, as shown in Figure 7. This demonstrates the importance and novelty of quality formation and control based on baking methods and technology parameters for baked adzuki beans.

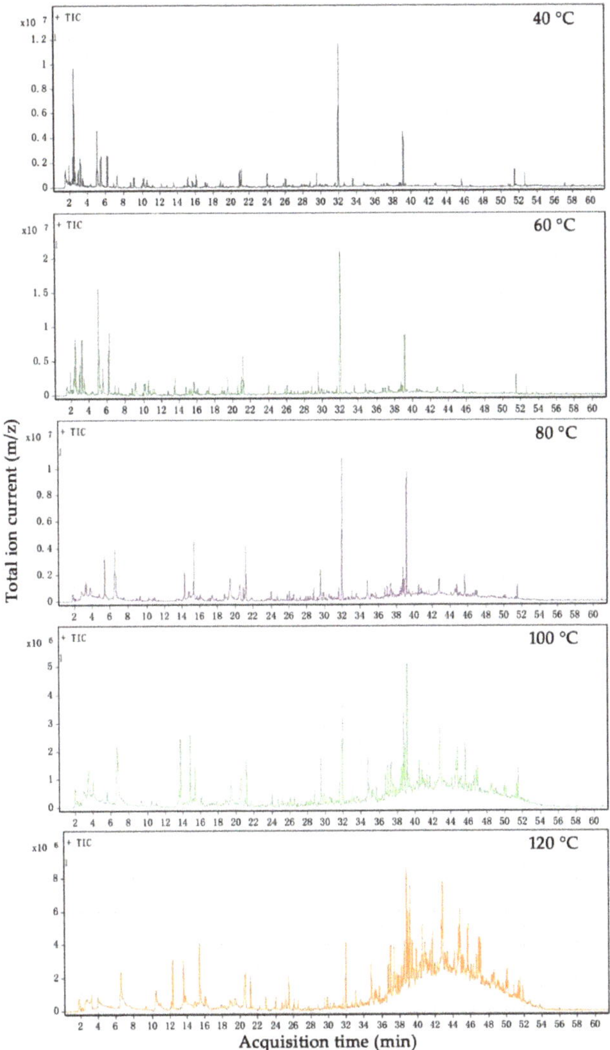

Figure 5. Total ion chromatogram (TIC) of baked beans under different temperatures.

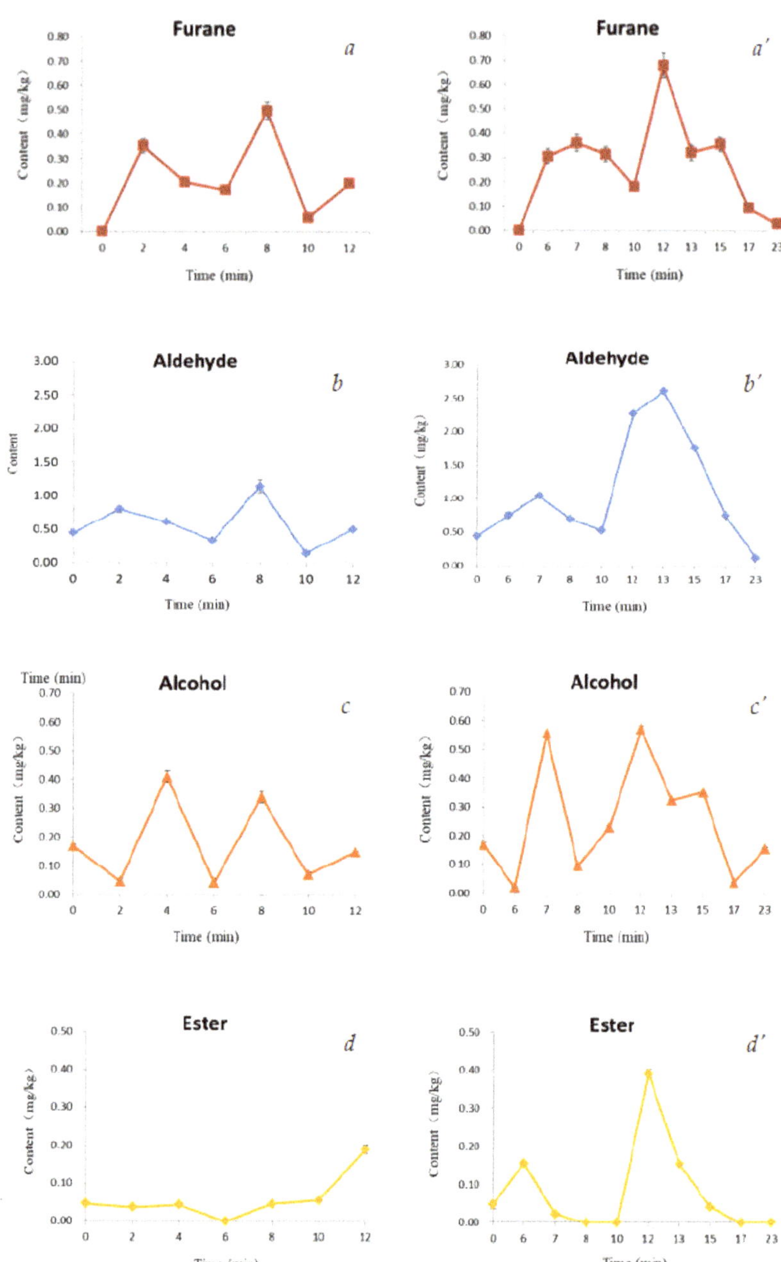

Figure 6. *Cont.*

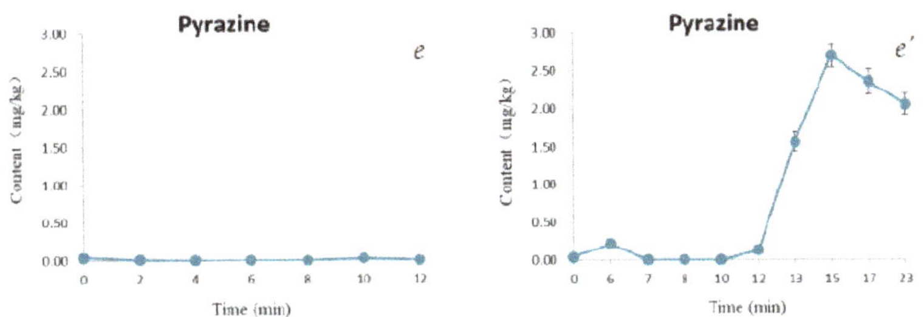

Figure 6. Changes in characteristic volatile compounds of baked beans during microwave-baking (*a–e*) and drum-roasting processes (*a'–e'*).

Figure 7. Baked adzuki bean, powder, and tune bean drinks.

4. Conclusions

The kinetics of acrylamide formation in adzuki beans under microwave baking are higher than those under drum roasting due to the greater kinetic coefficient in microwave baking. The critical temperature leading to this increase in acrylamide content is 116.5 °C, with a corresponding moisture content of 5.6% (*w.b.*) in microwave baking, and 91.6 °C, with a corresponding moisture content of 6.1% (*w.b.*). The starting point indicating the increase in acrylamide content is ∇E of 11.1, 3.6 under microwave baking and drum roasting, respectively. No statistically significant correlation was found for acrylamide content with a single color value of L^*, a^*, and b^*; however, the overall color of adzuki beans tends to decline with a rapidly rising acrylamide content, and the initial point of the formation of acrylamide content in baked beans may be evaluated based on its overall color ∇E level under the microwave-baking and drum-roasting methods. Characteristic volatile compounds in baked beans include furan, pyrazine, ketone, alcohols, aldehydes, esters, pyrroles, sulfocompound, phenols, and pyridine. The comparison of volatile formation in beans baked using drum roasting and microwave baking needs further investigation to evaluate flavor quality. This can provide guidance for controlling the acrylamide content in adzuki beans under microwave baking and drum roasting.

Supplementary Materials: The following are available online at https://www.mdpi.com/article/10.3390/foods10112762/s1, Table S1: Identified volatile compounds under microwave baking and drum roasting.

Author Contributions: Conceptualization, X.Y.; methodology, Z.L., Z.G. and Y.Z.; software, H.S.; validation, R.Z.; formal analysis, A.S.; investigation, S.L.; resources, T.L.; data curation, X.Z.; writing—original draft preparation, X.Y.; writing—review and editing, N.X., A.S. and Q.W.; supervision, Y.Z. and Q.W.; project administration, X.Y. All authors have read and agreed to the published version of the manuscript.

Funding: This research was funded by Leading Talent Reserve Leader Funding Project of Heilongjiang Province, grant number 2018 No. 384, Department of Human Resources and Social Security, Heilongjiang Province.

Institutional Review Board Statement: Not applicable.

Informed Consent Statement: Not applicable.

Data Availability Statement: The data presented in this study are available on request from the corresponding author.

Conflicts of Interest: The authors declare no conflict of interest.

References

1. Hu, H.; Liu, H.; Shi, A.; Liu, L.; Fauconnier, M.L.; Wang, Q. The Effect of microwave pretreatment on micronutrient contents, oxidative stability and flavor quality of peanut oil. *Molecules* **2019**, *24*, 62. [CrossRef]
2. Yu, S.; Hongkun, X.; Chenghai, L. Comparison of microwave assisted extraction with hot reflux extraction in acquirement and degradation of anthocyanin from powdered blueberry. *Int. J. Agric. Biol. Eng.* **2016**, *9*, 186–199.
3. Suhag, R.; Dhiman, A.; Deswal, G.; Thakur, D.; Sharanagat, V.S.; Kumar, K.; Kumar, V. Microwave processing: A way to reduce the anti-nutritional factors (ANFs) in food grains. *LWT* **2021**, *150*, 111960. [CrossRef]
4. Kruszewski, B.; Obiedziński, M.W. Impact of raw materials and production processes on furan and acrylamide contents in dark chocolate. *J. Agric. Food Chem.* **2020**, *68*, 2562–2569. [CrossRef] [PubMed]
5. Mesias, M.; Delgado-Andrade, C.; Holgado, F.; Morales, F.J. Acrylamide content in French fries prepared in food service establishments. *Food Sci. Technol.* **2019**, *100*, 83–91. [CrossRef]
6. Dibaba, K.; Tilahun, L.; Satheesh, N.; Geremu, M. Acrylamide occurrence in Keribo: Ethiopian traditional fermented beverage. *Food Control* **2018**, *86*, 77–82. [CrossRef]
7. Pérez-Nevado, F.; Cabrera-Bañegil, M.; Repilado, E.; Martillanes, S.; Martín-Vertedor, D. Effect of different baking treatments on the acrylamide formation and phenolic compounds in Californian-style black olives. *Food Control* **2018**, *94*, 22–29. [CrossRef]
8. Matoso, V.; Bargi-Souza, P.; Ivanski, F.; Romano, M.A.; Romano, R.M. Acrylamide: A review about its toxic effects in the light of Developmental Origin of Health and Disease (DOHaD) concept. *Food Chem.* **2019**, *283*, 422–430. [CrossRef]
9. Açar, Ö.Ç.; Gökmen, V. New approach to evaluate the risk arising from acrylamide formation in cookies during baking: Total risk calculation. *J. Food Eng.* **2010**, *100*, 642–648. [CrossRef]
10. Anese, M.; Quarta, B.; Peloux, L.; Calligaris, S. Effect of formulation on the capacity of l-asparaginase to minimize acrylamide formation in short dough biscuits. *Food Res. Int.* **2011**, *44*, 2837–2842. [CrossRef]
11. Santra, S.; Banerjee, A.; Das, B. Polycation charge and conformation of aqueous poly (acrylamide-co-diallyldimethylammonium chloride): Effect of salinity and temperature. *J. Mol. Struct.* **2021**, *1247*, 131292. [CrossRef]
12. Gil, M.; Ruiz, P.; Quijano, J.; Londono-Londono, J.; Jaramillo, Y.; Gallego, V.; Tessier, F.; Notario, R. Effect of temperature on the formation of acrylamide in cocoa beans during drying treatment: An experimental and computational study. *Heliyon* **2020**, *6*, e03312. [CrossRef]
13. Yıldız, H.G.; Palazoğlu, T.K.; Miran, W.; Kocadağlı, T.; Gökmen, V. Evolution of surface temperature and its relationship with acrylamide formation during conventional and vacuum-combined baking of cookies. *J. Food Eng.* **2017**, *197*, 17–23. [CrossRef]
14. Balagiannis, D.P.; Mottram, D.S.; Higley, J.; Smith, G.; Wedzicha, B.L.; Parker, J.K. Kinetic modelling of acrylamide formation during the finish-frying of french fries with variable maltose content. *Food Chem.* **2019**, *284*, 236–244. [CrossRef]
15. Nguyen, H.T.; Van der Fels-Klerx, H.J.; Van Boekel, M.A.J.S. Acrylamide and 5-hydroxymethylfurfural formation during biscuit baking. Part II: Effect of the ratio of reducing sugars and asparagine. *Food Chem.* **2017**, *2301*, 14–23. [CrossRef]
16. Arisseto, A.P.; Vicente, E.; Ueno, M.S.; Tfouni, S.A.V.; De Figueiredo Toledo, M.C. Furan levels in coffee as influenced by species, roast degree, and brewing procedures. *J. Agric. Food Chem.* **2011**, *59*, 3118–3124. [CrossRef] [PubMed]
17. Ding, S.Y.; Yang, J. The effects of sugar alcohols on rheological properties, functionalities, and texture in baked products—A review. *Trends Food Sci. Technol.* **2021**, *111*, 670–679. [CrossRef]
18. Bortolomeazzi, R.; Munari, M.; Anese, M.; Verardo, G. Rapid mixed mode solid phase extraction method for the determination of acrylamide in roasted coffee by HPLC–MS/MS. *Food Chem.* **2012**, *135*, 2687–2693. [CrossRef]
19. Chang, Y.; Zeng, X.; Sung, W.C. Effect of chitooligosaccharide and different low molecular weight chitosans on the formation of acrylamide and 5-hydroxymethylfurfural and Maillard reaction products in glucose/fructose-asparagine model systems. *LWT* **2020**, *119*, 108879. [CrossRef]
20. Sung, W.C.; Chang, Y.W.; Chou, Y.H.; Hsiao, H.I. The functional properties of chitosan-glucose-asparagine Maillard reaction products and mitigation of acrylamide formation by chitosans. *Food Chem.* **2018**, *243*, 141–144. [CrossRef]
21. Rannou, C.; Laroque, D.; Renault, E.; Prost, C.; Sérot, T. Mitigation strategies of acrylamide, furans, heterocyclic amines and browning during the Maillard reaction in foods. *Food Res. Int.* **2016**, *90*, 154–176. [CrossRef] [PubMed]
22. Zhang, G. Formation mechanism and risk assessments of acrylamide generated in heated foodstuffs. *J. Wuxi Univ. Light Ind.* **2003**, *22*, 91–99.
23. Van Boekel, M.A.J.S. Formation of flavour compounds in the Maillard reaction. *Biotechnol. Adv.* **2006**, *24*, 230–233. [CrossRef] [PubMed]

Article

Effect of Ultrasonic Induction on the Main Physiological and Biochemical Indicators and γ–Aminobutyric Acid Content of Maize during Germination

Liangchen Zhang [1], Nan Hao [2], Wenjuan Li [3], Baiqing Zhang [3], Taiyuan Shi [1], Mengxi Xie [1] and Miao Yu [1,*]

[1] Institute of Food and Processing, Liaoning Academy of Agricultural Sciences, Shenyang 110161, China; napoleon19831214@163.com (L.Z.); sty1965fightting@163.com (T.S.); moor1112@163.com (M.X.)

[2] Corn Research Institute, Liaoning Academy of Agricultural Sciences, Shenyang 110161, China; lnhn1981@163.com

[3] College of Food Science and Technology, Shenyang Normal University, Shenyang 110034, China; lwj20211111@163.com (W.L.); zbqfood@126.com (B.Z.)

[*] Correspondence: jannytiti@163.com; Tel./Fax: +86-159-9837-8968

Abstract: Research on the nutrient content of cereal grains during germination is becoming a hot topic; however, studies on germinated maize are still scarce. This study aimed to provide a technical reference and theoretical basis for the development of functional maize health foods and to expand the application of ultrasonic technology in the production of germinated grains. In this study, the germination rate of maize was used as the evaluation index, and the ultrasonic frequency, ultrasonic temperature, and induction time were selected as the influencing factors in orthogonal experiments to determine the optimal process parameters for ultrasonic induction of maize germination (ultrasonic frequency of 45 kHz, ultrasonic temperature of 30 °C, and ultrasonic induction time of 30 min). Based on this process, the effects of ultrasonic induction on the main physiological, biochemical, and γ–aminobutyric acid contents of maize during germination were investigated. The results showed that the respiration of the ultrasonic treated maize was significantly enhanced during germination, resulting in a 27% increase in sprout length, as well as a 4.03% higher dry matter consumption rate, and a 2.11% higher starch consumption rate. Furthermore, the reducing sugar content of germinated maize increased by 22.83%, soluble protein content increased by 22.52%, and γ–aminobutyric acid content increased by 30.55% after ultrasonic induction treatment. Throughout the germination process, the glutamate acid decarboxylase activity of the ultrasonically treated maize was higher than that of the control group, indicating that ultrasonication can promote maize germination, accelerate the germination process, and shorten the enrichment time of γ–aminobutyric acid in germinated maize. The results of this study can be applied to the production of γ–aminobutyric acid enrichment in germinated maize.

Keywords: ultrasonic; maize; germination; physiological and biochemical indicators; γ–aminobutyric acid

Citation: Zhang, L.; Hao, N.; Li, W.; Zhang, B.; Shi, T.; Xie, M.; Yu, M. Effect of Ultrasonic Induction on the Main Physiological and Biochemical Indicators and γ–Aminobutyric Acid Content of Maize during Germination. *Foods* **2022**, *11*, 1358. https://doi.org/10.3390/foods11091358

Academic Editor: Mohsen Gavahian

Received: 13 April 2022
Accepted: 6 May 2022
Published: 7 May 2022

Publisher's Note: MDPI stays neutral with regard to jurisdictional claims in published maps and institutional affiliations.

Copyright: © 2022 by the authors. Licensee MDPI, Basel, Switzerland. This article is an open access article distributed under the terms and conditions of the Creative Commons Attribution (CC BY) license (https://creativecommons.org/licenses/by/4.0/).

1. Introduction

Maize (*Zea mays* L.) is an annual cereal crop which produces kernels that are rich in nutrients, such as unsaturated fatty acids, glutathione, dietary fiber, vitamins, trace elements, protein, fat, and carotene [1]. However, maize protein has poor nutritional availability, with a utilization rate of only 6% in the food industry [2]. Modifying germination conditions can enhance their nutritional quality, reduce levels of antinutritional factors, and improve their taste [3]. Previous studies have confirmed that the γ–aminobutyric acid (GABA) content of brown rice, soybean, and fava bean increased significantly after germination [4]. GABA is an inhibitory neurotransmitter in mammalian cerebrospinal fluid that has a variety of health benefits [5]. GABA has been shown to inhibit cancer cell proliferation and stimulate apoptosis, regulate blood pressure and blood cholesterol, and reduce pain and anxiety [6]. Plant derived GABA products have received much attention for their safety and natural properties.

GABA is mainly from glutamate catalyzed by glutamate decarboxylase (GAD), and the activity of GAD is the main factor affecting the content of GABA in plants [7]. GAD is an endogenous enzyme in plant cells. Under environmental stimuli, such as hypoxia, darkness, ultraviolet light, salt stress, and mechanical vibration, its activity results in an increase in GABA content [4].

Ultrasonic waves refer to elastic waves propagated in gas, liquid, solid, and other media by mechanical vibration at a certain frequency. Ultrasonic waves can generate not only mechanical force but also thermal effects that cause various physical and chemical changes, and ultrasonic waves of appropriate intensity acting on biological tissues can affect plant metabolism and biochemical properties [8]. Ultrasonic waves have sterilizing effects which can eliminate bacteria on the surface of seeds, reduce bacterial infection of seeds, break the dormant period of seeds, and improve the vitality and germination rate of seeds [9]. Ultrasonic induction can accelerate the division and growth of plant cells and the protein synthesis of protoplasts by destroying the structure of biological macromolecules, increasing the pH value and Ca^{2+} concentration of cells, and changing the permeability of cell membranes to shorten the germination time of plant seeds [10]. Ultrasonic induction can stimulate the activity of endogenous enzymes in seeds; however, the activities of endogenous enzymes in seeds were found not to increase immediately after induction by ultrasonic waves but to increase significantly during germination compared with an untreated group [11]. Activation of endogenous enzymes is beneficial for the enrichment of functional factors in seeds [12]; studies have shown that when rice, soybean, mungbean, and other plant food raw materials are induced by ultrasonic waves and then germinated, the GABA content increases significantly [4]; the total phenolic content increased significantly in barley germinated after ultrasonic induction [8], while the content of resveratrol was significantly increased in peanuts germinated after ultrasonic induction [11]. The activation of endogenous enzymes induced by ultrasound can reduce or eliminate the concentration of some antinutrients, such as phytic acid and protein inhibitors, contained in grains and beans, increase their protein digestibility and bioavailability, and significantly improve nutritional quality [13,14].

Ultrasonic induction technology can improve the yield of germinated maize and improve its nutritional quality. However, the research on the artificial regulation of maize germination under stress has not been reported. In this work, we investigated the effect of ultrasonic induction treatment on maize germination rate, physiological and biochemical indicators, and GABA content. Additionally, the relationships between glutamic acid (Glu) content, Glutamate Decarboxylase (GAD) activity, and GABA content were investigated. This study aimed to provide a technical reference and theoretical basis for the development of functional maize health foods and to expand the application of ultrasonic technology in the production of germinated grains.

2. Materials and Methods

2.1. Materials and Reagents

For this work, the maize variety Liaodan 575 was provided by the Maize Research Institute of the Liaoning Academy of Agricultural Sciences. The maize seeds were harvested in October 2020, encapsulated in airtight containers, and stored in a sealed container at 4 °C. All chemicals and solvents used in this study were purchased from Sigma–Aldrich (St. Louis, MO, USA) or Solarbio (Beijing Solarbio Science & Technology Co., Ltd., Beijing, China).

2.2. Preparation for Maize Germination

2.2.1. Maize Soaking Treatment

The maize soaking treatment was assessed as described by Yu et al. [11], with minor modifications. A total of 300 g of maize seeds were weighed and sterilized by soaking in

1% by volume NaClO solution (Beijing Solarbio Science & Technology Co., Ltd., Beijing, China) for 30 min, rinsed with deionized water to restore neutral pH, and then soaked in an incubator for 6 h at 30 °C. Another 300 g of maize seeds were treated in the same way and then used as the control.

2.2.2. Ultrasonic Induction Treatment of Maize

The soaked maize samples were wrapped in a layer of sterile gauze and placed in an ultrasonic cleaner (KQ–300VDE, overall dimensions: 410 × 350 × 420 mm, tank dimensions: 300 × 240 × 180 mm, ultrasonic power: 250 W, heating power: 600 W, range of temperature: 20–80 °C ± 3°C, Kunshan Ultrasonic Instrument Co., Ltd., Kunshan, China) for ultrasonic induction treatment. We utilized three different ultrasonic frequencies: 28 kHz, 45 kHz, and 100 kHz. Additionally, induction times of 10, 20, and 30 min were tested, and temperatures of 30 °C, 35 °C, and 40 °C were utilized. $L_9 3^3$ orthogonal testing was carried out to identify the optimum ultrasonic induction treatment. The experimental design is shown in Table 1.

Table 1. Factors and levels of the orthogonal experiment.

Levels	Factors		
	Ultrasound Frequencies/KHz	Ultrasound Time/min	Ultrasonic Temperature/°C
1	28	10	30
2	45	20	35
3	100	30	40

2.2.3. Maize Germination Treatment

A single layer of soaked maize seeds was placed in Petri dishes and kept at a constant temperature of 30 °C and 90% relative humidity for 96 h. Portions of the maize seeds were removed at 0 h, 12 h, 24 h, 36 h, 48 h, 60 h, 72 h, 84 h, and 96 h during germination, rinsed three times with distilled water, and freeze-dried to a moisture fraction of less than 10%. The maize seeds were then stored at 4 °C before testing.

2.3. Methods

2.3.1. Determination of Germination Rate and Sprout Length

The germination rate was calculated as described previously [15]. Twenty-five maize samples were randomly selected at different time points and sprout lengths were measured using a vernier caliper (500–196–30, Mitutoyo, Tokyo, Japan).

2.3.2. Determination of Respiration Intensity

The respiratory intensity was measured as described by He et al. [16], with minor modifications. A total of 25 maize seeds were randomly selected at different time points, placed in a 40 mm × 80 mm jar, and measured with an infrared gas analyser (GXH–3051, Jun Fang Li Hua Technology Research Institute, Beijing, China) at 20 °C. Deoxygenated air was utilized as the carrier gas and measurements were taken for one hour. The respiration intensity was expressed as the mass of CO_2 released from 1 g of maize respired for 1 h in mg/g·h.

2.3.3. Determination of Dry Matter Content

Seven maize seeds were selected and weighed to determine their fresh weight (M), then dried at 105 °C for 15 min, followed by baking at 80 °C to a constant weight (m). The dry matter content of the maize was then taken as m/M.

2.3.4. Determination of Starch and Reducing Sugar Content

The starch content of maize was determined using the polarimetry method [17] and the reducing sugar content was determined using the 3,5–dinitrosalicylic acid method [18].

2.3.5. Determination of Soluble Protein

Soluble protein content was determined by the Bradford method [19].

2.3.6. Determination of γ–Aminobutyric Acid and Free Amino Acid Content

The determination method for GABA and free amino acid contents was assessed according to the method described by Chen et al. [20] and Shen et al. [21], with some modifications. A 1.0 g sample of germinated maize was crushed and placed in a glass tube containing 5 mL of deionized water. This was then put into an ultrasonic apparatus (PRESET–SW–12H, Sonoswiss, Switzerland) at 40 °C and 45 kHz for 30 min. The sonicated sample was centrifuged in a centrifuge at $8000\times g$ for 5 min and the supernatant was collected. The precipitate was then extracted again as described above. The supernatants of the two extractions were combined and a 3 mL sample of the combined supernatants was mixed with 7 mL of 100% ethanol, stored at 4 °C overnight, and centrifuged at $8000\times g$ for 10 min. The supernatant was collected and dried under nitrogen (DN–12A, Bilang, Shanghai, China), the dried material was dissolved with 1 mL of 0.02 M HCl (Beijing Solarbio Science & Technology Co., Ltd., Beijing, China), and centrifuged at $8000\times g$ for 10 min. The supernatant contained GABA and free amino acid extracts, which were then were filtered through a 0.45 μm syringe filter (0.45 μm, Shanghai Jinlan Instrument Manufacturing Co. Ltd., Shanghai, China), and an automatic amino acid analyzer (L–8900, Hitachi, Tokyo, Japan) was used to measure GABA and free amino acid content.

2.3.7. Determination of Glutamate Decarboxylase Activity

The determination method of GAD activity was assessed according to the method described by Khwanchai et al. [22], with some modifications. A 0.7 g sample of crushed maize was mixed with 3.2 mL of potassium phosphate buffer (0.05 M, pH 5.8). The potassium phosphate buffer contained 2 mM ethylenediaminetetraacetic acid (Beijing Solarbio Science & Technology Co., Ltd., Beijing, China), 2 mM β–mercaptoethanol (Beijing Solarbio Science & Technology Co., Ltd., Beijing, China), and 0.2 mM pyridoxal phosphate (Beijing Solarbio Science & Technology Co., Ltd., Beijing, China). The mixture was centrifuged at $1000\times g$ for 20 min at 4 °C, and the resulting supernatant was used as the GAD extract. A 200 μL sample of GAD extract was mixed with 100 μL of 1% Glu (Sigma–Aldrich, St. Louis, MO, USA) and incubated for 2 h at 40 °C, pH 5.8, then the enzyme was inactivated at 90 °C for 10 min in a water bath and the GABA content was measured. The determination method of GABA content was described in Section 2.3.6. One unit of GAD activity was defined as the production of 1 μmol GABA from Glu per min at 40 °C.

2.3.8. Statistical Analyses

All the experiments were performed in triplicate and the results are reported as the means ± standard deviation. Values were analyzed by one-way ANOVA, followed by Duncan's multiple range tests using Statistical Product Service Solutions 17.0 (SPSS Inc., Chicago, IL, USA), and a difference was considered significant when the *p*-value was < 0.05.

3. Results and Discussion

3.1. Selection of the Ultrasonic Induction Treatment Process for Germinating Maize

The germination rates of maize treated by different ultrasonic processes are shown in Table 2. The results for the ultrasound group showed that the highest germination rate was achieved at an ultrasonic frequency of 45 kHz and a temperature of 30 °C for a duration of 30 min, with a germination rate of 90.00 ± 1.63%, while the germination rate of maize without ultrasonic induction treatment was 79.50 ± 2.00%. The germination rate of maize after ultrasonic induction treatment was increased by 10.5%. As germination rate is the key to germinating maize preparation, these process parameters were chosen for the ultrasonic induction treatment process for germinating maize and provided a good basis for subsequent experiments. Similar results have been found in other crops, including

Oryza sativa L. [23], *Capparis spinosa* L. [24], *Giycine max* L. [25], *Triticum aestivum* L. [26], and *Cicer arietinum* L. [27].

Table 2. Orthogonal experimental table $L_9 3^3$ and the experimental results for each index.

Factors Number	Factors and Levels			Germination Rate/%
	Ultrasound Frequencies/KHz	Ultrasound Time/min	Ultrasonic Temperature/°C	
1	28	10	30	85.50 ± 3.79 b
2	28	20	35	83.50 ± 2.25 cd
3	28	30	40	84.50 ± 1.91 c
4	45	10	35	86.00 ± 2.58 b
5	45	20	40	86.50 ± 1.91 b
6	45	30	30	90.00 ± 1.63 a
7	100	10	40	83.00 ± 2.58 d
8	100	20	30	83.50 ± 3.00 cd
9	100	30	35	82.50 ± 1.00 d

Different letters indicate significant differences at $p < 0.05$.

The influence of ultrasonic waves on maize germination rate is mainly due to the mechanical vibration and thermal effect of the cavitation effect. The mechanical vibration generated by an ultrasonic wave can accelerate the rate at which water enters maize seed and help the seed to end its dormant state. The thermal effect generated by ultrasonication will accelerate the decomposition and dissolution of maize seeds and provide sufficient nutrients for seed germination. The efficiency of this thermal effect is several times higher than that of ordinary heating. Ultrasound is also a form of energy, which converts electrical energy into kinetic energy in the form of ultrasonic vibration and transmits it to the cells of the seed, stimulates the enhancement of life movement, and accelerates the physiological and biochemical reactions inside the seed [28]. Ultrasonic induction improves the germination rate of maize seeds through biological, chemical and physical methods.

3.2. Effects of Ultrasonic Induction Treatment on Physiological and Biochemical Indicators during Maize Germination

3.2.1. Effect of Ultrasonic Induction Treatment on Sprout Length during Maize Germination

As shown in Figure 1, the growth pattern of maize in the ultrasound group was significantly better than that of the control group. The fibrous roots of maize in the ultrasound group had already grown at 60 h after germination, while the fibrous roots of maize in the control group only appeared visibly at 96 h after germination, which visually proves that ultrasound induction can promote maize germination.

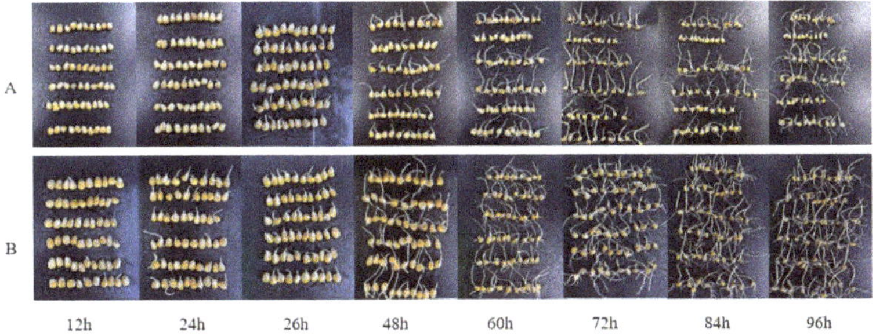

Figure 1. Different stages of germinated maize. (**A**) Control group. (**B**) Ultrasound group.

As shown in Figure 2A, the sprout length of maize in the ultrasound group was significantly higher than that in the control group from 24 h to 96 h after germination ($p < 0.05$). The sprout length of maize in the ultrasound group was 44.45 ± 1.06 mm at the end of germination, which was 1.27 times longer than that in the control group. Ding et al. [28] found that treating long-grain de-hulled rice (*Oryza sativa* L.) at a 25 kHz ultrasonic frequency for 5 min also significantly increased sprout length.

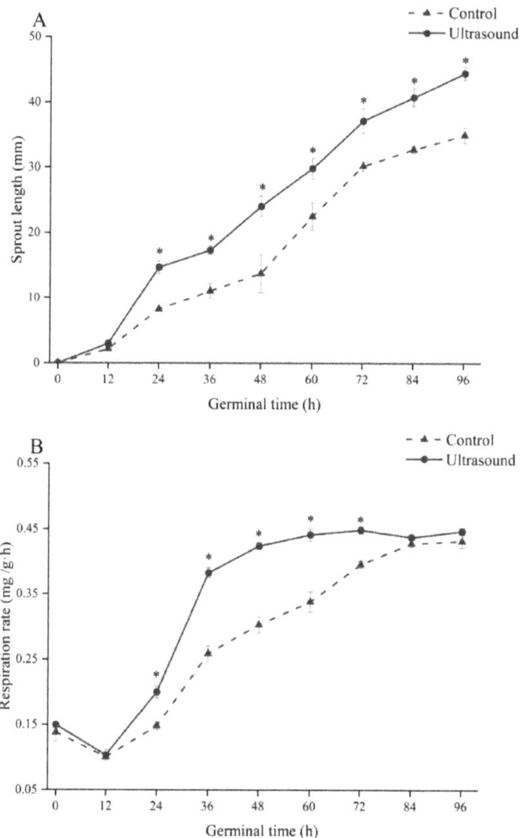

Figure 2. The effect of ultrasound induction treatment on sprout length (**A**) and respiration rate (**B**) during maize germination. * Indicates a significant difference ($p < 0.05$) between the ultrasound and control groups.

Ultrasonic induction treatment has also been demonstrated to stimulate cell growth, accelerate protein synthesis in protoplasts, accelerate the induction of plant cell division, promote the growth of plant seed scions, and shorten the germination period of seeds [29,30]. Cavitation generated by ultrasonic induction treatment shatters seed coats and thereby markedly reduces resistance to water diffusion. This not only promotes the passage of water molecules through cell walls and increases the transfer rate of nutrients but also enables the embryo to freely absorb excess water, thereby promoting embryo development and growth [31].

3.2.2. Effect of Ultrasonic Induction Treatment on the Respiratory Rate of Germinating Maize

As shown in Figure 2B, the respiratory intensity of maize in both groups decreased initially, followed by a gradual increase as germination progressed. However, the respiration rate of the ultrasound group increased more strongly compared to the control and was positively correlated with sprout length (r = 0.969, $p < 0.05$).

Enhanced respiration of maize seeds is one of the most significant physiological changes in the germination process; the enhancement of respiration in the ultrasound group also proved that ultrasonic induction can promote the growth and metabolism of maize seeds. The respiratory intensity of maize in both groups in this experiment started to increase after 12 h of germination, which is consistent with trends reported in rice, wheat, and other cereal crops [26,32]. The respiratory intensity of maize was weakened and then increased during the 0–24 h stage, which is related to the predominance of anaerobic respiration in the early germination of maize [33].

3.2.3. Effect of Ultrasonic Induction Treatment on Dry Matter Content of Maize during Germination

As shown in Figure 3A, the dry matter content in both the control and ultrasound groups showed a similar trend over the course of 96 h. At 96 h, the dry matter content of the ultrasound group was 49.20% of the dry matter content before germination (0.31 ± 0.003 g/g), while that of the control group was 53.23% of the dry matter content before germination (0.33 ± 0.01 g/g).

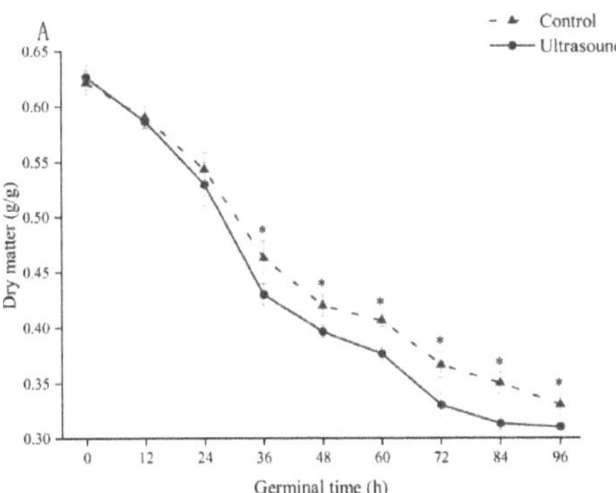

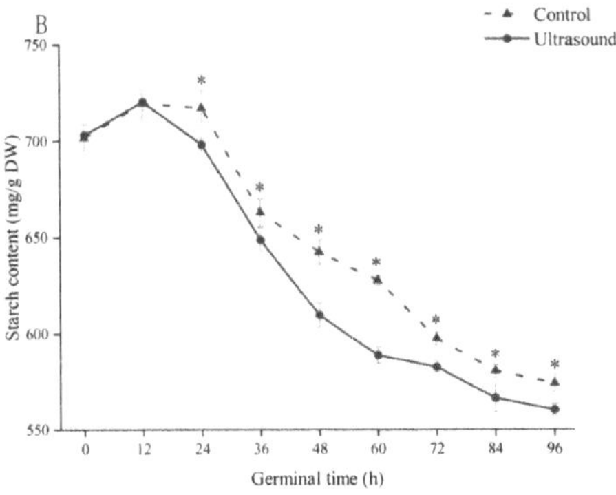

Figure 3. *Cont.*

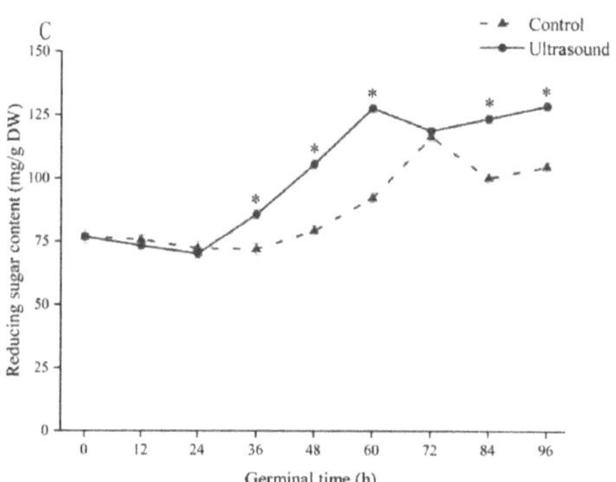

Figure 3. The effect of ultrasound induction treatment on dry matter (**A**), starch content (**B**), and reducing sugar content (**C**) during maize germination. * Indicates a significant difference ($p < 0.05$) between the ultrasound and control groups.

The dry matter content in the ultrasound group was significantly negatively correlated with respiration intensity (R = -0.987, $p < 0.05$). The increase in respiration during maize germination was accompanied by the rapid consumption of organic matter [34], which explains the increased rate of dry matter reduction seen in the ultrasound group compared to the control group.

3.2.4. Effects of Ultrasonic Induction Treatment on the Starch and Reducing Sugar Contents of Germinating Maize

As shown in Figure 3B, changes in the starch contents of both groups were similar as germination progressed, with a slow initial increase followed by a continuous decrease. The starch content of the ultrasound group was significantly lower than that of the control group from 24 h to 96 h of germination. At 96 h of germination, the starch content of the control group was 574.04 $\pm$ 4.41 mg/g, with a starch consumption rate of 18.21%, while the starch content of the ultrasound group was 560.21 $\pm$ 2.97 mg/g, with a starch consumption rate of 20.32%. The increase in starch content in the early stages of germination in both groups was correlated with the decrease in dry matter content due to water absorption by the maize, and the continuous decrease in starch content in the middle and late stages of germination was likely due to the activation of endogenous enzymes and hydrolysis of starch [3]. Hu et al. [35] found that maize showed a 4.74 fold increase in α–amylase activity, a 16.82 fold increase in β–amylase activity, and a 14.41 fold increase in total amylase activity during germination.

The starch consumption rate of maize in the ultrasound group was 2.12% higher than that in the control group at the end of germination ($p < 0.05$), which may be related to the ability of ultrasonic induction treatment to increase the endogenous enzyme activity of plant cells and promote cell growth and biosynthesis [36]. Ultrasound has multiple effects on the hydrolysis of starch during the germination of cereal seeds, including altering the active site and configuration of enzymes present in the grain, affecting synthesis and the conversion of enzymes during germination, and changing the microstructure of starch, leading to changes in enzyme sensitivity [37].

As shown in Figure 3C, both control and ultrasound groups showed similar trends in reducing sugar throughout the germination process. Within 0–12 h of maize germination, the content of reducing sugar did not change significantly; after 12 h of germination, the soluble sugar content started to decrease, likely due to growth of the radicle [38], the activation of endogenous enzymes, and consumption of small sugar molecules [39,40]. The fluctuation in soluble sugar content during the germination process in both groups indicated that the synthesis and metabolism of soluble sugars took place simultaneously. The reducing sugar content of maize in the ultrasound group was significantly higher than that in the control group from 36 to 60 h and from 84 to 96 h of germination, with the peak occurring at 60 h of germination, 12 h earlier than that in the control group. The reducing sugar content in the ultrasound group at the end of germination was 128.5 ± 1.68—an increase of 22.83% compared to the control group. This indicates that ultrasonic induction treatment was able to promote carbohydrate metabolism during maize germination, and this promotion effect primarily manifested in the middle and late stages of maize germination.

Reducing sugar in maize mainly comes from the hydrolysis of starch. Starch is hydrolyzed into maltose under the action of amylase, and maltose is hydrolyzed into glucose by maltase. Under the action of germination and ultrasonic induction, the enzyme activity involved in starch hydrolysis is greatly enhanced and is increased by the reducing sugar content in maize during germination. Xia et al. [23] found that brown rice treated with ultrasound showed a more pronounced decrease in starch content and an increase in reducing sugars compared to untreated samples. In addition to enhancing the activity of amylases during seed germination, ultrasound can increase the concentration of substrates for starch hydrolysis reactions by disrupting starch–protein and starch–lipid complexes, thereby accelerating the rate of starch metabolism [41]. These factors also explain why the reducing sugar content of the maize in the ultrasound group was higher than that in the control group during the germination of the maize.

3.2.5. Effect of Ultrasonic Induction Treatment on the Soluble Protein Content of Germinating Maize

As shown in Figure 4A, the peak soluble protein content of maize in the ultrasound group appeared at 60 h of germination, which was 36 h earlier than that of the control group. During germination, the soluble protein content of maize in the ultrasound group was significantly higher than that in the control group. At the end of germination, the soluble protein content of maize in the ultrasound group was 23.94 ± 0.29 mg/100 g, which was 22.52% higher than that in the control group.

The trend of soluble protein content change during the germination of maize in the two groups was different. In the early stage of maize germination, the soluble protein content increased and its source was the degradation of stored proteins; with the conversion of soluble forms and the synthesis of new proteins, the increase in soluble protein content during germination is presumably due to the formation and completion of the enzyme system, which can be regarded as the result of the dissolution of stored proteins under the action of proteases and the synthesis of new protein substances [42]. The rapid increase in the soluble protein content of maize in the ultrasound group during the early stages of germination suggests that the ultrasonic induction treatment accelerated the completion of this process and that the soluble protein content of maize in the ultrasound group stabilized after reaching its peak, probably because the ultrasonic induction treatment brought forward the morphological establishment of the seedlings and balanced protein synthesis and consumption [43]. The reason for this is that, under the action of ultrasound, a variety of physical mechanical effects are generated between the cell wall of plant cells and the liquid medium and these mechanical effects change the physical structure and integrity of plant cells [33,44].

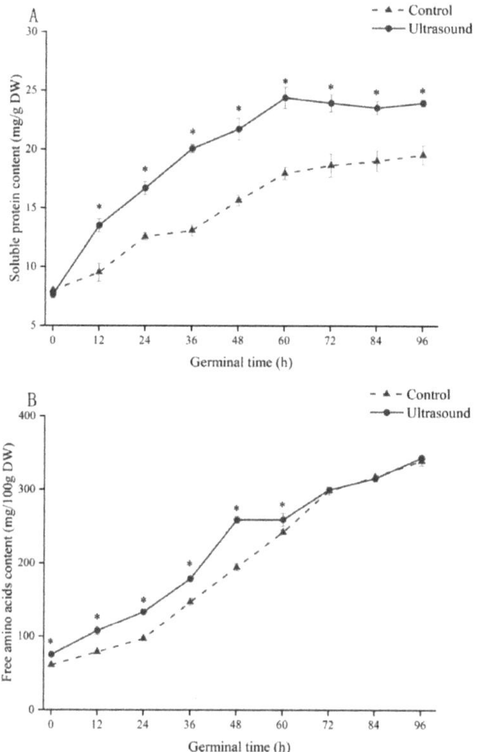

Figure 4. The effect of ultrasound induction treatment on soluble protein content (**A**) and free amino acid content (**B**) during maize germination. * Indicates a significant difference ($p < 0.05$) between the ultrasound and control groups.

3.2.6. Effect of Ultrasonic Induction Treatment on the Free Amino Acid Content of Germinating Maize

As shown in Figure 4B, the free amino acid content of maize in both the control and ultrasound groups continued to increase during germination, and the free amino acid content of maize in the ultrasound group was significantly higher than that in the control group from 0 to 60 h of maize germination ($p < 0.05$). The free amino acid content of maize in the ultrasound group increased from 75.13 ± 2.74 mg/100 g to 343.71 ± 2.94 mg/100 g, indicating that ultrasound promotes protein hydrolysis in maize. Similar findings were obtained by Ding et al. [45], who applied ultrasound to germinating red rice. Our results indicated that ultrasound treatment did not affect free amino acid content in the late stages of germination, which may be related to the concentration of soluble protein in maize at this stage.

During germination, a large number of biological enzymes are activated and released, causing seeds to undergo a series of complex chemical and physiological changes. Proteins are hydrolyzed into small peptides and amino acids by the action of endogenous enzymes, and the products of these hydrolysis reactions are involved in either catabolism or anabolism to produce new proteins. Ultrasonic induction accelerates this process by changing the activity of endogenous enzymes, and ultrasonic waves can destroy the binding structures of protein and starch, change the spatial structures of proteins, increase the concentrations of protein hydrolysis substrates, and accelerate the generation of free amino acids in maize during germination.

3.3. Effect of Ultrasonic Induction Treatment on γ–Aminobutyric Enrichment in Germinating Maize

3.3.1. Effect of Ultrasonic Induction Treatment on the Glutamic Acid Content of Germinating Maize

As shown in Figure 5A, the Glu content in maize continued to increase as germination time increased. The Glu content in the ultrasound group was 10.58–29.26% higher than the control group from 0 to 60 h. The Glu content in the ultrasound group reached a maximum value of 48.91 ± 0.25 mg/100 g at 60 h of maize germination. In the present study, maize Glu content continued to increase during maize germination, while Xu et al. [46] found that Glu content continued to decrease with germination time during the pre-germination period of naked oats, and Ma et al. [47] found that soybean Glu content decreased significantly after germination. Other studies have suggested that the changes in protein and amino acid contents follow species-dependent patterns [48]. The increase in Glu content in both the control and ultrasound groups slowed down in the later stages of germination due to the increased respiratory intensity and the increased dry matter consumption that caused increased Glu consumption. Similar results were found by Kamjijam et al. [49], who examined the Glu content of white and colored rough rice during germination.

Glu acts as an important substrate for intracellular amino acid metabolism, is directly involved in the assimilation and isomerization of ammonia, and is consumed during the synthesis of other amino acids [22,39]. Free Glu is mainly derived from three pathways: proteolysis, the glutamine synthetase cycle, and GABA transaminase reactions. The conversion pathway of Glu is mainly utilized to generate precursors for GABA synthesis [50]. The change in Glu content in the ultrasound group indicates that ultrasonic induction treatment can accelerate the rate of Glu synthesis and the conversion rate of Glu.

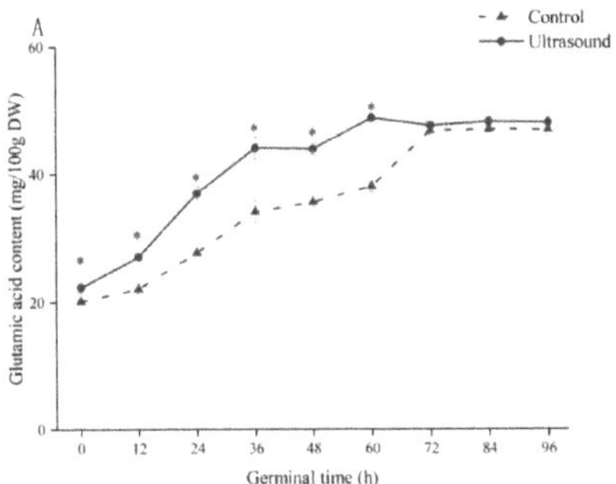

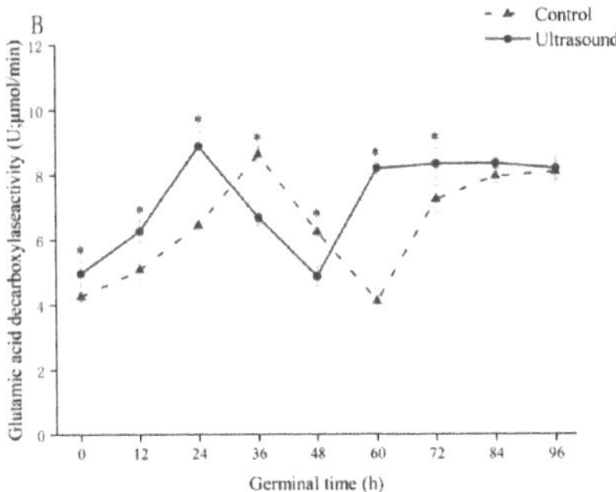

Figure 5. *Cont.*

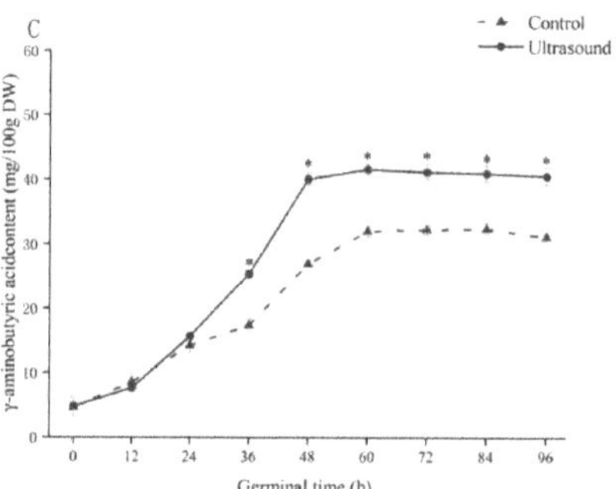

Figure 5. The effect of ultrasonic induction treatment on glutamic acid content (**A**), glutamic acid decarboxylase activity (**B**), and γ–aminobutyric acid content (**C**) during maize germination. * Indicates a significant difference ($p < 0.05$) between the ultrasound and control groups.

3.3.2. Effect of Ultrasonic Induction Treatment on the Activity of Glutamate Decarboxylase in Germinating Maize

As shown in Figure 5B, in both groups, GAD activity initially increased, then decreased, and eventually began increasing again during germination, resulting in two peaks of GAD activity. The first peak of GAD activity was 8.63 ± 0.24 U at 36 h after germination in the control group, while the first peak of GAD activity was 8.89 ± 0.47 U at 24 h after germination in the ultrasound group. The changes in GAD activity in the ultrasound group occurred earlier compared to those in the control group, indicating that ultrasonic induction treatment could promote the GAD activation process.

The peak of GAD activity in the two groups in this study occurred when the Glu content increased most rapidly, and the Glu content increased at a slower rate during the same time period when GAD activity decreased. GAD activity fluctuates in a wide range during grain germination [45], and GAD activity is regulated by the concentration of Glu [22]. GAD is a Ca^{2+}/calmodulin-dependent enzyme with a calmodulin-binding region [51], and ultrasonic induction treatment promotes the degradation of substances around the plant cell wall, increasing intracellular Ca^{2+} concentration and thus GAD activity [45].

3.3.3. Effect of Ultrasonic Induction Treatment on γ–Aminobutyric Content in Germinating Maize

As shown in Figure 5C, the GABA content in both groups continued to increase as germination progressed, reaching a peak of 32.44 ± 0.87 mg/100 g in the control group at 84 h of germination and 41.58 ± 0.81 mg/100 g in the ultrasound group at 60 h. Towards the end of germination, the GABA content of maize in the ultrasound group was 40.60 ± 1.39 mg/100 g, which was 8.42 times higher than the GABA content of ungerminated maize and 30.55% higher than the GABA content of maize in the control group.

GABA is an important intermediate product of the tricarboxylic acid cycle (TCA cycle) branch, which is obtained from GAD catalyzed decarboxylation of Glu in plants. GABA is converted to succinic semialdehyde by transamination and then further converted to succinic acid, thus entering the TCA cycle [52]. The rapid increase in GABA in the early stages of germination in both groups was likely due to the increase in Glu concentration as well as the increase in GAD activity. As germination time progressed, the rate of GABA proliferation began to slow down after the peak of GABA content in both groups, which slowed the increase in Glu and decreased GAD activity at this stage, directly limiting the increase in GABA content (Figure 5A,B).

From 0 to 12 h of germination, the GABA content of maize in the ultrasound group was lower than that of the control group, likely due to the ultrasonic induction treatment accelerating GABA leaching [53]. From 36 to 96 h of germination, the GABA content of maize in the ultrasound group was significantly higher than that in the control group, which may be attributed to the promotion of protein hydrolysis and the activation of endogenous enzymes by ultrasound. Endogenous enzymes are thought to be activated relatively slowly during seed germination [54], and ultrasound can accelerate this process through cavitation. As well as thermal and mechanical effects, ultrasonic induction enhances GAD activity in maize. At the same time, the activation of protease and the destruction of protein molecules induced by ultrasonic waves accelerate the decomposition of protein in maize, so that the concentration of Glu, the precursor of GABA, is maintained at a high concentration; therefore, ultrasonic induction promotes the reaction of GABA generation by increasing the enzyme activity and increasing the substrate concentration, resulting in GABA proliferation in maize. Ultrasound has been shown to significantly increase the GABA content in plants by causing damage to cell interiors and lowering cytoplasmic pH, thereby favoring Glu decarboxylation over GABA transamination [54]; therefore, ultrasonic induction can reduce or slow down GABA metabolism in maize during germination.

Liu et al. [55] found that the GABA content of brown rice treated with ultrasound at 40 KHz power for 30 min was 39.76% higher after germination compared with that of germinated brown rice without ultrasound treatment. GABA content accumulation was also found to be increased more by low-frequency compared to high-frequency ultrasound. As the ultrasound frequency increases, localized high temperature zones may be generated, leading to partial degradation of GABA and a decrease in GABA content [56].

4. Conclusions

This study showed that ultrasonic induction treatment significantly increased germination rate, sprout length, and respiration in maize. These changes may be related to the accelerated metabolism of maize by ultrasonic induction treatment, resulting in accelerated consumption of dry matter and starch and significant increases in soluble proteins, free amino acids, Glu, GABA content, and GAD activity. The GABA content and GAD activity of maize treated with ultrasound were higher compared to the control group throughout the germination process, indicating that ultrasonic induction treatment could promote the germination of maize, accelerate the process of germination, and shorten the time of GABA enrichment in germinating maize. The results of this study can be applied to increase GABA content in germinating maize.

Author Contributions: Data curation, L.Z. and W.L.; formal analysis, W.L.; funding acquisition, M.Y.; investigation, B.Z.; methodology, L.Z.; project administration, M.Y.; resources, N.H. and T.S.; validation, N.H. and B.Z.; writing—original draft, L.Z.; writing—review and editing, M.X. and M.Y. All authors have read and agreed to the published version of the manuscript.

Funding: Shenyang Science and Technology Bureau (RC190134). Fundamental Research Funds of Liaoning Academy of Agricultural Sciences (2021-HBZ-1010).

Acknowledgments: The authors gratefully acknowledge the technical assistance of the food nutrition and quality safety team from LAAS and their allowing the authors to use the necessary instruments, as well as their guidance in carrying out the experiments. This research was funded by the "Fundamental Research Funds of Liaoning Academy of Agricultural Sciences" (2021-HBZ-1010) and the "Science and Technology Innovation Program for Talents in Shenyang" (RC190134).

Conflicts of Interest: The authors declare no conflict of interest.

References

1. Shiferaw, B.; Prasanna, B.M.; Hellin, J.; Bänziger, M. Crops that feed the world 6. Past successes and future challenges to the role played by maize in global food security. *Food Secur.* **2011**, *3*, 307–327. [CrossRef]
2. Salika, R.; Riffat, J. Abiotic stress responses in maize: A review. *Acta Physiol. Plant.* **2021**, *43*, 130. [CrossRef]
3. Cho, D.H.; Lim, S.T. Germinated brown rice and its bio–functional compounds. *Food Chem.* **2016**, *196*, 259–271. [CrossRef] [PubMed]
4. Wang, J.; Ma, H.; Wang, S. Application of Ultrasound, Microwaves, and Magnetic Fields Techniques in the Germination of Cereals. *Food Sci. Technol. Res.* **2019**, *25*, 489–497. [CrossRef]

5. Liao, W.C.; Wang, C.Y.; Shyu, Y.T.; Yu, R.C.; Ho, K.C. Influence of preprocessing methods and fermentation of adzuki beans on γ–aminobutyric acid (GABA) accumulation by lactic acid bacteria. *J. Funct. Foods* **2013**, *5*, 1108–1115. [CrossRef]
6. Zhang, Q.; Xiang, J.; Zhang, L.; Zhu, X.; Evers, J.; Wopke, V.; Duan, L. Optimizing soaking and germination conditions to improve gamma–aminobutyric acid content in japonica and indica germinated brown rice. *J. Funct. Foods* **2014**, *10*, 283–291. [CrossRef]
7. Bouché, N.; Fromm, H. GABA in plants: Just a metabolite? *Trends Plant Sci.* **2004**, *9*, 110–115. [CrossRef]
8. Yaldagard, M.; Mortazavi, S.A.; Tabatabaie, F. Application of ultrasonic waves as a priming technique for accelerating and enhancing the germination of barley seed: Optimization of method by the taguchi Approach. *J. Inst. Brew.* **2008**, *114*, 14–21. [CrossRef]
9. Chiu, K.Y. Ultrasonication–enhanced microbial safety of sprouts produced from selected crop species. *Oncology* **2015**, *88*, 120–126.
10. Qin, Y.-C.; Lee, W.-C.; Choi, Y.-C.; Kim, T.-W. Biochemical and physiological changes in plants as a result of different sonic exposures. *Ultrasonics* **2003**, *41*, 407–411. [CrossRef]
11. Yu, M.; Liu, H.; Shi, A.; Liu, H.; Wang, Q. Preparation of resveratrol–enriched and poor allergic protein peanut sprout from ultrasound treated peanut seeds. *Ultrason. Sonochem.* **2016**, *28*, 334–340. [CrossRef]
12. Xia, Q.; Green, B.D.; Zhu, Z.; Li, Y.; Gharibzahedi, S.M.T.; Roohinejad, S.; Barba, F.J. Innovative processing techniques for altering the physicochemical properties of wholegrain brown rice (*Oryza sativa* L.)—Opportunities for enhancing food quality and health attributes. *Crit. Rev. Food Sci.* **2019**, *59*, 3349–3370. [CrossRef] [PubMed]
13. Cui, L.; Pan, Z.; Yue, T.; Atungulu, G.G.; Berrios, J. Effect of Ultrasonic Treatment of Brown Rice at Different Temperatures on Cooking Properties and Quality. *Cereal Chem.* **2010**, *87*, 403–408. [CrossRef]
14. Hasan, M.M.; Yun, H.-K.; Kwak, E.-J.; Baek, K.-H. Preparation of resveratrol–enriched grape juice from ultrasonication treated grape fruits. *Ultrason. Sonochem.* **2014**, *21*, 729–734. [CrossRef] [PubMed]
15. Xia, Q.; Wang, L.; Li, Y. Exploring high hydrostatic pressure–mediated germination to enhance functionality and quality attributes of wholegrain brown rice. *Food Chem.* **2018**, *249*, 104–110. [CrossRef] [PubMed]
16. He, Y.; Fan, G.; Wu, C.; Kou, X.; Li, T.; Tian, F.; Gong, H. Influence of packaging materials on postharvest physiology and texture of garlic cloves during refrigeration storage. *Food Chem.* **2019**, *298*, 125019. [CrossRef]
17. Mccleary, B.V.; Solah, V.; Gibson, T.S. Quantitative measurement of total starch in cereal flours and products. *J. Cereal Sci.* **1994**, *20*, 51–58. [CrossRef]
18. Saqib, A.A.N.; Whitney, P.J. Differential behaviour of the dinitrosalicylic acid (DNS) reagent towards mono– and di–saccharide sugars. *Biomass Bioenerg.* **2011**, *35*, 4748–4750. [CrossRef]
19. Bradford, M.M. A rapid and sensitive method for the quantification of microgram quantities of protein utilizing the principle of protein–dye binding. *Anal. Biochem.* **1976**, *71*, 248–254. [CrossRef]
20. Chen, H.; Chang, H.; Chen, Y.; Hung, C.; Lin, S.; Chen, Y. An improved process for high nutrition of germinated brown rice production: Low–pressure plasma. *Food Chem.* **2016**, *191*, 120–127. [CrossRef]
21. Shen, S.; Wang, Y.; Li, M.; Xu, F.; Chai, L.; Bao, J. The effect of anaerobic treatment on polyphenols, antioxidant properties, tocols and free amino acids in white, red, and black germinated rice (*Oryza sativa* L.). *J. Funct. Foods* **2015**, *19*, 641–648. [CrossRef]
22. Khwanchai, P.; Chinprahast, N.; Pichyangkura, R.; Chaiwanichsiri, S. Gamma–aminobutyric acid and glutamic acid contents, and the GAD activity in germinated brown rice (*Oryza sativa* L.): Effect of rice cultivars. *Food Sci. Biotechnol.* **2014**, *23*, 373–379. [CrossRef]
23. Xia, Q.; Tao, H.; Li, Y.; Pan, D.; Cao, J.; Liu, L.; Zhou, X.; Barba, F.J. Characterizing physicochemical, nutritional and quality attributes of wholegrain *Oryza sativa* L. subjected to high intensity ultrasound–stimulated pre–germination. *Food Control* **2020**, *108*, 106827. [CrossRef]
24. Rinaldelli, E. Effect of ultrasonic waves on seed germination of *Capparis spinosa* L. as related to exposure time, temperature, and gibberellic acid. *Adv. Hortic. Sci.* **2000**, *14*, 182–188.
25. Lo Porto, C.; Ziuzina, D.; Los, A.; Boehm, D.; Palumbo, F.; Favia, P.; Tiwari, B.; Bourke, P.; Cullen, P.J. Plasma activated water and airborne ultrasound treatments for enhanced germination and growth of soybean. *Innov. Food Sci. Emerg.* **2018**, *49*, 13–19. [CrossRef]
26. Guimarães, B.; Polachini, T.C.; Augusto, P.E.D.; Telis–Romero, J. Ultrasound–assisted hydration of wheat grains at different temperatures and power applied: Effect on acoustic field, water absorption and germination. *Chem. Eng. Process.* **2020**, *155*, 108045. [CrossRef]
27. Goussous, S.J.; Samarah, N.H.; Alqudah, A.M.; Othman, M.O. Enhancing seed germination of four crop species using an ultrasonic technique. *Exp. Agric.* **2010**, *46*, 231–242. [CrossRef]
28. Ding, J.; Hou, G.G.; Dong, M.; Xiong, S.; Zhao, S.; Feng, H. Physicochemical properties of germinated dehulled rice flour and energy requirement in germination as affected by ultrasound treatment. *Ultrason. Sonochem.* **2018**, *41*, 484–491. [CrossRef]
29. Ampofo, J.O.; Ngadi, M. Ultrasonic assisted phenolic elicitation and antioxidant potential of common bean (*Phaseolus vulgaris*) sprouts. *Ultrason. Sonochem.* **2020**, *64*, 104974. [CrossRef]
30. Lorimer, T.; Paniwnyk, L.; Lorimer, J.P. The uses of ultrasound in food technology. *Ultrason. Sonochem.* **1996**, *3*, 253–260.
31. Yaldagard, M.; Mortazavi, S.A.; Tabatabaie, F. Influence of ultrasonic stimulation on the germination of barley seed and its alpha–amylase activity. *Afr. J. Biotechnol.* **2008**, *7*, 2465–2471.
32. Kaur, H.; Gill, B.S. Changes in physicochemical, nutritional characteristics and ATR–FTIR molecular interactions of cereal grains during germination. *J. Food Sci. Technol.* **2021**, *58*, 2313–2324. [CrossRef] [PubMed]

33. Naguib, D.M.; Abdalla, H. Metabolic status during germination of nano silica primed zea mays seeds under salinity stress. *J. Crop Sci. Biotechnol.* **2019**, *22*, 415–423. [CrossRef]

34. Soufan, W.; Okla, M.K. Effect of salt stress on germination, growth and yield of maize (*Zea mays* L. CV. Ghota–1). *J. Pure Appl. Microbiol.* **2014**, *8*, 185–194.

35. Hu, X.; Jiang, X.; Hwang, H.; Liu, S.; Guan, H. Promotive effects of alginate–derived oligosaccharide on maize seed germination. *J. Appl. Phycol.* **2004**, *16*, 73–76. [CrossRef]

36. Wu, F.; Chen, H.; Yang, N.; Wang, J.; Duan, X.; Jin, Z.; Xu, X. Effect of germination time on physicochemical properties of brown rice flour and starch from different rice cultivars. *J. Cereal Sci.* **2013**, *58*, 263–271. [CrossRef]

37. Wang, D.; Yan, L.; Ma, X.; Wang, W.; Zou, M.; Zhong, J.; Ding, T.; Ye, X.; Liu, D. Ultrasound promotes enzymatic reactions by acting on different targets: Enzymes, substrates and enzymatic reaction systems. *Int. J. Biol. Macromol.* **2018**, *119*, 453–461. [CrossRef]

38. Helland, M.H.; Wicklund, T.; Narvhus, J.A. Effect of germination time on α–amylase production and viscosity of maize porridge. *Food Res. Int.* **2002**, *35*, 315–321. [CrossRef]

39. Lu, T.-C.; Meng, L.-B.; Yang, C.-P.; Liu, G.-F.; Liu, G.-J.; Ma, W.; Wang, B.-C. A shotgun phosphoproteomics analysis of embryos in germinated maize seeds. *Planta* **2008**, *228*, 1029–1041. [CrossRef]

40. Young, T.E.; Juvik, J.A.; De Mason, D.A. Changes in carbohydrate composition and α–amylase expression during germination and seedling growth of starch–deficient endosperm mutants of maize. *Plant Sci.* **1997**, *129*, 175–189. [CrossRef]

41. Yu, J.; Engeseth, N.J.; Feng, H. High intensity ultrasound as an abiotic elicitor—effects on antioxidant capacity and overall quality of romaine lettuce. *Food Bioprocess Tech.* **2015**, *9*, 262–273. [CrossRef]

42. Zhang, Q.; Liu, Y.; Yu, Q.; Ma, Y.; Gu, W.; Yang, D. Physiological changes associated with enhanced cold resistance during maize (Zea mays) germination and seedling growth in response to exogenous calcium. *Crop Pasture Sci.* **2020**, *71*, 529–538. [CrossRef]

43. Yang, H.; Gao, J.; Yang, A.; Chen, H. The ultrasound–treated soybean seeds improve edibility and nutritional quality of soybean sprouts. *Food Res. Int.* **2015**, *77*, 704–710. [CrossRef]

44. Li, H.; Yu, J.; Ahmedna, M.; Goktepe, I. Reduction of major peanut allergens Ara h 1 and Ara h 2, in roasted peanuts by ultrasound assisted enzymatic treatment. *Food Chem.* **2013**, *141*, 762–768. [CrossRef] [PubMed]

45. Ding, J.; Ulanov, A.V.; Dong, M.; Yang, T.; Nemzer, B.V.; Xiong, S.; Zhao, S.; Feng, H. Enhancement of gama–aminobutyric acid (GABA) and other health–related metabolites in germinated red rice (*Oryza sativa* L.) by ultrasonication. *Ultrason.Sonochem.* **2018**, *40*, 791–797. [CrossRef]

46. Xu, J.; Hu, Q.; Duan, J.; Tian, C. Dynamic changes in gamma–aminobutyric acid and glutamate decarboxylase activity in oats (*Avena nuda* L.) during steeping and germination. *J. Agric. Food Chem.* **2010**, *58*, 9759–9763. [CrossRef] [PubMed]

47. Ma, M.; Zhang, H.; Xie, Y.; Yang, M.; Tang, J.; Wang, P.; Yang, R.; Gu, Z. Response of nutritional and functional composition, anti–nutritional factors and antioxidant activity in germinated soybean under UV–B radiation. *LWT Food Sci. Technol.* **2020**, *118*, 108709. [CrossRef]

48. Ghumman, A.; Kaur, A.; Singh, N. Impact of germination on flour, protein and starch characteristics of lentil (Lens culinari) and horsegram (*Macrotyloma uniflorum* L.) lines. *LWT Food Sci. Technol.* **2016**, *65*, 137–144. [CrossRef]

49. Kamjijam, B.; Bednarz, H.; Suwannaporn, P.; Jom, K.N.; Niehaus, K. Localization of amino acids in germinated rice grain: Gamma–aminobutyric acid and essential amino acids production approach. *J. Cereal Sci.* **2020**, *93*, 102958. [CrossRef]

50. Shelp, B.J.; Bown, A.W.; Mclean, M.D. Metabolism and functions of gamma–aminobutyric acid. *Trends Plant Sci.* **1999**, *4*, 446–452. [CrossRef]

51. Wu, C.-H.; Dong, C.-D.; Kumar Patel, A.; Rani Singhania, R.; Yang, M.-J.; Guo, H.-R.; Kuo, J.-M. Characterization of waste cell biomass derived glutamate decarboxylase for in vitro gamma–aminobutyric acid production and value–addition. *Bioresour. Technol.* **2021**, *337*, 125423. [CrossRef] [PubMed]

52. Kinnersley, A.M.; Turano, F.J. Gamma aminobutyric acid (gaba) and plant responses to stress. *Crit. Rev. Plant Sci.* **2000**, *19*, 479–509. [CrossRef]

53. Carrera, C.; Ruiz–Rodriguez, A.; Palma, M.; Barroso, C.G. Ultrasound–assisted extraction of amino acids from grapes. *Ultrason. Sonochem.* **2015**, *22*, 499–505. [CrossRef]

54. Ding, J.; Hou, G.G.; Nemzer, B.V.; Xiong, S.; Dubat, A.; Feng, H. Effects of controlled germination on selected physicochemical and functional properties of whole–wheat flour and enhanced gamma–aminobutyric acid accumulation by ultrasonication. *Food Chem.* **2018**, *243*, 214–221. [CrossRef]

55. Liu, R.; He, X.; Shi, J.; Nirasawa, S.; Tatsumi, E.; Li, L.; Liu, H. The effect of electrolyzed water on decontamination, germination and γ–aminobutyric acid accumulation of brown rice. *Food Control* **2013**, *33*, 1–5. [CrossRef]

56. Ding, J.; Johnson, J.; Chu, Y.F.; Feng, H. Enhancement of gamma–aminobutyric acid, avenanthramides, and other health–promoting metabolites in germinating oats (*Avena sativa* L.) treated with and without power ultrasound. *Food Chem.* **2019**, *283*, 239–247. [CrossRef] [PubMed]

 *foods*

Article

Discrete Element Simulation Study of the Accumulation Characteristics for Rice Seeds with Different Moisture Content

Jinwu Wang [1], Changsu Xu [1], Xin Qi [1], Wenqi Zhou [1] and Han Tang [1,2,*]

1 College of Engineering, Northeast Agricultural University, Harbin 150030, China; jinwuw@neau.edu.cn (J.W.); ChangsuXu@neau.edu.cn (C.X.); XinQi@neau.edu.cn (X.Q.); zwq@neau.edu.cn (W.Z.)
2 Key Laboratory of Crop Harvesting Equipment Technology of Zhejiang Province, Jinhua Polytechnic, Jinhua 321007, China
* Correspondence: tanghan@neau.edu.cn; Tel.: +86-0451-55190950

Abstract: To study the accumulation characteristics of rice seeds with different moisture content, an accurate model of rice seeds was established by 3D scanning technology. The accumulation state of rice seeds by the "point source" accumulation method was analyzed by proportioning and measuring the simulation parameters with different moisture content. The accumulation process was simulated at 10.23%, 14.09%, 17.85%, 21.77%, 26.41% and 29.22% moisture content, respectively. The velocity and force state of the seeds were visually analyzed by using the accumulation process with a moisture content of 29.22%. The accumulation process was divided into four stages according to the velocity characteristics of the seeds. The average force and kinetic energy of the rice seeds outside the cylinder were obtained, and the average force of the rice seeds outside the cylinder was proved to be the direct cause of the velocity change during the accumulation process. The mechanical characteristics of rice seeds in the quasi-static accumulation stage were partitioned and systematically analyzed. The force distribution of the "central depression" structure of rice seeds with a moisture content of 10.23%, 14.09% and 17.85% on the horizontal surface appeared. The higher the moisture content of rice seeds, the more likely the typical "circular" force structure appeared, and the more uniformly the force on the horizontal surface was distributed in the circumference direction.

Keywords: natural repose angle; point source; velocity characteristics; mechanical characteristics; distribution

Citation: Wang, J.; Xu, C.; Qi, X.; Zhou, W.; Tang, H. Discrete Element Simulation Study of the Accumulation Characteristics for Rice Seeds with Different Moisture Content. *Foods* **2022**, *11*, 295. https://doi.org/10.3390/foods11030295

Academic Editors: P. J.A. Sobral and Ioan Cristian Trelea

Received: 5 December 2021
Accepted: 20 January 2022
Published: 22 January 2022

Publisher's Note: MDPI stays neutral with regard to jurisdictional claims in published maps and institutional affiliations.

Copyright: © 2022 by the authors. Licensee MDPI, Basel, Switzerland. This article is an open access article distributed under the terms and conditions of the Creative Commons Attribution (CC BY) license (https://creativecommons.org/licenses/by/4.0/).

1. Introduction

The accumulation is one of the most common states of existence during the production, storage and processing of rice seeds. It is a highly dynamic and variable system with no viscous connection between seeds [1]. As a complex nonlinear system, the seed medium responds to various perturbations by "self-organizing" with changes in external factors, and its unevenly distributed internal chain-like mechanical structure allows it to exhibit many singularities like both solid and liquid [2]. Agricultural engineering and food engineering are the most widely used fields of grain systems. Grain accumulation is involved in the storage of materials, subsequent processing and the development of related equipment. The research on the accumulation characteristics of grain systems has important basic scientific significance and engineering application value [3].

Due to the dual behavior of solid and liquid, there are interactions between grain and granary before accumulation, such as blockage, arching, funnel flow and unstable flow [4,5]. The interaction between grain and granary affects the occurrence of accumulation, and the blockage problem has been widely concerned by many scholars. Wan et al. [6] studied the influence of orifice shape on particle velocity through the combination of simulation and experiment. Zaki et al. [7] explored the effects of different hopper shapes on grain discharge speed and efficiency, and established the relationship between particle speed and flow at the outlet of the silo. The lower the particle velocity and flow rate at the outlet of the silo, the easier the blockage problem is. Liu et al. [8] analyzed the relationship between silos with different orifice sizes and arch structure, and established the orifice size prediction model of curved fabric buckets. Xiao et al. [9] explored that the arch structure of particle flow at the outlet is the main cause of blockage from the perspective of particle dynamics. Ahmadi et al. [10] studied that the friction coefficient of particles has an important influence on the size formation of arch structure. The above research focuses on the interaction between grain particles and granaries, but there is little research on the dynamic characteristics of the grain itself in the process of unloading. The dynamic and mechanical characteristics of particles in the process of stacking are of great significance to guide the safe storage of grain. After the formation of the accumulation, the particle system will show a certain static and stable state, in which the "central depression" phenomenon is one of the key elements of the study. The maximum normal force at the bottom of the accumulation does not always occur at the center of the accumulation but may occur at the periphery of the center of the accumulation [11,12]. Horabik et al. [13] studied that the pressure of particles with different shapes on the bottom presented different distribution states. The phenomenon of "central depression" was directly manifested by significant differences in the distribution of contact forces. The contact force between particles was strongly correlated with the frictional characteristics of the surface of the particles [14]. Shi et al. [15] explored the effects of particle sliding friction and rolling friction on the accumulation process. In addition, Cao et al. [16] showed that the tangential force was higher when the bottom surface roughness was larger, and the corresponding "central depression" of the normal force was more obvious. Horabik et al. [17] studied the effects of stress distribution and contact friction on the "central depression" phenomenon. Carlevaro et al. [18] studied the formation of force chains and the arching mechanism in the accumulation of the particles and showed that the structure of the arching force chain supported the seed accumulation and carried most of the acting forces in the accumulation.

The non-transparency of the particle system led to the fact that the mechanical characteristics of the accumulation cannot be directly observed through tests. Meanwhile, the large amount of the particle material in engineering cannot be directly analyzed and determined, which largely limited the development of bulk mechanics. With the progressive development of computer simulation technology and similarity theory, the discrete element method has become an effective numerical simulation method to study the behavior of particle systems [19]. In the field of agricultural engineering and food engineering, the discrete element method is widely used in solving the movement characteristics of materials in machines, such as the screening status of grain particles in combine harvesters [20], gas and solid coupling simulation test [21], and the dynamic changes of grain in the rice milling machine [22]. In the discrete element numerical simulation, accurately obtaining the physical parameters of materials is the key to effectively exploring the particle motion and mechanical properties. The physical parameters mainly include material properties and interaction properties [23]. The different moisture content will lead to the change of physical properties of materials, which directly affects the accuracy of simulation results [24].

The spherical seeds with a single physical characteristic were mostly used by the discrete element simulation process. The physical characteristic of grain crops was changed significantly as their surface irregularities and different moisture content, which can cause variability in the mechanical characteristic during the accumulation process [25]. In this paper, the simulation model was accurately established based on rice grain by 3D scanning technology, and the numerical values of simulation parameters were explored through experiments with different moisture content. The accumulation state and mechanical properties of rice grains under different moisture content were studied by the discrete element method.

2. Materials and Methods

2.1. Parameter Determination

The characteristics reflected by rice with different moisture content are mainly manifested in the differences of physical parameters such as the density, modulus of elasticity, Poisson's ratio, etc., [26]. Taking the rice grains of "Longjing 29" widely planted in the main rice producing areas of China as the research object, the 1000 grains were selected to measure and analyze their triaxial dimensions by using vernier calipers (Ningbo Deli tools Co., Ltd., Ningbo, China). The grain length is mainly distributed between 5–6 mm, the width is mainly distributed between 1–2 mm, and the thickness is mainly distributed between 1–2 mm. The average values of the triaxial dimensions are 5.63 mm, 2.05 mm and 1.80 mm, respectively.

The moisture content of rice seeds during the harvesting period was concentrated in the range of 10.23–29.22%. To achieve the test range of moisture content for the ideal rice seeds, the undamaged rice seeds of 1000 were artificially calibrated by the water supply method [27]. Each group of parameters was repeated three times, and the results were averaged. The main parameters of rice grain include moisture content [28], density [29], elastic modulus [30], static friction coefficient and dynamic friction coefficient [31], collision recovery coefficient [32] and Poisson's ratio [33]. The parameter measuring methods are reported in Supplementary Materials (Table S1).

The rice seeds always interact with mechanical parts during production, storage and processing. Therefore, the mechanical contact material was set to be Q235 steel with the following physical characteristics: the density of 7850 kg/m^3, the modulus of elasticity of 212 MPa and the Poisson's ratio of 0.31. The results of the physical parameters of the rice seeds with different moisture content and the contact mechanical parameters with Q235 steel are shown in Table 1.

Table 1. Mean values (±standard error) of measuring results of parameters at different moisture contents of rice seeds.

Parameters	10.23% Moisture Content	14.09% Moisture Content	17.85% Moisture Content	21.77% Moisture Content	26.41% Moisture Content	29.22% Moisture Content
Density/kg·m^{-3}	910 ± 6.93 [e]	966 ± 7.00 [d]	990 ± 9.54 [cd]	1005 ± 7.94 [c]	1197 ± 4.00 [b]	1253 ± 32.19 [a]
Modulus of elasticity/MPa	252 ± 11.53 [a]	220 ± 13.11 [b]	211 ± 4.00 [b]	208 ± 3.61 [b]	170 ± 5.29 [c]	102 ± 4.58 [d]
Poisson's ratio	0.38 ± 0.01 [a]	0.37 ± 0.00 [ab]	0.35 ± 0.01 [bc]	0.33 ± 0.01 [c]	0.30 ± 0.01 [d]	0.25 ± 0.03 [e]
Static friction coefficient *	0.28 ± 0.03 [d]	0.30 ± 0.00 [cd]	0.32 ± 0.02 [bc]	0.32 ± 0.01 [bc]	0.35 ± 0.01 [b]	0.44 ± 0.03 [a]
Dynamic friction coefficient *	0.015 ± 0.001 [c]	0.017 ± 0.001 [c]	0.026 ± 0.003 [b]	0.026 ± 0.002 [b]	0.028 ± 0.001 [ab]	0.031 ± 0.001 [a]
Collision recovery coefficient *	0.61 ± 0.02 [a]	0.58 ± 0.01 [ab]	0.56 ± 0.01 [b]	0.43 ± 0.03 [c]	0.38 ± 0.00 [d]	0.30 ± 0.01 [e]
Static friction coefficient **	0.44 ± 0.04 [c]	0.48 ± 0.04 [c]	0.51 ± 0.01 [c]	0.56 ± 0.02 [b]	0.60 ± 0.01 [ab]	0.63 ±0.00 [a]
Dynamic friction coefficient **	0.028 ± 0.002 [d]	0.030 ± 0.001 [d]	0.031 ± 0.000 [d]	0.037 ± 0.002 [c]	0.042 ± 0.001 [b]	0.050 ± 0.003 [a]
Collision recovery coefficient **	0.55 ± 0.03 [a]	0.52 ± 0.02 [a]	0.48 ± 0.01 [b]	0.45 ± 0.00 [b]	0.33 ± 0.02 [c]	0.27 ± 0.03 [d]

Different letters within the same row indicate a statistically significant difference ($p < 0.05$); * represents the measured value between rice grain and Q235 steel; ** represents the measured value between rice grain and rice grain.

The effects of different moisture content on physical characteristics were evaluated by SPSS statistics 22.0 (SPSS, Chicago, IL, USA). Analysis of variance (ANOVA) was performed with Duncan's multiple comparison to assess the significance level with a threshold of $p < 0.05$. The results were expressed as the mean $\pm$ standard error. Each replicate ($n = 3$) was taken as a random effect in this model.

2.2. Discrete Element Numerical Simulation Scheme

The overall structure of the rice seeds is complex. To accurately establish the model of the rice seeds, the rice seeds with approximate average values of the three axes dimensions were selected (5.58 mm, 2.05 mm and 1.83 mm). The Reeyee X5 3D laser scanner (Nanjing Weibu 3D Technology Co., Ltd., Nanjing, China) was used to extract the 3D geometric feature parameters of the rice seeds. A 3D laser scanner projected gratings onto the surface of the rice grain. According to the shape of the fringes changing with the curvature, the spatial coordinates of each point of the surface were accurately calculated by using the orientation method and the triangulation method. The 3D point cloud data were generated. Shading, denoising, point cloud registration, point cloud triangulation, merging and model correction were carried out in turn. Using automated reverse engineering software to convert scan data into an accurate digital model, the ideal 3D scanning model of rice grain was obtained. This 3D laser-scanned rice grain model was imported into EDEM and filled with spherical element particle polymerization [34]. The process of discrete element modeling of rice seeds is reported in Supplementary Materials (Figure S1).

The surface of rice seeds is smooth without adhesion. To accurately represent the contact characteristics of rice seeds, the Hertz–Mindlin (no-slip) model was used to construct the contact between rice seeds and rice seeds and between rice seeds and steel plates [35].

Place the cylinder vertically on the horizontal plane (the cylinder and the horizontal plane are made of Q235 steel). The diameter of the cylinder is 50 mm and the height is 150 mm. The Z-axis direction was selected as the direction of gravity with reference to the constructed model coordinates, and its value was set to -9.81 m/s^2. The "falling rain method" was used to make the rice seeds accumulate naturally in the cylinder with zero initial velocity. The total number of rice seeds was set to 4000. When the rice seeds were completely settled in the cylinder, it started to lift the cylinder at a speed of 0.05 m/s. Until the rice seeds in the cylinder were naturally scattered into an accumulation under the action of gravity, the simulation was stopped after the rice seeds state of the accumulation was completely stationary and stable. The Rayleigh time step was determined jointly by the software based on parameters such as the seed radius and the density [36]. The Rayleigh time step was finally determined to be 20% and the total simulation time for each group was set to 3 s. The simulations of six groups for rice seeds were conducted with different moisture content, and this method was used for each group in this paper.

2.3. Comparative Verification of Accumulate Test

To verify the accuracy of the established discrete element model of rice seeds before simulation, the natural rest angle was measured by comparison between EDEM simulation and bench test. For each group, the 4000 rice seeds of "Long Japonica 29" with different moisture content were selected as the test varieties. The cylinder and horizontal surface of the same size and material were selected, and the same speed was used to lift the cylinder to ensure the consistency between EDEM simulation and bench test.

To accurately determine the natural rest angle of rice seeds, the accumulation of rice seeds in three-dimensional space was photographed, and the MATLAB software was used to process the image noise, grayscale and binarization [37].

3. Results and Discussion

3.1. Results of Comparative Verification for Accumulate Test

The unilateral contour curve of rice seed accumulation was extracted and the ratio of the vertical to horizontal pixel values of the curve was used to determine the magnitude of the natural rest angle. The results of the bench test and simulation test of natural rest angle for rice seeds with different moisture content are shown in Supplementary Materials (Figure S2).

The natural rest angle of rice seeds with different moisture content was analysed (Figure 1). The natural rest angle increased gradually with the increase of moisture content. This was mainly due to the fact that the higher the moisture content, the higher the friction between seeds, resulting in poor mobility of the population. The overall results of the comparison between the natural rest angle bench test and the simulation test for rice seeds with different moisture content had certain errors, and the maximum error was 2.45%.

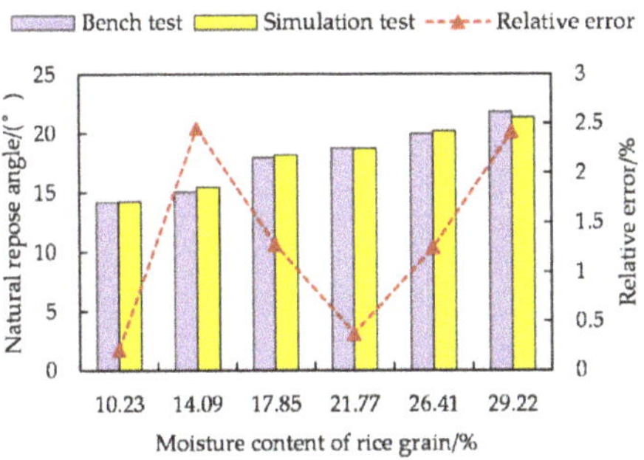

Figure 1. Natural rest angle of rice seeds with different moisture content.

This was mainly due to the fact that the rice seed metamodel consists of multiple small spheres aggregated to increase the surface area of rice seeds, and the frictional force at each contact part of adjacent seeds increased accordingly in the free accumulation process. Meanwhile, the overall shape of rice seeds in the bench test showed some variability. Within the allowable range of error, it showed that the discrete element model of the rice seed was accurate and effective, which laid a foundation for the study of mechanical characteristics of rice seed in the later stage of the accumulation.

3.2. Velocity Distribution in the Accumulation Process of Rice Seeds

To explore the velocity distribution of rice seeds in the accumulation process, the accumulation process of rice seeds with 29.22% moisture content as an example was taken. The velocity distribution in the accumulation process was deeply analyzed (Figure 2a).

The seeds gathered at the mouth of the cylinder and formed a column of material flow out from the bottom at the initial stage of accumulation for rice seeds (0–0.5 s). When the seeds reached the horizontal surface of the substrate, the seeds gradually accumulated in a cone shape at the center of the substrate. The speed of the inner layer of seeds was small to form the inner static support structure, and the outer layer of seeds had a larger speed and gradually stayed on the surface of the cone to expand outward at this time. This stage was called the substrate support forming stage. With the increase of time (0.5–0.9 s), the basal accumulation area of rice seeds kept expanding, and the number of outer seeds with higher velocity increased. This was mainly due to the continuous accumulation of the seeds layer, and the continuous seeds flow still fell heavily under the action of the gravity in the internal cylinder, which had a very strong disturbance effect on the seeds that were about to accumulate at rest. The seeds were forced to continue to roll with the flow of seeds. This stage was called the expansion accumulation stage, which was characterized by large seeds speed as a whole and rapid accumulation shape. The rice seeds had completely fallen out of the cylinder at the moment of 0.9–1.3 s, and the rice seeds in the accumulation process had no constraints and limitations of the cylinder. The rice seeds were freely scattered and shaped under the action of gravity, at which time the speed of the rice seeds gradually decreased. This stage was called the free accumulation stage. The accumulation process of the rice seeds was completed and the morphology of the accumulation no longer changes after 1.3 s. The accumulation of the rice seeds at this stage had completely stabilized. This stage was called the quasi-static accumulation stage.

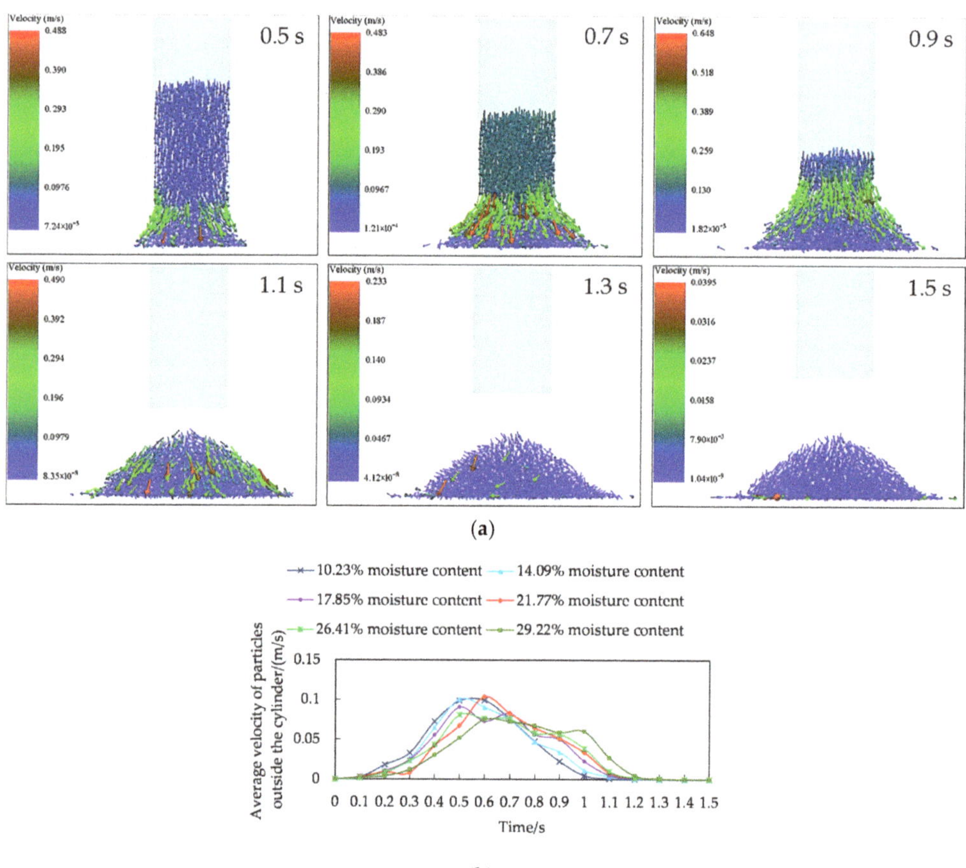

(a)

(b)

Figure 2. Characteristics of rice seeds accumulation velocity with the moisture content of 29.22%: (**a**) velocity distribution in the accumulation process; (**b**) average velocity of the rice seeds outside the cylinder at different moments.

To continuously quantify the velocity distribution of rice seeds with different moisture content during the accumulation process, the average velocity of the seeds outside the cylinder (the seeds formed the accumulation) at different moments was compared and analyzed with the accumulation time as the horizontal coordinate and the average velocity of the seeds as the vertical coordinate (Figure 2b). The average velocity of rice seeds with different moisture content showed a trend of increasing first and then decreasing during the accumulation process. When the moisture content was 10.23%, the average speed of rice seeds first rose to the maximum value. With the increase of moisture content, the average speed rosed to the maximum value in turn. The higher the moisture content, the later the average speed rosed to the maximum value. When the average speed of rice seeds reached the maximum value, the average speed of rice seeds with a moisture content of 10.23% firstly decreased and quickly reduced to 0. With the increase of moisture content, the average speed decreased to 0 in turn. The higher the moisture content, the later the time of the average speed decreased to the lowest value shifted. This may be due to the fact that as the moisture content increased, the friction on the surface of the rice seeds increased. Both the friction with the inner wall of the cylinder and the friction with the seeds produced a certain hysteresis effect, resulting in a longer time between the base-supported forming stage and the expanded accumulation stage in the accumulation process, and a corresponding lag between the free accumulation stage and the quasi-static accumulation stage.

3.3. Analysis of the Mechanical Characteristics of Rice Seeds during Accumulation

To further analyze the force situation during the accumulation of rice seeds at different moments, the accumulation process of rice seeds with a moisture content of 29.22% was taken as an example, and the force state of the inner and outer layers of the accumulation of rice seeds was visually analyzed by longitudinal section along the diameter direction of the accumulation of rice seeds. The force distribution in the accumulation process was deeply analyzed (Figure 3a).

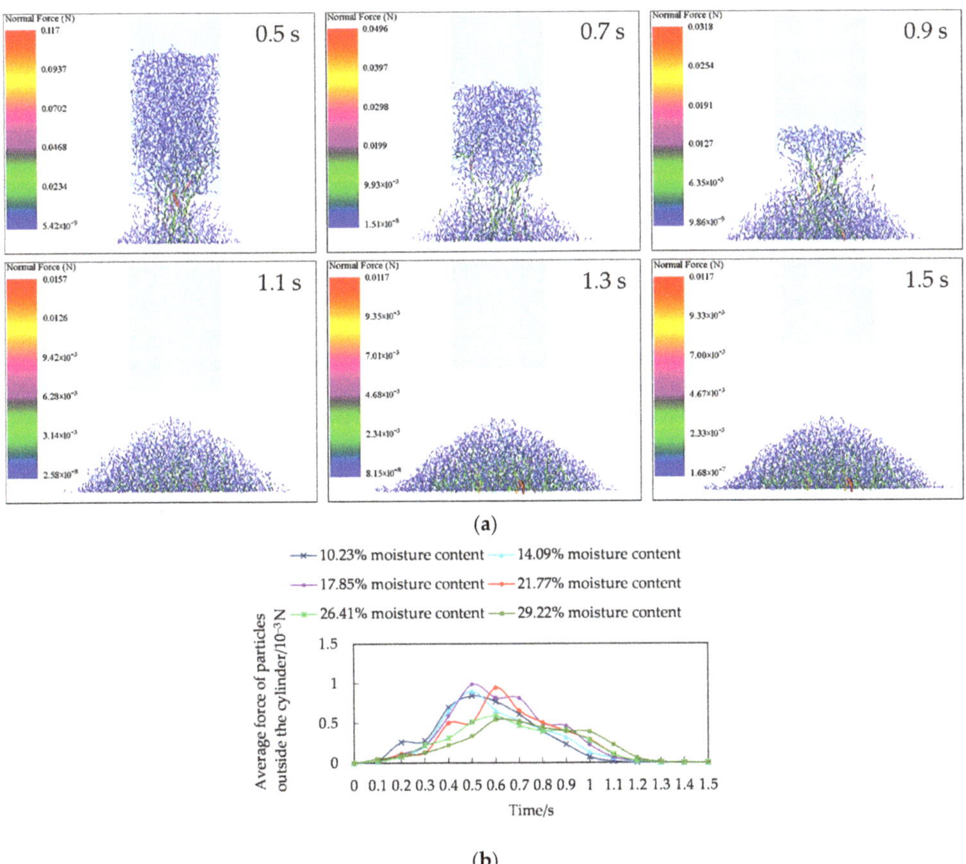

Figure 3. Force of rice seeds accumulation with 29.22% moisture content: (**a**) force distribution; (**b**) average force of the rice seeds outside the cylinder at different moments.

In the substrate support forming stage, the inter-seed force was small and relatively uniform in the outer layer of the accumulation, while the local force between the seeds was larger in the inner part of the accumulation. A few seeds were supporting the weight of the whole accumulation and gradually forming the accumulation form.

When the rice seeds entered the expansion accumulation stage, the maximum force between the seeds gradually decreased with time and still occurred inside the accumulation. This was mainly due to the fact that the inside seeds were not only subjected to their own gravity, but also to the pressure of the seeds falling from the inside cylinder.

When the rice seeds entered the free accumulation stage, the maximum force between the seeds continued to decrease, but the maximum force occurred in the area of the inner bottom of the seed accumulation in contact with the horizontal surface. When the rice seeds entered the quasi-static accumulation stage, the maximum force between the seeds no longer changed, and the maximum force area still occurred at the bottom of the inner layer where the seeds were in contact with the horizontal surface. The force gradually decreased in a radial pattern from the inner layer to the outer layer of the seeds. The force in the outer layer was the smallest, which was mainly due to the fact that the outer layer of seeds was free only by its own gravity and the support force of the seed accumulation. At this time, the overall force on the accumulation reached equilibrium, and the accumulation state no longer changes.

To quantitatively analyze the distribution of accumulation forces on rice seeds with different moisture content during the accumulation process, and also to study the influencing factors on the variation of accumulation speed of rice seeds with different moisture content, the forces on seeds outside the cylinder (seeds forming the accumulation) were compared and analyzed with the accumulation time as the horizontal coordinate and the average force on seeds as the vertical coordinate at different times (Figure 3b). The average force of rice seeds with different moisture content showed a trend of increasing first and then decreasing during the accumulation process. At 10.23% of moisture content, the average force of rice seeds first rose to the maximum value. As the moisture content increased, the average force rose to the maximum value. The higher the moisture content, the later the average force rose to the maximum value. After the average force of rice seeds reached the maximum value, the average force of rice seeds with a moisture content of 10.23% firstly decreased and soon reduced to 0. With the increase of moisture content, the average force decreased to 0 in turn. The higher the moisture content, the time that the average force decreased to the lowest value shifted later. The changing trend of the average force of the seeds outside the cylinder at different times was consistent with the changing trend of the average velocity. This indicated that the force during the accumulation of the seeds was the direct cause of the change of their velocity.

To reveal the change of motion posture of the rice seeds with different moisture content during the accumulation process from the perspective of energy, the average kinetic energy of the rice seeds outside the cylinder was analyzed (Figure 4). As can be seen from Figure 4a,c, the overall translational kinetic energy was greater than the rotational kinetic energy during the accumulation of rice seeds, indicating that the sliding motion posture took up a larger proportion in the accumulation process. The average translational kinetic energy and average rotational kinetic energy of rice seeds with different moisture content showed a trend of first increasing and then decreasing, and the higher the moisture content, the longer the time of increasing and decreasing of translational kinetic energy and rotational kinetic energy.

The phenomenon was mainly due to the accumulation of rice seeds in the early stage of substrate support forming stage, and most of the rice seeds as a skeleton structure scattered in the bottom of the horizontal surface by comprehensive analysis of the speed and force during the accumulation of rice seeds. With the time increased, a large number of rice seeds formed the accumulation pattern and filled the spaces between the seeds, and the rice seeds became more prone to sliding or rolling. When there were no rice seeds down in the cylinder, no new rice seeds collided and pressed each other, the motion of the seeds was no longer active, and the average translational kinetic energy and average rotational kinetic energy gradually decreased after the free accumulation stage to the quasi-static accumulation stage. As can be seen from Figure 4b,d, the higher the moisture content, the higher the energy in the process of seed accumulation and the longer the time of kinetic energy decayed. This is explained from the side that the higher the moisture content, the longer the time of increase and decrease of the translational kinetic energy and rotational kinetic energy.

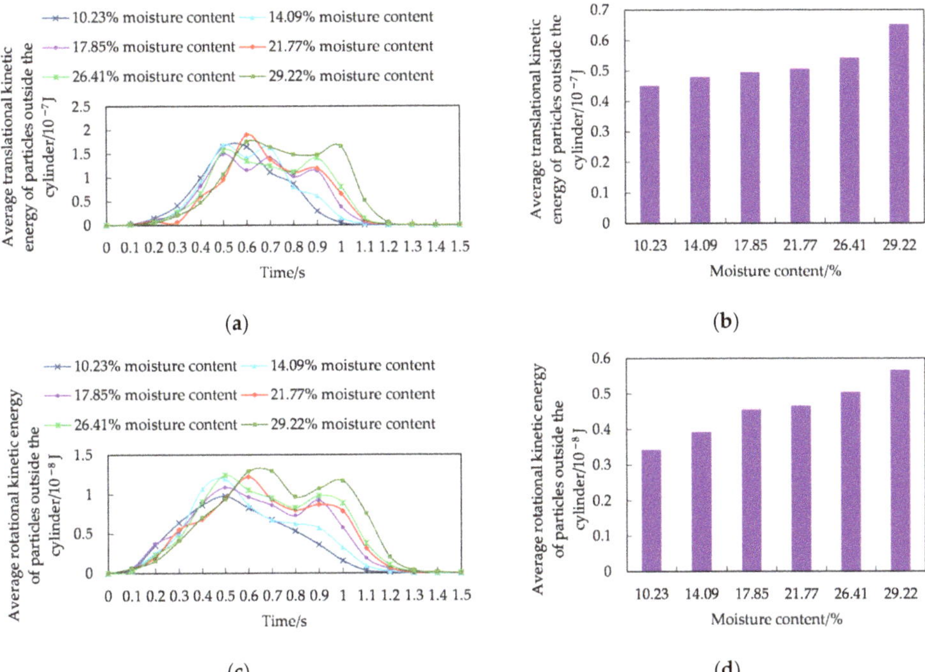

Figure 4. Average kinetic energy of rice seeds outside the cylinder: (**a**) average translational kinetic energy of rice seeds outside the cylinder at different times; (**b**) average translational kinetic energy of rice seeds outside the cylinder at the accumulation process; (**c**) average rotational kinetic energy of rice seeds outside the cylinder at different times; (**d**) average rotational kinetic energy of rice seeds outside the cylinder at the accumulation process.

3.4. Analysis of Mechanical Characteristics of Rice Seeds in Quasi-Static Accumulation Stage

The rice seed system showed a certain static and stable state on the macroscopic level after the accumulation formation. However, the mechanical characteristics of the seed-to-seed and the influence law of the seed accumulation on the mechanical characteristics of the horizontal surface were still unclear. To analyze the mechanical characteristics of rice seeds in the quasi-static accumulation stage in detail, the seed accumulation was uniformly partitioned in a circular shape. The average value of the force on the seeds within a cylinder with a radius of 6 mm arranged in the center of the accumulation was the value of radial distance 0. The radial distance of 10 mm represented the average value of the force on the seeds within a circular cylinder with an inner diameter of 4 mm and an outer diameter of 16 mm, and so on. The partitions with radial distances of 0, 10, 20, ..., 140 mm were divided. The partitioning method is reported in Supplementary Materials (Figure S3a).

The variation of the normal contact force between the rice seeds along the radial distance was analysed (Figure 5a). The normal contact force between rice seeds with different moisture content was negative, indicating that the direction of contact force was downward. With the increase of radial distance, the normal contact force gradually decreased and reached 0 at the radial distance of 140 mm. This was mainly due to the fact that the amount of rice seeds in the area gradually decreased with the increase of radial distance, which led to the gradual decrease of the contact force between the seeds.

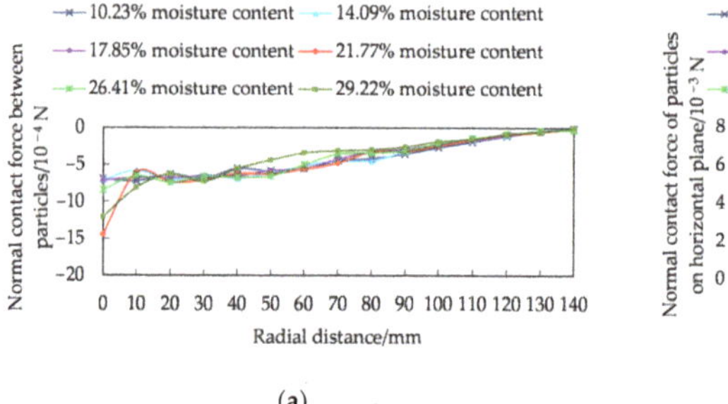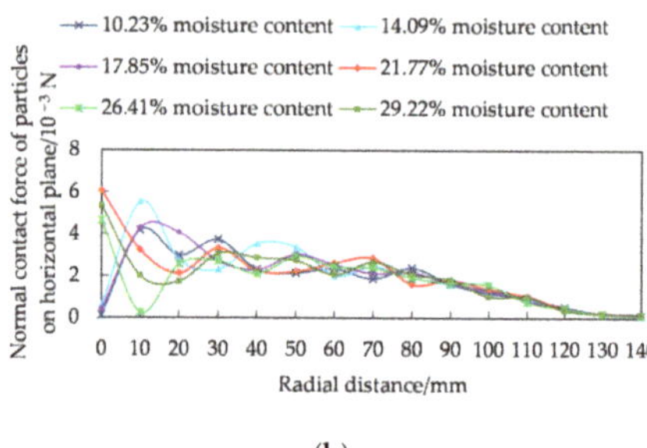

(a) (b)

Figure 5. Analysis of mechanical characteristics of rice seeds in quasi-static accumulation stage: (**a**) normal contact force between rice seeds, (**b**) normal contact force on horizontal surface.

The contact force between the seed accumulation in the quasi-static accumulation stage was transferred to the bottom of the seeds through the force chain. The normal stress distribution at the bottom of the granular pile has always been a hot issue in the study of granular mechanics.

In this paper, the normal contact force on the horizontal surface of rice seed accumulation with different moisture content in radial position was extracted (Figure 5b). The normal contact force of rice seed accumulation with 10.23%, 14.09% and 17.85% moisture content to the horizontal surface was not the maximum value at the central part (radial distance of 0), but with the increased radial distance, the maximum value of normal contact force to the horizontal surface appeared at the position of the radial distance of 10 mm. The force distribution of the "central depression" structure was shown for rice seed accumulation with 10.23%, 14.09% and 17.85% moisture content. With the increased radial distance, the normal contact force on the horizontal surface gradually decreased, in which the second maximum peak of normal contact force occurred at the radial distance of 30 mm for 10.23% moisture content. The second peak of normal contact force occurred at the radial distance of 40 mm for 14.09% moisture content rice seeds. The 17.85% moisture content rice seeds at the radial distance of the second peak of normal contact force were observed at a radial distance of 50 mm. It showed that the peak of the second normal contact force became larger and larger from the center point as the moisture content increased. With the increased radial distance, the normal contact force on the horizontal surface gradually decreased and decreased to 0 at the radial distance of 140 mm. The normal contact force of the rice seed accumulation with 10.23%, 14.09% and 17.85% moisture content on the horizontal surface showed a maximum value at the center. Although the force value fluctuated with the increase of radial distance, the overall trend gradually decreased and reduced to 0 at the radial distance of 140 mm. The force distribution of the structure was no "central depression".

The trend of normal contact force between rice grains was different from that of rice grains on the horizontal plane, which may be related to the fact that the shape of rice seeds and the form of interaction between seeds change the magnitude of the force and the form of force transfer. This changed the normal contact force acting on the horizontal plane, which was studied in depth by Zhao et al. [38].

To show the distribution state of the contact force of the seed accumulation on the horizontal surface more clearly, the circular force at the bottom of the horizontal surface was divided into 8 discrete cells in a clockwise direction with the simulation X-axis as the horizontal zero point, and the center points along the radial direction were taken from the center of the circle of 14, 28, 42, 56, 70 mm on each discrete cell. The normal force of the bottom seed accumulation in the cell with the length of 14 mm and the width of 6 mm was taken at the center point. The normal force of the bottom seeds in the cell with a length of 14 mm and width of 6 mm was taken as the normal force of the horizontal plane with radial distance of 25, 56, 84, 112, 140 mm. The partitioning method is reported in Supplementary

Materials (Figure S3b). The normal force distribution of the horizontal surface was analysed (Figure 6).

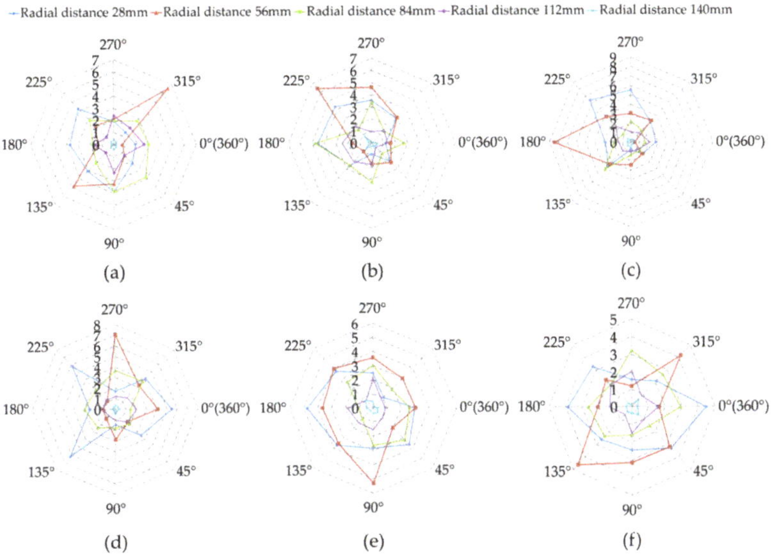

Figure 6. Normal force distribution on horizontal surface (unit: 10^{-3} N): (**a**–**f**) indicate the action of moisture content of 10.23%, 14.09%, 17.85%, 21.77%, 26.41%, 29.22% on the horizontal surface, respectively.

The distribution state of the force on the horizontal surface of rice seeds with different moisture content basically showed the same law. The distribution of the forces on the radial distance of 28 mm, 84 mm, 112 mm and 140 mm was relatively uniform, among which the force on the radial distance of 140 mm was the smallest. However, the force distribution fluctuated greatly at the radial distance of 56 mm, and there were extreme values. With the increased moisture content, the force at different radial distances gradually tended to a regular shape, indicating that the higher the moisture content of rice seeds, the more likely to appear the typical "ring force" structure, and the more uniformly the force on the horizontal surface was distributed in the circumferential direction. It was mainly due to the fact that the higher the moisture content of rice seeds, the larger the natural rest angle, the easier the seeds to move downward along the accumulation in the process of forming the accumulation, resulting in a certain "beat tight" effect along the surrounding area. Finally, a "ring force" structure with the cumulative effect of layers of seeds formed. When the moisture content was small, the natural rest angle formed by the accumulation of seeds was small, and the "beat tight" effect was not fully developed. Therefore, the "ring force" structure was not easy to appear.

Rice is one of the most important food crops in the world. The production, processing and storage of rice are of great significance to world economic stability and food safety [39]. This study focused on the physical characteristics of rice grains with different moisture content. The research results can be applied to the control and design of large grain unloading equipment and food processing receiving equipment. With the increase of moisture content, the time of substrate support forming stage and expansion accumulation stage becomes longer. That is, the rice grains with high moisture content cannot quickly enter the free accumulation stage, and the blockage problem is easy to occur. Therefore, when the unloading equipment is used to deal with rice with high moisture content, it is necessary to design a larger unloading port or raise a larger height to facilitate the discharge of rice grains. The greater the moisture content, the longer the kinetic energy attenuation time, the longer the time to reach the stable accumulation stage, and the longer the time required in the grain unloading stage. Therefore, in the automatic production line of food processing, when processing rice with large moisture content, the equipment needs to be set for a longer time to ensure the stability of grain unloading.

In addition, in the quasi-static accumulation stage, the pressure of low moisture content rice on the bottom has the characteristics of central depression. Therefore, it is necessary to reinforce the bottom plate at the non-central position at the bottom of the horizontal receiving tray of storage equipment or food processing equipment. The higher the moisture content, the greater the natural repose angle, the uniform stress in the circumferential direction of the bottom, and the smaller the surface area of the horizontal receiving tray. The rice grains with moisture content of 10.23%, 14.09% and 17.85% have the structure of central depression. For the food processing machinery with the accumulation center as the feed inlet, the blockage problem will occur. Therefore, the arc bottom groove and vibration are used to interfere with the stability of the central depression structure, so as to eliminate the arch effect of the indirect contact force of the particles and ensure the smooth entry of the grains into the feed inlet [12].

Rice seeds form the accumulation in a variety of ways, including "rainfall", "dumping" and "point source" [17]. This paper studied the velocity and mechanical characteristics of rice seeds with different moisture content in the accumulation process and the mechanical characteristics of the quasi-static accumulation stage by taking the classical "point source" type as an example, showing that the accumulation characteristics of rice seeds with different moisture content were not the same, and not all rice seeds showed a significant "central depression" law in the normal force on the horizontal surface, which may be closely related to the physical characteristics of rice seeds. Rice with high moisture content needs drying treatment during long-term storage, otherwise, it is easy to lead to microbial development and mycotoxin formation. In the future, we will introduce different varieties of rice seeds as research objects to study the micromechanical characteristics of "long grain", "medium grain" and "short grain", and study the main factors influencing the mechanical distribution law of "central depression", so as to further improve the system of research on the mechanical characteristics of rice seed accumulation. At the same time, the mechanism of microbial development and fungal formation during accumulation with different moisture content will be carried out to explore the changes of rice grains after accumulation from a chemical point of view. Combined with this study, a systematic and complete research system on the safe storage of rice grains will be formed from both physical and chemical aspects.

4. Conclusions

This study accurately established the model of rice seeds based on 3D scanning technology, simulated the accumulation state of rice seeds by "point source" accumulation method by discrete element method, and obtained the velocity, force, kinetic energy and force distribution under the accumulation state of rice seeds with different moisture content. The main conclusions are as follows:

(1)　According to the velocity characteristics of rice seeds, the accumulation process can be divided into four stages: substrate support forming stage, expansion accumulation stage, free accumulation stage and quasi-static accumulation stage. The average velocity of rice seeds with different moisture content showed a trend of increasing and then decreasing during the accumulation process. The force during the accumulation of seeds was the direct cause of its velocity change.

(2)　The average translational kinetic energy and the average rotational kinetic energy of rice seeds with different moisture content showed a trend of increasing and then decreasing, and the higher the moisture content, the greater the energy in the process of seed accumulation and the longer the kinetic energy decayed.

(3)　The force distribution of the "central depression" structure of the action of rice seeds with the moisture content of 10.23%, 14.09% and 17.85% on the horizontal surface.

The higher the moisture content of rice seeds, the more likely the typical "ring force" structure appeared, and the more evenly the force on the horizontal surface was distributed in the circumferential direction.

(4) When rice grains with 10.23% moisture content formed accumulation, the kinetic energy decay time was the shortest and the speed was the fastest; In the quasi-static accumulation stage, the force of rice grain with 21.77% water content on the horizontal plane is the most uniform, and it was the least prone to blockage when unloading at the accumulation center. This study provides a reference for the design and development of food processing equipment and safe storage equipment.

Supplementary Materials: The following supporting information can be downloaded at: https://www.mdpi.com/article/10.3390/foods11030295/s1, Table S1: Parameters measuring methods; Figure S1: The process of discrete element modeling of rice seeds; Figure S2: The results of bench test and simulation test of natural rest angle for rice seeds with different moisture content: (a1–a6) are the results of natural rest angle of bench test, respectively, and its moisture contents are 10.23%, 14.09%, 17.85%, 21.77%, 26.41%, and 29.22%, respectively. (b1–b6) are the result of the natural rest angle of EDEM simulation, and its moisture content are 10.23%, 14.09%, 17.85%, 21.77%, 26.41%, and 29.22%, respectively. The red dotted line is the primary fitting curve; Figure S3: Rice seed accumulation partitioning diagrams: (a) and (b) are the two partitioning methods respectively.

Author Contributions: Conceptualization, J.W. and C.X.; methodology, C.X.; software, X.Q. and W.Z.; validation, C.X. and H.T.; investigation, X.Q. and H.T.; resources, J.W. and H.T.; data curation, H.T. and C.X.; writing—original draft preparation, H.T. and C.X.; writing—review and editing, J.W., W.Z. and H.T.; visualization, C.X.; supervision, H.T.; funding acquisition, J.W. and H.T. All authors have read and agreed to the published version of the manuscript.

Funding: This work was supported by the National Natural Science Foundation of China (NSFC), grant number: 31901414; the Program on Industrial Technology System of National Rice (CN), grant number: CARS-01-48; and Key Laboratory of Crop Harvesting Equipment Technology of Zhejiang Province, grant number: 2021KZ01.

Institutional Review Board Statement: Not applicable.

Informed Consent Statement: Not applicable.

Data Availability Statement: All data are presented in this article in the form of figures and tables.

Acknowledgments: We thank Yanan Xu and Zuodong Fu for their contributions to the software applications.

Conflicts of Interest: The authors declare no conflict of interest.

References

1. Han, Y.; Zhao, D.; Chu, Y.; Zhen, J.; Li, G.; Zhao, H.; Jia, F. Breakage behaviour of single rice particles under compression and impact. *Adv. Powder Technol.* **2021**, *32*, 4635–4650. [CrossRef]
2. Hesse, R.; Krull, F.; Antonyuk, S. Prediction of random packing density and flowability for non-spherical particles by deep convolutional neural networks and Discrete Element Method simulations. *Powder Technol.* **2021**, *393*, 559–581. [CrossRef]
3. Horabik, J.; Molenda, M. Parameters and contact models for DEM simulations of agricultural granular materials: A review. *Biosyst. Eng.* **2016**, *147*, 206–225. [CrossRef]
4. Zheng, Q.J.; Yu, A.B. Finite element investigation of the flow and stress patterns in conical hopper during discharge. *Chem. Eng. Sci.* **2015**, *129*, 49–57. [CrossRef]
5. Tian, T.; Su, J.; Zhan, J.; Geng, S.; Xu, G.; Liu, X. Discrete and continuum modeling of granular flow in silo discharge. *Particuology* **2018**, *36*, 127–138. [CrossRef]
6. Wan, J.; Wang, F.; Yang, G.; Zhang, S.; Wang, M.; Lin, P.; Yang, L. The influence of orifice shape on the flow rate: A DEM and experimental research in 3D hopper granular flows. *Powder Technol.* **2018**, *335*, 147–155. [CrossRef]
7. Zaki, M.; Siraj, M.S. Study of a flat-bottomed cylindrical silo with different orifice shapes. *Powder Technol.* **2019**, *354*, 641–652. [CrossRef]
8. Liu, H.; Han, Y.; Jia, F.; Xiao, Y.; Chen, H.; Bai, S.; Zhao, H. An experimental investigation on jamming and critical orifice size in the discharge of a two-dimensional silo with curved hopper. *Adv. Powder Technol.* **2021**, *32*, 88–98. [CrossRef]
9. Xiao, Y.; Han, Y.; Jia, F.; Liu, H.; Li, G.; Chen, P.; Meng, X.; Bai, S. Research on clogging mechanisms of bulk materials flowing trough a bottleneck. *Powder Technol.* **2021**, *381*, 381–391. [CrossRef]

10. Ahmadi, A.; Seyedi Hosseininia, E. An experimental investigation on stable arch formation in cohesionless granular materials using developed trapdoor test. *Powder Technol.* **2018**, *330*, 137–146. [CrossRef]

11. Cheng, X.; Zhang, Q.; Shi, C.; Yan, X. Model for the prediction of grain density and pressure distribution in hopper-bottom silos. *Biosyst. Eng.* **2017**, *163*, 159–166. [CrossRef]

12. Liu, K.; Xiao, Z.; Wang, S. Development of arching and silo wall pressure distribution in storage and discharging state based on discrete element analysis. *Nongye Gongcheng Xuebao/Trans. Chin. Soc. Agric. Eng.* **2018**, *34*, 277–285. [CrossRef]

13. Horabik, J.; Molenda, M. Distribution of static pressure of seeds in a shallow model silo. *Int. Agrophys.* **2017**, *31*, 167–174. [CrossRef]

14. Gallego, E.; Fuentes, J.M.; Wiącek, J.; Villar, J.R.; Ayuga, F. DEM analysis of the flow and friction of spherical particles in steel silos with corrugated walls. *Powder Technol.* **2019**, *355*, 425–437. [CrossRef]

15. Shi, L.; Yang, X.; Zhao, W.; Sun, W.; Wang, G.; Sun, B. Investigation of interaction effect between static and rolling friction of corn kernels on repose formation by dem. *Int. J. Agric. Biol. Eng.* **2021**, *14*, 238–246. [CrossRef]

16. Cao, R.H.; Cao, P.; Lin, H.; Zhang, K.; Tan, X.W. Particle flow analysis of direct shear tests on joints with different roughnesses. *Yantu Lixue/Rock Soil Mech.* **2013**, *34*, 456–463.

17. Horabik, J.; Parafiniuk, P.; Molenda, M. Discrete element modelling study of force distribution in a 3D pile of spherical particles. *Powder Technol.* **2017**, *312*, 194–203. [CrossRef]

18. Carlevaro, C.M.; Pugnaloni, L.A. Arches and contact forces in a granular pile. *Eur. Phys. J. E* **2012**, *35*, 44. [CrossRef]

19. Berger, R.; Kloss, C.; Kohlmeyer, A.; Pirker, S. Hybrid parallelization of the LIGGGHTS open-source DEM code. *Powder Technol.* **2015**, *278*, 234–247. [CrossRef]

20. Ma, Z.; Li, Y.; Xu, L. Discrete-element method simulation of agricultural particles' motion in variable-amplitude screen box. *Comput. Electron. Agric.* **2015**, *118*, 92–99. [CrossRef]

21. Wang, Z.; Liu, M. Semi-resolved CFD–DEM for thermal particulate flows with applications to fluidized beds. *Int. J. Heat Mass Transf.* **2020**, *159*, 120150. [CrossRef]

22. Li, A.; Han, Y.; Jia, F.; Zhang, J.; Meng, X.; Chen, P.; Xiao, Y.; Zhao, H. Examination milling non-uniformity in friction rice mills using by discrete element method and experiment. *Biosyst. Eng.* **2021**, *211*, 247–259. [CrossRef]

23. Wang, Q.; Mao, H.; Li, Q. Modelling and simulation of the grain threshing process based on the discrete element method. *Comput. Electron. Agric.* **2020**, *178*, 105790. [CrossRef]

24. Markauskas, D.; Ramírez-Gómez, Á.; Kačianauskas, R.; Zdancevičius, E. Maize grain shape approaches for DEM modelling. *Comput. Electron. Agric.* **2015**, *118*, 247–258. [CrossRef]

25. Bhushan, B.; Raigar, R.K. Influence of moisture content on engineering properties of two varieties of rice bean. *J. Food Process Eng.* **2020**, *43*, e13507. [CrossRef]

26. Ebrahimifakhar, A.; Yuill, D. Inverse estimation of thermophysical properties and initial moisture content of cereal grains during deep-bed grain drying. *Biosyst. Eng.* **2020**, *196*, 97–111. [CrossRef]

27. Liu, J.; Liu, Y.; Wang, A.; Dai, Z.; Wang, R.; Sun, H.; Strappe, P.; Zhou, Z. Characteristics of moisture migration and volatile compounds of rice stored under various storage conditions. *J. Cereal Sci.* **2021**, *102*, 103323. [CrossRef]

28. Zhang, B.; Qian, C.; Jiao, J.; Ding, Z.; Zhang, Y.; Cui, H.; Liu, C.; Feng, L. Rice moisture content detection method based on dielectric properties and SPA-SVR algorithm. *Nongye Gongcheng Xuebao/Trans. Chin. Soc. Agric. Eng.* **2019**, *35*, 237–244. [CrossRef]

29. Constant, M.; Coppin, N.; Dubois, F.; Artoni, R.; Lambrechts, J.; Legat, V. Numerical investigation of the density sorting of grains using water jigging. *Powder Technol.* **2021**, *393*, 705–721. [CrossRef]

30. Yang, Z.; Sun, J.; Guo, Y. Effect of moisture content on compression mechanical properties and frictional characteristics of millet grain. *Nongye Gongcheng Xuebao/Trans. Chin. Soc. Agric. Eng.* **2015**, *31*, 253–260. [CrossRef]

31. Jiang, M.; Chen, G.; Liu, C.; Liu, W.; Zhang, Z. Effects of moisture content on elastic-plastic properties of bulk wheat. *Nongye Gongcheng Xuebao/Trans. Chin. Soc. Agric. Eng.* **2020**, *36*, 245–251. [CrossRef]

32. Wang, J.; Xu, C.; Xu, Y.; Wang, Z.; Qi, X.; Wang, J.; Zhou, W.; Tang, H.; Wang, Q. Influencing Factors Analysis and Simulation Calibration of Restitution Coefficient of Rice Grain. *Appl. Sci.* **2021**, *11*, 5884. [CrossRef]

33. Qiu, S.; Yuan, X.; Guo, Y.; Cui, Q.; Wu, X.; Zhang, Z. Effects of variety and moisture content on mechanical properties of millet. *Nongye Gongcheng Xuebao/Trans. Chin. Soc. Agric. Eng.* **2019**, *35*, 322–326. [CrossRef]

34. Wu, M.; Cong, J.; Yan, Q.; Zhu, T.; Peng, X.; Wang, Y. Calibration and experiments for discrete element simulation parameters of peanut seed particles. *Nongye Gongcheng Xuebao/Trans. Chin. Soc. Agric. Eng.* **2020**, *36*, 30–38. [CrossRef]

35. Romuli, S.; Karaj, S.; Müller, J. Discrete element method simulation of the hulling process of Jatropha curcas L. fruits. *Biosyst. Eng.* **2017**, *155*, 55–67. [CrossRef]

36. Horabik, J.; Wiącek, J.; Parafiniuk, P.; Bańda, M.; Kobyłka, R.; Stasiak, M.; Molenda, M. Calibration of discrete-element-method model parameters of bulk wheat for storage. *Biosyst. Eng.* **2020**, *200*, 298–314. [CrossRef]

37. Wei, H.; Tang, X.; Ge, Y.; Li, M.; Saxén, H.; Yu, Y. Numerical and experimental studies of the effect of iron ore particle shape on repose angle and porosity of a heap. *Powder Technol.* **2019**, *353*, 526–534. [CrossRef]

38. Zhao, L.L.; Zhao, Y.M.; Liu, C.S.; Li, J. Discrete element simulation of mechanical properties of wet granular pile. *Wuli Xuebao/Acta Phys. Sin.* **2014**, *63*, 034501. [CrossRef]

39. Chan, Y.J.; Lu, W.C.; Lin, H.Y.; Wu, Z.R.; Liou, C.W.; Li, P.H. Effect of rice protein hydrolysates as an egg replacement on the physicochemical properties of flaky egg rolls. *Foods* **2020**, *9*, 245. [CrossRef]

Article

Effects of Radio Frequency Pretreatment on Quality of Tree Peony Seed Oils: Process Optimization and Comparison with Microwave and Roasting

Zhi Wang, Chang Zheng, Fenghong Huang, Changsheng Liu *, Ying Huang and Weijun Wang

Oil Crops Research Institute of the Chinese Academy of Agricultural Sciences, Oil Crops and Lipids Process Technology National & Local Joint Engineering Laboratory, Key Laboratory of Oilseeds Processing, Ministry of Agriculture, Hubei Key Laboratory of Lipid Chemistry and Nutrition, Wuhan 430062, China; 82101195248@caas.cn (Z.W.); zhengchang@caas.cn (C.Z.); huangfenghong@oilcrops.cn (F.H.); huangying01@caas.cn (Y.H.); Wangweijun@caas.cn (W.W.)

* Correspondence: liuchangsheng@caas.cn; Tel.: +86-027-86827874

Abstract: In this study, we explored the technical parameters of tree peony seeds oil (TPSO) after their treatment with radio frequency (RF) at 0 °C–140 °C, and compared the results with microwave (MW) and roasted (RT) pretreatment in terms of their physicochemical properties, bioactivity (fatty acid tocopherols and phytosterols), volatile compounds and antioxidant activity of TPSO. RF (140 °C) pretreatment can effectively destroy the cell structure, substantially increasing oil yield by 15.23%. Tocopherols and phytosterols were enhanced in oil to 51.45 mg/kg and 341.35 mg/kg, respectively. In addition, antioxidant activities for 2,2-diphenyl-1-picrylhydrazyl (DPPH) and ferric-reducing antioxidant power (FRAP) were significantly improved by 33.26 μmol TE/100 g and 65.84 μmol TE/100 g, respectively ($p < 0.05$). The induction period (IP) value increased by 4.04 times. These results are similar to those of the MW pretreatment. The contents of aromatic compounds were significantly increased, resulting in improved flavors and aromas (roasted, nutty), by RF, MW and RT pretreatments. The three pretreatments significantly enhanced the antioxidant capacities and oxidative stabilities ($p < 0.05$). The current findings reveal RF to be a potential pretreatment for application in the industrial production of TPSO.

Keywords: tree peony seed oil; heating pretreatment; microstructure; volatile compounds; bioactive compounds; oxidative stability

Citation: Wang, Z.; Zheng, C.; Huang, F.; Liu, C.; Huang, Y.; Wang, W. Effects of Radio Frequency Pretreatment on Quality of Tree Peony Seed Oils: Process Optimization and Comparison with Microwave and Roasting. *Foods* **2021**, *10*, 3062. https://doi.org/10.3390/foods10123062

Academic Editors: Yiannis Kourkoutas and Alessandra Bendini

Received: 13 September 2021
Accepted: 18 November 2021
Published: 9 December 2021

Publisher's Note: MDPI stays neutral with regard to jurisdictional claims in published maps and institutional affiliations.

Copyright: © 2021 by the authors. Licensee MDPI, Basel, Switzerland. This article is an open access article distributed under the terms and conditions of the Creative Commons Attribution (CC BY) license (https://creativecommons.org/licenses/by/4.0/).

1. Introduction

The peony (*Paeonia Suffruticosa*) has been a popular traditional ornamental and medicinal plant in China for 2000 years [1]. More recently, tree peony has been widely cultivated for the high oil yield rate of its seeds (>27%) and because the protein content of the meal reaches 32.44% [2]. The high level of interest in tree peony seed oil (TPSO) might be explained by its abundant unsaturated fatty acids (UFAs, >90%), high proportions of α-linolenic acid (C18:3, ALA, >40%) and richness in sterols (55.00%), γ-tocopherol (average 362.98 mg/kg) and phenolic compounds (ranging from 49.16 to 130.59 mg/kg) [3–5]. These nutrition compounds have many physiological functions, which can exert antidiabetic, antihyperlipidemic, serum cholesterol lowering and antitumor effects [6–8].

Oil is usually extracted from seeds by pressing, solvent, aqueous enzymatic, supercritical CO_2 extraction, and microwave-assisted methods. The most commonly used methods to extract oils are solvent extraction (SE) and hot pressing. SE is the most efficient method, and produces less residual oil but is environmentally unfriendly. Oil extraction by hot-pressing is simpler, safer, and has fewer steps than SE [9], but these two methods produce poor quality oil due to high processing temperatures and a relatively high number of processing steps. Its high cost on an industrial scale has limited the development of aqueous enzymatic and supercritical CO_2 extraction processing. Cold pressing has been developed given the demand for natural green food, and is used in rapeseed oil [9], camellia seed oil, and linseed oil [10], to effectively preserve the nutrients in oils. However, the maximum oil yield of TPSO through cold pressing is only 23.11% [11]. Suitable heating pretreatments should provide efficient technologies for producing high-quality oils, and have various effects such as the promotion of oil solubility, increasing the level of bioactive components, and improving flavor. Currently studied technologies include autoclave, cooking, oven-heating, roasting (RT), steam explosion, microwave (MW) and radio frequency (RF) [12].

RF and MW are electromagnetic-field-based heating technologies. Compared to RT (convection heating), dielectric heating can penetrate the inside of seeds and provide fast and volumetric heating effects. In contrast to MW (2450 MHz), RF (27.12 MHz) has a longer wavelength, which can hasten the rate of drying, increase the quantity, and ensure heating is more uniform [13]. RT, as a traditional technology, is a simple and accessible pretreatment that is generally performed at temperatures of 100–200 °C, with a high-temperature and a long time process. A previous study revealed that RT treatment compromises several of the physicochemical properties of chia seeds [14]. Excessive RT heating of TPSO produces high contents of benzoic acid [15], which results in poor oil quality. Microwave heating, which was considered to be a new high-efficiency and promising technology has been widely used in many processes, including in the processing of rapeseed and camellia seed, to improve the texture and the release of nutrient components and enhance its antioxidant effects [16,17]. Longer nutrient retention was observed in the rapid conversion of electromagnetic energy to thermal energy in MW heating than for conventional RT [18]. RF heating is mainly used to control pests and fungi in diverse grains, nuts and beans, and is widely used for heating, drying dis-infestations, and pasteurization [19]. Recently, RF was proven to be able to extract pectin, phenolic compounds, and anthocyanins [20]. Ruange Lana et al. [21] showed that RF heating could be an efficient method for heterogeneous food matrices. Conversely, there is no information in the literature on the application of RF as a pretreatment of oilseeds to enhance the quality of TPSO.

Previous studies have mostly focused on exploring the influence of different varieties and genotypes of tree peony seeds (TPS) on oil quality [8,22,23]. However, the product of oil processing technology has not been discussed, and few researchers have studied the effect of heating pretreatment on the quality of TPSO. Therefore, the objectives of this study were to optimize the operating conditions involved in the RF assisted TPS extraction of TPSO in achieving efficient extraction, and to compare the physicochemical properties, bioactivity, volatile compounds and antioxidant activity of TPSO extracted by RF, MW and RT using the same operating conditions.

2. Materials and Methods

2.1. Materials and Reagents

Fresh tree peony seeds were obtained from Heze (Shandong, China). The moisture content was 4.78%. The adjustment of moisture content of 8% was decided after the preliminary experimental study of the effect of different moisture (6%, 7%, 8%, 9%, 10%) (Supplement Table S1) on oil extracted from the seeds at RF 140 °C, for which it was found that 8% led to a higher oil yield.

The 5α-cholestane (purity ≥ 97%), α-tocopherol, β-tocopherol, γ-tocopherol, δ-tocopherol, 37 fatty acid methyl ester mixtures standard, BSTFA + TMCS, squalene, DPPH, TPTZ, $FeCl_3$, KBr, Trolox and p-anisidine were purchased from Sigma-Aldrich (St. Louis, MO, USA). Chromatographic grade reagents were purchased from Merck (Darmstadt, Germany). The analytical reagents were purchased from Sinopharm Chemical Reagent Co., Ltd. (Beijing, China).

2.2. Pretreatment and Preparation of the Samples

Radio-frequency pretreatment (with hot-air) was conducted using the GJD-2B-27II-JY (Huashi Jiyuan high-frequency equipment Co., Ltd., Hebei, China). TPS (weight 500 ± 0.5 g) were evenly spread in 150 mm-in-diameter Pyrex Petri dishes, with electrode gaps of 70 mm, and a cavity temperature set to 80 °C. The seeds were treated with different RF temperatures (0 °C, 80 °C, 100 °C, 120 °C and 140 °C).

For the microwave treatment, 500 g of seed-kernels was placed in 8 Pyrex Petri dishes (9 cm diameter) and conditions were set at 800 W, 5 min (140 $\pm$ 5 °C) and a frequency of 2450 MHz inside a microwave (Model Mars, 800 W, CEM, Charlotte, NC, USA).

For the roasting treatment, TPS (500 g, kernel) on aluminum foil were oven-roasted at 140 °C for 20 min. Each condition was repeated for triplicate tests.

2.3. Cold Pressing

TPS was pressed with a cold-press machine (CA59, Monchengladbach, Germany), and centrifuged at 10,000 rpm for 20 min (Sorvall Stratos, Thermo Fisher Co., Ltd., Waltham, MA, USA). All samples were placed in amber glass bottles and were sealed and stored in a refrigerator (4 $\pm$ 2 °C). The oil yield was calculated following the method of Cong et al. [24].

2.4. Fourier-Transform Infrared Spectroscopy (FTIR)

The sample was mixed with potassium bromide (KBr) at a mass ratio of 1:30 for tableting. The infrared spectra of the gels were detected from 4000 to 400 cm^{-1} at a 4 cm^{-1} resolution using the Fourier transform infrared (FT-IR) spectra, and recorded on a TENSOR 27 FTIR spectrometer (Bruker, Karlsruhe, Germany).

2.5. Transmission Electron Microscopy (TEM)

The seed sample was cut to a size of less than 1 mm^3. The sample was completely immersed in a 2.5% glutaraldehyde solution and 0.1 M phosphate buffer solution and fixed for 4 h. A 0.1 M phosphate buffer solution was used for washing for 15 min for 3 repetitions, then fixed in a 1% osmium acid solution for 1–1.5 h. This was repeated with a 0.1 M phosphate buffer solution, with washing performed 3 times, for 15 min each. The sample was dehydrated with an acetone solution at gradient concentrations. Then, the sample was embedded in Epon 812 resin. A pure acetone and an embedding solution were placed at room temperature for 0.5 h, and 37 °C for 1.5 h, respectively. Finally, the sample was cured at 37, 45, or 60 °C for 24 h. A 70 mm slice was obtained using a Reichert-Jung UL TRACUT E ultrathin slicer (Leica, Vienna, Austria), and was dyed with a lead citrate solution and uranyl acetate for 15 min, then observed under a JEM 1200 transmission electron microscope (JEOL, Tokyo, Japan).

2.6. Determination of Physical and Chemical Properties

The acid value (AV) was measured according to GB 5009.229-2016 [25]. For the determination of color, we used a PFXi-Series automatic lovbind tintometer (PFXI 995, German) to evaluate the CIE (L*, a*, b*) of samples. The following parameters were determined: lightness (L*), redness+/greenness (a*), and yellowness+/blueness (b*). The total phenolic content (TPC) was determined using the method described by Cong et al. [26]. The results are expressed as gallic acid equivalents (mg GAE/100 g). The carotenoid content of the oils was determined according to the MPOB (2005) standard [27].

2.7. Determination of the Tocopherols Content

Tocopherols contents were measured according to the AOCS Official Method Ce 8-89 [28]. The HPLC conditions were the same as those reported by Cong et al. [26].

2.8. Determination of Fatty Acids Profile

The fatty acids content was determined according to GB5009.168-2016 [25]. Briefly, we placed three drops of TPSO in a 10 mL centrifuge tube, adding 0.5 mL (0.5 mol/L) sodium methoxide and 2.5 mL n-hexane, after which the supernatant was injected into a GC (GC7890A, Agilent, CA, USA) equipped with a flame ionization detector (FID) to quantify fatty acid methyl esters. The column was DB-FFAP (30 m $\times$ 250 µm id $\times$ 0.25 µm thickness). The temperatures of the injector and detector were both set to 300 °C, the flow rate of nitrogen was 1.5 mL/min with a split ratio of 80:1. The column temperature was maintained at 210 °C for 9 min, then it was increased by 20 °C/min to 250 °C and maintained for 10 min.

2.9. Determination of the Phytosterols and Squalene

For saponifying 200 mg of oil, we added 0.1 mL of 1 mg/mL 5α-cholestane as the internal standard, then 2 mL of 2 mol/L KOH-Ethanol was added to the sample, and the oil samples were placed in a saponified bath at 70 °C for 60 min. After saponification was completed, we added 1.5 mL of distilled water and 1 mL of hexane to the sample. The extraction was repeated three times. Following this, the mixture was centrifuged at 4000 rpm for 10 min. Then, we transferred the sample to a 2 mL injection bottle. After the removal of hexane by nitrogen blowing, 100 μL of BSTFA + TMCS was added to the glass tube containing the injection bottle and the mixture was allowed to react in an oven at 105 °C for 15 min. Finally, we used n-hexane to increase the volume to 1 mL for analysis.

A 1 μL derivative sample, with a resolution ratio of 20:1, was injected into an Agilent GC7890A and 7683B series equipped with an FID, and a silica gel capillary column (DB-5MS, 30 m × 0.25 mm × 0.25 μm; Agilent, Santa Clara, CA, USA). The flow rate was 1.2 mL/min, and helium was used as the carrier gas. The temperature of the injector and detector were both set to 290 °C. The column was kept at 100 °C for 1 min. The column temperature was increased from 60 °C to 290 °C at a rate of 40 °C/min and maintained for 15 min. The sterol was identified via a comparison with the retention time of the standard, and quantified based on the relative peak area of the internal standard. The GC-MS ion source temperature was set to 290 °C, and the transmission line was set to 290 °C in the SIM monitoring mode. An electron impact mass spectrum was generated at 70 eV. The scanning range was m/z 50–500.

2.10. Sensory Evaluation

The sensory evaluation was carried out by 8 trained members between the ages of 25 and 40 from the Institute of Oil Crops Research. Each team member received one week of TPSO assessment training. TPSOs were produced through different pretreatments methods. The team members compared the aroma parameters (roasted, nuts, fat, green, sweetness and pungency), and assigned a score to each sample based on the odor intensity for a sensory evaluation of the aroma of peony oil, with a score range of 0–10, where 0–3 indicates that the odor is very weak and difficult to distinguish, 3–7 means that the odor is clear and the intensity is low, and 8–10 means that the odor is clear and strong. At the same time, the overall scent degree was measured using a score range of 0–10 points.

2.11. Determination of Volatile Compounds

Volatile compounds were determined by headspace solid-phase micro-extraction (HS-SPME) according to Xiao et al. [29]. The condition of extraction and the GC-MS conditions were slightly different. The oil was equilibrated at 60 °C for 20 min through an automatic SPME module (Agilent, CTC G6500, Santa Clara, CA, USA) and the volatile compounds were automatically extracted for 30 min.

The Agilent 7890A GC and 5975C series mass spectrometers (GC-MS, Agilent, Santa Clara, CA, USA) were used to identify the volatiles in the TPSO extracted by the three different pretreatment methods. The extracted compounds were separated in polar DB-WAX (30 m × 320 μm × 0.25 μm). The working conditions were 250 °C in split mode with an inlet temperature of 1:10. The carrier gas was helium (99.999%), and the flow rate was 1.5 mL/min. The operating conditions were as follows: heating at 40 °C for 2 min, heating at 5 °C/min to 200 °C, and maintained for 2 min, then heating at 8 °C/min to 250 °C, and finally, maintaining this temperature for 5 min. The GC-MS transfer line was set to 280 °C; an electron impact mass spectrum was generated at 70 eV. The scanning range was *m/z* 38–400. All compounds were identified according to the NIST 2017 (Qual $\geq$ 80) mass spectrometry library and reliable standards.

2.12. Determination of the Oxidative Stabilities of the Oils

The peroxide value (POV) was measured according to GB 5009.227–2016 [24]. For K_{232} (CD) and K_{270} (CT), absorbance at 232 (K_{232}) and 270 (K_{270}) nm was measured according to the ISO standard 3656:2002 [30]. P-anisidine value (P-AV) was measured according to the AOCS Official Method Cd18-90 [31]. The induction period (IP) was determined according to the method of Azadmard et al. [16].

2.13. Determination of the Antioxidant Activity of the Oils

The 2,2-diphenyl-1-picrylhydrazyl (DPPH) and ferric-reducing antioxidant power (FRAP) were measured following the method of Cong et al. [26]. The DPPH and FRAP results are expressed as micromoles of Trolox equivalents per 100 g of sample (μmol TE/100 g).

2.14. Statistical Analysis

The experiment data were expressed as the mean value $\pm$ standard deviation from the triplicate tests (3 replicates per treatment). SPSS 20.0 (SPSS Inc., Chicago, IL, USA) was used to perform the ANOVA and Duncan's multiple range test ($p < 0.05$). All figures were drawn in Origin 9.0 software (Origin Lab Corp., Northampton, England).

3. Results

3.1. Different Pretreatments Seed Microstructure (FTIR and TEM)

Figure 1 displays the main peaks of the FTIR spectrum and the TPS of the different temperatures of RF groups. The FTIR spectrum can be classified into 7 groups: (1) 3000–3600 cm^{-1}, (2) 3282 cm^{-1}, (3) 2800–3000 cm^{-1}, (4) 1745 cm^{-1}, (5) 1645–1647 cm^{-1}, (6) 1500–1600 cm^{-1} (7) 1246 cm^{-1}. The wave number range of 3000–3600 cm^{-1} was a result of the broadband center and was attributed to potential hydrogen bond interactions (OH and NH groups) and hydrogen bond stretching vibrations of cis-olefinic double-bonds. The existence of peaks in the wave number range of 3282 cm^{-1} indicates the band assignments of the ν (=C–H) and OH groups vibrating strongly. The existence of peaks in the wave number range of the lipid band area 2800–3000 cm^{-1} indicates a characteristic of the stretching C–H bond in methyl groups. The 1745 cm^{-1} peak is associated with C=O bonding, conjugated bonds of triglycerides, bending and the stretching vibrations of aliphatic groups of triglycerides. The observed peaks in 1645–1647 cm^{-1}, 1500–1600 cm^{-1}, 1246 cm^{-1} are probably due to the existence of amide I (C=O bonding), amide II, amide III in the structure, respectively [32,33]. The visual investigation of the IR spectra of TPS treated by different temperatures does not show any marked difference. While comparing peak intensities, a slight change was noticed in certain IR spectral peaks of TPS of different temperatures. FTIR spectroscopy was applied to investigate the inter molecular interactions in protein and polysaccharide matrices [32]. The TPS at the RF of different temperatures showed a decrease in the intensity of peak at 1745 cm^{-1} compared to those that were untreated. The reason could be related to the polysaccharides (carbonyl group) of the seed sample decreasing with increasing temperatures. Similar results were observed in the Fourier transform infrared spectrum of chia seed flours [34]. The intensity of the peaks at 1645 cm^{-1} and 2800–3000 cm^{-1} decreased in the samples, which may be due to a decrease in the triglycerides and protein contents. Zhong et al. [35] reported similar results of a reduction in lipid and protein contents (%) after RF treatment. The peaks in the FTIR spectrum could also be related to the creation of melanoidins during the RF heating process [33].

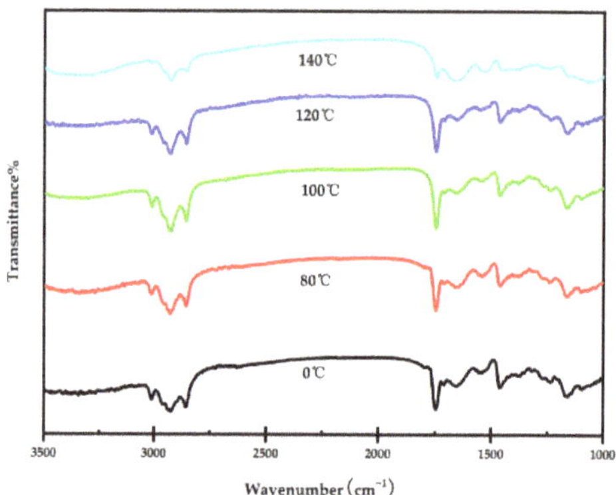

Figure 1. FTIR spectra of untreated seeds (0 °C) and of the different radio frequency pretreatment seed samples treated at a temperature of 80, 100, 120 and 140 °C.

Figure 2 shows the TEM results of the internal structure of the TPS, of that which was untreated, RF140 °C, MW140 °C and RT140 °C. The results show that the destruction of the cell structure formed irregular cavities and evident porosity after heating through RF, MW and RT pretreatments. Compared to those that were untreated, those treated by RF TPS showed structural changes in the protein and lipid bodies in the cells, and no intact cells were observed. The cellular bodies were ruptured to release the oil, which gathered into droplets and sufficiently interacted with oil, with the droplets finally coalescing. Wroniak et al. [36] suggested that MW can reduce the moisture content and increase cell rupture during pressing, ultimately achieving the objective of increasing oil extractability. The reason for the increase in oil yield was that MW has high-frequency electromagnetic waves, which can be quickly converted into heat energy, leading to cell lysis [17]. The principle of RF was similar to that of MW. RF pretreatment at 140 °C could also damage the cell structure. According to the TEM images, the cell wall was highly likely to be severely damaged, as a result of which the oil would be released from damaged cells to achieve a high free oil yield. All three pretreatment technologies can, to some extent, destroy the cell structure.

3.2. Physicochemical Properties of Tree Peony Seed Oil

In this study, the oil content of the seed was 34.55%. The physical and chemical properties of TPS treated by RF at different temperatures (80, 100, 120 and 140 °C) are presented in Table 1. The effects on the oil yield were firstly investigated at different temperatures of RF. The oil yield significantly increased after different temperatures ($p < 0.05$), ranging from 15.57% (untreated) to 30.80% (RF at 140 °C), which was 1.98 times higher than that of the untreated TPS. As the temperature increased in RF treatments, the acid value (AV), TPC, and content of total carotenoids increased (Table 1). More specifically, the acid value (AV) ranged from 4.07 ± 0.04 (mg/g) (for untreated) to 5.88 ± 0.01 (mg/g) (at RF treatments140 °C), the total phenolic content ranged from 4.78 ± 0.01 (mg GAE/100 g) (for untreated) to 54.12 ± 0.06 (mg GAE/100 g) (RF treatments at 140 °C), and the β-carotenoid content ranged from 0.41 ± 0.02 (mg/kg) (for untreated) to 0.62 ± 0.14 (mg/kg) (at RF treatments140 °C). Both AV and TPC increased drastically at 120 °C, the initial acid value of TPSO was relatively high, and the deacidification process needed to be considered.

Then, we compared the physicochemical properties of TPSO extracted by different pretreatment (RF, MV, RT) methods at 140 °C. The oil yield treated with RF, MW and RT increased by 15.23%, 16.32% and 14.71%, respectively. The reason for this finding is that the cell membrane and cell wall of the seed easily rupture, so oil was released after heating treatment. Hu et al. [37] studied MV-assisted oil extraction from tea seed, and the oil yield was increased by 7.13% compared to the oil that was untreated. This result was consistent with the data in this paper. RF proved to be an effective pretreatment method to increase the oil output rate. The L* was significantly decreased at 140 °C for all three pretreatments ($p < 0.05$). Ling et al. [38] used different pistachios pretreatments, whereby the L* of oil was darker after MW and RT. A similar trend was observed in this study in the L* value. The a* value of TPSO treated at different RF temperatures increased significantly, darkening the sample (reddish). The darkening was probably due to caramelization and Maillard reactions and other chemical reactions [39]. Among the three different pretreatments methods, RT resulted in the most significant change, followed by MW and RF. Carotenoid particularly affected the b* value (yellow pigment) of vegetable oil (significantly correlated at the 0.05 level (r = 0.541). After RF, MW and RT, the TPCs increased significantly by 11.42, 13.4 and 12.9 times, respectively. This may be due to the bond between polyphenols and protein being broken.

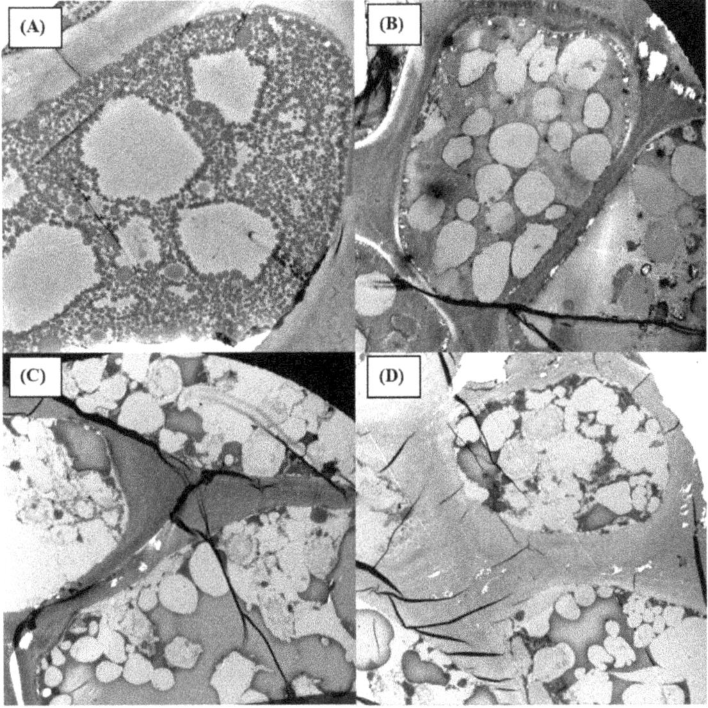

Figure 2. TEM observations of seeds subjected to different pretreatments. (**A**) untreated; (**B**) RF (radio frequency) at 140 °C; (**C**) MW (microwave) at 140 °C; (**D**) RT (roasted) at 140 °C.

Table 1. Acid values, colour, oil yield, total phenolic content and total carotenoids of TPSO.

Treatment	Acid Value (mg/g)	L	a*	b*	Oil Yield (%)	TPC (mgGAE/100 g)	Total Carotenoids (mg/kg)
Untreated	4.07 ± 0.04 [a]	64.81 ± 0.23 [a]	−4.22 ± 0.13 [a]	91.52 ± 0.40 [e]	15.57 ± 013 [a]	4.78 ± 0.01 [a]	0.41 ± 0.02 [a]
RF80 °C	5.14 ± 0.01 [b]	65.20 ± 0.11 [a]	−1.24 ± 0.13 [b]	91.42 ± 0.62 [e]	25.59 ± 0.11 [b]	28.78 ± 0.21 [b]	0.47 ± 0.02 [a]
RF100 °C	5.21 ± 0.00 [b]	65.82 ± 0.44 [a]	2.73 ± 0.52 [c]	91.19 ± 0.28 [e]	25.58 ± 0.07 [b]	33.86 ± 1.57 [c]	0.47 ± 0.02 [a]
RF120 °C	6.31 ± 0.03 [c]	65.03 ± 0.29 [ab]	6.53 ± 0.08 [d]	95.48 ± 0.35 [d]	27.61 ± 0.05 [c]	50.62 ± 2.85 [d]	0.47 ± 0.02 [a]
RF140 °C	6.87 ± 0.01 [d]	64.32 ± 0.29 [b]	7.67 ± 0.33 [e]	104.34 ± 0.33 [b]	30.79 ± 0.11 [d]	54.12 ± 0.06 [e]	0.62 ± 0.15 [a]
MW140 °C	6.65 ± 0.03 [d]	63.48 ± 0.50 [c]	11.79 ± 0.33 [f]	101.26 ± 0.24 [c]	31.88 ± 0.13 [e]	64.31 ± 0.01 [f]	1.35 ± 0.18 [b]
RT140 °C	6.91 ± 0.02 [d]	62.32 ± 0.27 [d]	12.10 ± 0.01 [f]	105.69 ± 0.07 [a]	30.27 ± 0.21 [d]	61.88 ± 0.93 [f]	0.71 ± 0.02 [a]

Values represent the mean ± SD of triplicates ($p < 0.05$). Different small letters indicate significant differences ($p < 0.05$) within a row. CIE L*a*b* color system (L* luminosity; a* red/green coordinate; b* yellow/blue coordinate); TPC, total phenolic content.

3.3. Fatty Acid Compositions, Tocopherols and Phytosterols in TPSO

The bioactive properties of TPSO, produced by the different temperatures of RF, were investigated, and the results of MW and RT at 140 °C were compared with the untreated oil obtained from untreated seed samples. The fatty acid composition, the tocopherol and phytosterol contents of TPSO produced through the three heating treatments of TPS are presented in Table 2. The most abundant fatty acids were palmitic (16:0), oleic (18:1), linoleic acids (18:2) and α-linolenic acid (18:3). The fatty acid content remained at 5.64–5.73% (16:0), 22.32–22.91% (18:1), and 24.01–24.65% (18:2) and 45.27–46.06% (18:3), respectively. The results showed that unsaturated fatty acids (UFAs) were above 92%, of which the content of α-linolenic acid (ALA, C18:3) was the highest, at above 45%. ALA is an essential fatty acid that cannot be synthesized by humans and must be obtained through consumption. TPSO shows potential as a resource. The α-linoleic acid levels were slightly reduced in oil after RF, MW, RT pretreatments at 140 °C, decreasing by 1.73%, 0.40% and 0.69%, respectively. The saturated fatty acids (SFA, 16:0 and 18:0) contents were 7.17–7.85% for TPSOs. We observed no significant difference in the 0.05 levels of UFAs and SFAs after RF, MV, RT pretreatments at 140 °C. Different types of UFAs in TPSO can lower blood cholesterol and are closely related to human health [40].

Table 2. Fatty acid composition, and tocopherols, and phytosterols contents in TPSOs obtained after different pretreatments (RF, RT, MW).

Fatty Acids (%)							
	Untreated	RF80 °C	RF100 °C	RF120 °C	RF140 °C	MW140 °C	RT 140 °C
C16:0	5.640 ± 0.003 a	5.640 ± 0.091 a	5.635 ± 0.008 a	5.648 ± 0.009 ab	5.695 ± 0.003 ab	5.646 ± 0.014 ab	5.731 ± 0.006 b
C16:1	0.334 ± 0.020 a	0.385 ± 0.011 ab	0.365 ± 0.098 ab	0.424 ± 0.030 ab	0.381 ± 0.018 ab	0.463 ± 0.062 b	0.372 ± 0.034 ab
C18:0	1.316 ± 0.045 a	1.275 ± 0.065 a	1.242 ± 0.008 a	1.297 ± 0.004 a	1.282 ± 0.018 a	1.256 ± 0.000 a	1.300 ± 0.008 a
C18:1	22.687 ± 0.078 bc	22.580 ± 0.296 abc	22.608 ± 0.004 abc	22.824 ± 0.093 bc	22.911 ± 0.018 c	22.327 ± 0.105 a	22.590 ± 0.042 abc
C18:2	24.011 ± 0.380 a	24.104 ± 0.062 ab	24.655 ± 0.064 c	24.291 ± 0.025 abc	24.414 ± 0.041 bc	24.426 ± 0.078 bc	24.258 ± 0.002 ab
C18:3	46.066 ± 0.501 c	45.919 ± 0.333 bc	45.445 ± 0.012 ab	45.467 ± 0.081 abc	45.270 ± 0.027 a	45.883 ± 0.107 bc	45.750 ± 0.079 abc
SFA	7.288 ± 0.028 a	7.298 ± 0.166 a	7.246 ± 0.081 a	7.368 ± 0.035 a	7.357 ± 0.033 a	7.365 ± 0.076 a	7.403 ± 0.036 a
UFA	92.763 ± 0.042 b	92.602 ± 0.025 a	92.707 ± 0.080 ab	92.585 ± 0.037 a	92.595 ± 0.033 a	92.635 ± 0.076 a	92.598 ± 0.035 a
Tocopherols(mg/kg)							
α-Tocopherols	51.76 ± 0.78 d	49.21 ± 0.79 bcd	44.92 ± 4.39 ab	43.64 ± 0.48 a	45.96 ± 1.44 abc	49.99 ± 0.20 cd	44.57 ± 0.97 a
γ-Tocopherols	714.61 ± 4.66 a	717.64 ± 2.85 ab	733.58 ± 12.50 bc	748.56 ± 0.39 cd	778.76 ± 11.45 e	761.75 ± 2.96 d	750.44 ± 1.13 d
σ-Tocopherols	6.40 ± 0.05 b	6.26 ± 0.03 ab	6.37 ± 0.05 b	6.60 ± 0.03 b	9.51 ± 0.28 c	5.63 ± 0.49 a	6.32 ± 0.42 b
Total	785.26 ± 10.28 a	795.43 ± 24.55 a	820.24 ± 16.29 a	781.70 ± 27.64 a	836.71 ± 13.19 a	820.02 ± 3.50 a	803.61 ± 1.68 a
Phytosterols(mg/kg)							
Squalene	13.82 ± 0.30 a	14.45 ± 0.47 ab	14.36 ± 0.59 ab	15.22 ± 0.17 bc	15.62 ± 0.58 cd	16.58 ± 0.57 d	15.16 ± 0.25 bc
Campesterol	30.30 ± 0.48 a	30.63 ± 0.54 ab	30.98 ± 0.36 ab	30.78 ± 0.76 ab	32.02 ± 0.21 bc	32.87 ± 0.39 c	33.035 ± 1.23 c
γ-Stigmasterol	8.61 ± 0.85 a	9.90 ± 0.47 b	9.73 ± 0.57 b	10.43 ± 0.43 b	10.30 ± 0.02 b	10.62 ± 0.11 b	10.04 ± 0.24 b
β-Sitosterol	1454.75 ± 4.16 a	1558.20 ± 4.92 bc	1532.77 ± 4.74 ab	1575.57 ± 2.76 bcd	1645.67 ± 3.03 d	1628.88 ± 0.43 cd	1590.28 ± 3.72 bcd
Fucosterol	677.24 ± 0.97 a	686.41 ± 3.37 abc	680.31 ± 2.48 ab	701.03 ± 2.13 cd	705.75 ± 1.31 d	707.64 ± 2.22 d	694.68 ± 1.21 bcd
Δ-5-Avenasterol	117.98 ± 2.29 a	126.84 ± 2.93 a	159.79 ± 2.17 b	164.56 ± 1.90 bc	167.36 ± 1.51 bc	171.29 ± 0.98 c	164.67 ± 1.89 bc
Δ-7-Avenasterol	33.73 ± 1.32 a	34.69 ± 0.14 ab	38.04 ± 0.74 c	36.64 ± 0.48 bc	37.40 ± 1.15 c	38.14 ± 0.71 c	38.18 ± 0.86 c
24-Methylenecycloartan-3β-ol	119.37 ± 2.26 a	129.19 ± 0.37 b	158.25 ± 2.18 c	178.93 ± 0.98 de	172.74 ± 1.38 d	191.16 ± 2.01 f	182.13 ± 1.85 e
Total	2455.79 ± 44.04 a	2590.29 ± 32.59 b	2624.21 ± 32.86 bc	2713.14 ± 14.16 cd	2786.85 ± 77.75 d	2797.14 ± 7.99 d	2728.16 ± 24.78 cd

Values are means ± standard deviations, $n = 3$ replicates per treatment. Different superscript letters within the same row indicate significant differences (one-way ANOVA and Duncan's test, $p \leq 0.05$). RF, radio frequency; MW, microwave; RT, roasting; SFAs, saturated fatty acids; UFAs, unsaturated fatty acids.

The contents of tocopherols in oil are shown in Table 2. The most abundant tocopherol in the TPSO samples was γ-tocopherol, followed by α- and δ-tocopherols for which the contents were relatively low. Firstly, at different RF temperatures, findings revealed that as the RF temperature increased, the amount of total tocopherol increased significantly, at 6.5% higher than the initial value. More specifically, the γ-tocopherol content varied between 714.61 mg/kg (untreated) and 778.76 mg/kg (RF at 140 °C). The δ-tocopherol content was between 6.40 mg/kg (untreated) and 9.51 mg/kg (RF at 140 °C). In addition, the α-tocopherols content was slightly reduced in TPSO, determined to be between 45.96 mg/kg (RF at 140 °C) and 51.76 mg/kg (control). Additionally, by preheating using the RF, MW and RT pretreatments, at 140 °C, the γ-tocopherol content was significantly increased to 778.76, 761.74 and 750.44 mg/kg, respectively. However, the same technology caused a significant ($p < 0.05$) reduction in α-tocopherol contents by 5.80, 1.78 and 7.2 mg/kg, respectively. There was no significant difference ($p < 0.05$) in the total content. Tocopherols contents in TPSO were high, which could prevent the high contents of UFAs from oxidation, γ-tocopherol showed a better antioxidant ability, decreasing the risk of cancer, and performing anti-inflammatory activities [41,42].

A statistical analysis showed that the contents of the seven phytosterols and squalene TPSOs produced by different pretreatments varied significantly ($p < 0.05$). The phytosterol with the highest content of TPSO was β-sitosterol (1454.75 ± 4.16 mg/kg), followed by fucosterol, which was a unique plant sterol in TPSO, since a similar distribution was rarely reported in other vegetable oils. The phytosterols compounds in this study were, to a limited extent, similar to the findings reported by Chang et al. [5]. Of the different RF temperatures, the highest squalene content was 15.62 mg/kg at 140 °C. The content of β-sitosterol in the different groups ranged from 1454.75 mg/kg to 1645.67 mg/kg, fucosterol ranged from 677.24 mg/kg to 705.75 mg/kg, which increased by 190.92 mg/kg and 28.51 mg/kg respectively, followed by Δ-5-avenasterol and 24-methylenecycloartan-3β-ol, which increased from 4.83% to 6.04% and 4.89% to 6.23% respectively. After three different pretreatments at 140 °C, the total phytosterol content in RF, MV and RT oil increased by 13.48%, 13.90%, 11.09% (compared to untreated samples), respectively, which proved that the three pretreatment methods can increase the extraction content. For individual phytosterols, the level of β-sitosterol, when treated by RF, MW and RT, increased significantly (compared to untreated samples) by 13.12%, 11.97% and 9.32%, respectively. The β-sitosterol and fucosterol in RF oil (1645.67 mg/kg and 705.75 mg/kg, respectively) were significantly higher than in RT oil (1590.28 mg/kg and 694.68 mg/kg, respectively). However, the increase in the total content of phytosterols for the three pretreatments of TPSO were not significant. In conclusion, a longer nutrient retention period was produced by the rapid conversion of electromagnetic energy to thermal energy in MW and RF heating than for the conventional RT, which was consistent with the data reported by Al Juhaimi et al. [18].

3.4. Sensory Characteristics and Volatile Compounds of TPSO

Volatile components are also important indicators for evaluating vegetable oils [29,43]. The volatile components of TPSO were significantly enhanced through RF, MW and RT treatment at 140 °C (Supplementary Table S2), increases were determined for mainly aldehydes, ketones, esters, and heterocyclic compounds. In the untreated TPSO, 3-octanol and 1-octen-3-ol, (Z)-2-penten-1-ol were detected. According to previous studies, these compounds only appear in untreated oil and are easily destroyed at a high temperature, mainly due to the thermal decomposition of linoleic acid methyl hydrogen peroxide [44]. Figure 3 shows that the total amount of volatile components was lowest with RF treatment; more heterocyclic substances were produced by MW and RT produced more ketone substances. The most dominant heterocyclic substances were 4,7-dimethylbenzofuran and 2,6-dimethylpyrazine, which were significantly increased under RF, MW and RT treatment at 140 °C ($p < 0.05$). Previous research indicated that pyrazines contribute to the roasted aroma [15]. Some alcohols, phenol, and acids were also detected in the TPSO sample of different treatments. The presence of benzyl alcohol and maltol increased with RT and RF treatment. RF and MW were conducive to an increase in lipids, especially γ-butyrolactone (caramel, sweet). Xiao et al. [29] revealed that volatile components in camellia seed oils mainly occurred due to the oxidation of lipase. Furthermore, 2-Furanmethanol increased significantly after pretreatment and was the main alcohol after processing. Benzoic acid increased significantly with RT, providing an unfavorable contribution to the flavor, which was described as pungent (Table S2). The level of benzoic acid in TPSO was very high, and found to be the active constituent, providing the acaricidal activity of the peony root bark [45]. Acetic acid also increased significantly for RF, MV, RT at 140 °C, this substance was also detected in palm kernel oil and Camellia seed oil. It is generated at an early stage in the Maillard reactions [46], and in this study we observed that the content increased most dramatically after RT treatment at 140 °C.

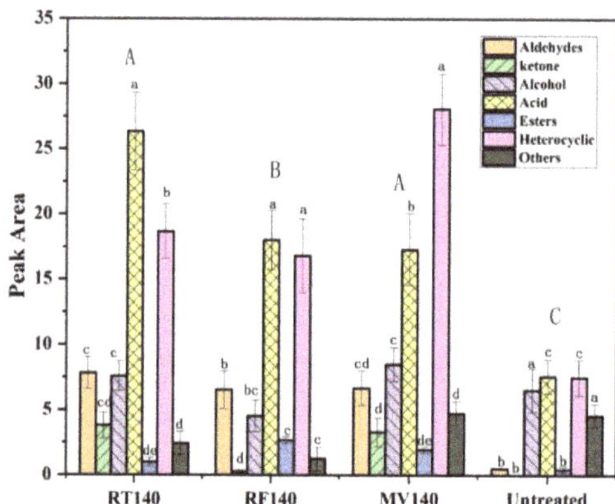

Figure 3. The content of volatile compounds treated by RT, RF, MV pretreatments at 140 °C. Different superscript letters indicate significant differences within the same pretreatment (lowercase, a) and different pretreatments (uppercase, A), respectively (one-way ANOVA and Duncan's test, $p \leq 0.05$). RF, radio frequency; MW, microwave; RT, roasting.

According to Figure 4, the TPSO scored higher as being grassy and nutty, which are classified as positive sensory attributes, and pungency was reported as an aftertaste in TPSO. This may be caused by the slight oxidation of TPSO due to a high α-linolenic acid level ($\geq$45.53%), and by benzoic acid and other phenolics in TPSO. After RF, MW, and RT treatments at 140 °C, the roasted aroma was found to be strong, the overall aroma increased, the grassy aroma was masked, and nutty and oily characteristics weakened. This also verifies the results of GC-MS measurements, as the heterocyclic substances considerably increased after three treatments. At the same temperature, the volatile aroma component of the RF sample was weaker than that of MW and RT, whereas the scent of the MW samples was gentler than that of the RT samples. This may be due to the differences in molecular vibration [12,21]. The actual mechanism for these differences is unclear.

3.5. Effect of Pretreatment on the Antioxidant Capacities and Oxidative Stabilities of Oils

Firstly, the antioxidant capacities of TPSO were evaluated through the free radicals' scavenging ability (DPPH and FRAP). Table 3 shows that the in vitro antioxidant properties of TPSO, after being treated with different RF temperatures, significantly increased. The extract of oil produced by the untreated seed was able to remove NO·free radicals and the removal rate was very small (DPPH) compared to other vegetable oils [4,14,18]. The DPPH and FRAP scavenging ability of the untreated sample were 2.65 and 27.69 μmol TE/100 g, respectively, which significantly increased (33.26–26.40 μmol TE/100 g DPPH and 65.84–82.54 μmol TE/100 g FRAP) with different RF temperature treatments. Radical scavenging increased significantly after different RF temperatures. The antioxidant effect was significantly increased with RF120 °C compared to the untreated sample (from 2.65 to 23.82 μmol TE/100 g). To better compare the different thermal based pressing methods under same operating conditions, antioxidation was tested during the pressing process, the results for which are shown in Table 3. The DPPH and FRAP values of TPSO, regardless of the use of RF, MW or RT heating pretreatments at 140 °C, increased compared to the values for the untreated groups. FRAP and DPPH scavenging abilities increased by 2.37, 2.98 and 2.42 times/12.55, 12.22 and 9.96 times, respectively. RF, MW and RT treatments produced excellent antioxidant properties in the resulting oi. These results may be caused by the substantial increase in the content of TPC [47]. FRAP and DPPH were associated with reducing the power of an antioxidant, and their increase in the TPSO implies an increased ability to quench free radicals in a system [48]. Therefore, increased radical scavenging could be explained by the possible formation of reducing compounds.

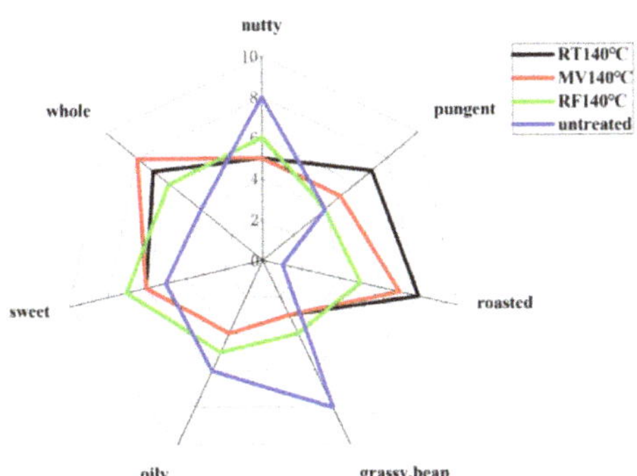

Figure 4. Sensory evaluation treated by RT, RF, MV pretreatments at 140 °C. RF, radio frequency; MW, microwave; RT, roasting.

Table 3. The content of antioxidant capacities and oxidation stability index of TPSO.

	FRAP (umol TE/100 g)	DPPH (umol TE/100 g)	CD	CT	POV (mmol O_2/kg)	P-AV	IP(h)
Untreated	27.69 ± 1.05 [a]	2.65 ± 0.13 [a]	1.24 ± 0.005 [a]	0.16 ± 0.003 [a]	0.29 ± 0.01 [a]	1.20 ± 0.11 [a]	1.07 ± 0.12 [a]
RF80 °C	42.56 ± 0.88 [b]	3.81 ± 0.31 [a]	1.31 ± 0.013 [a]	0.16 ± 0.005 [a]	0.35 ± 0.01 [a]	1.23 ± 0.08 [a]	1.21 ± 0.08 [a]
RF100 °C	79.41 ± 1.86 [c]	5.51 ± 0.45 [b]	1.49 ± 0.021 [a]	0.19 ± 0.009 [a]	0.71 ± 0.01 [b]	1.20 ± 0.12 [a]	2.14 ± 0.08 [b]
RF120 °C	89.61 ± 0.69 [d]	23.82 ± 0.37 [c]	1.65 ± 0.025 [a]	0.24 ± 0.003 [a]	1.07 ± 0.01 [c]	1.18 ± 0.05 [a]	3.07 ± 0.11 [c]
RF140 °C	93.53 ± 2.36 [e]	35.91 ± 0.15 [e]	2.21 ± 0.048 [b]	0.23 ± 0.007 [a]	1.42 ± 0.01 [d]	1.26 ± 0.11 [a]	4.33 ± 0.13 [d]
MW140 °C	110.23 ± 2.04 [f]	35.04 ± 0.05 [e]	1.35 ± 0.027 [a]	0.16 ± 0.002 [a]	1.13 ± 0.01 [c]	1.49 ± 0.05 [a]	4.53 ± 0.15 [d]
RT140 °C	94.59 ± 1.46 [e]	29.05 ± 0.46 [d]	1.86 ± 0.008 [b]	0.25 ± 0.007 [a]	1.54 ± 0.01 [d]	1.43 ± 0.03 [a]	3.11 ± 0.10 [c]

Values are means ± standard deviations, $n = 3$. Different superscript letters within the same row indicate significant differences (one-way ANOVA and Duncan test, $p \leq 0.05$). DPPH, 2,2-diphenyl-1-picrylhydrazyl; FRAP, ferric-reducing antioxidant power; CD, conjugated diene; CT, conjugated trienes; POV, peroxide value; P-AV, p-anisidine value; IP, induction period.

The correlation analysis results of its main bioactive compound (Table 4) show that DPPH, FRAP and the main bioactive compounds in TPSO at RT, MW, RF at 140 °C were significantly correlated at a 0.01 level or 0.05 level. Different heating treatments increased the bioactive compounds and enhanced the oxidation stability of TPSO. Regardless of the pretreatment, positive correlations were observed between the antioxidant activity, and the levels of γ-tocopherol, β-sitosterol, fucosterol, squalene and TPC. More specifically, DPPH was found to be highly correlated with TPC (r = 0.955, $p < 0.05$), γ-tocopherol content, (r = 0.965, $p < 0.05$), β-sitosterol levels (r = 0.996, $p < 0.05$), fucosterol content (r = 0.925, $p < 0.01$), and squalene content (r = 0.902, $p < 0.01$) suggesting that most of the antioxidant activity could be attributed to these components. In addition, FRAP was highly correlated ($p < 0.05$) with TPC (r = 0.986) and β-sitosterol (r = 0.960), followed by γ-tocopherol, fucosterol and squalene (r = 0.932–0.895; $p < 0.01$). In general, highly positive correlations between bioactive properties and the antioxidant capacities of raw samples and those treated using an RF, MV or RT pretreatment at 140 °C demonstrated the excellent contribution of these compounds to the bioactivity and functional potentials of TPSO.

POV, CD, CT, and P-AV were determined as indicators of the oxidation of lipids during the treatment process. The primary oxidation products, POV and CD, increased with an increasing RF heating temperature, from 0.29 to 0.42 mmol/kg and from 1.24 to 1.65, respectively. An increase in temperature caused the production of hydrogen peroxide, which increases the hydrolytic rancidity [10]. The POV of TPSO of untreated and heat-treated seeds ranged between 0.29 and 1.54 mmolO$_2$/kg, and the values significantly differ from each other. In contrast, oxidation stabilization effect of MW pretreatment was more pronounced. The increase in POV may be due to the high-temperature oxidation used in the heating treatment. However, Ghafoor et al. [49] studied the POV of poppy seeds, and the oil had similar values to those obtained in this study, but its antioxidant capacity (DPPH) and TPC content were reduced after RT and MV pretreatment. Their results contradict those of this study, which may be a result of differing materials and technology conditions. Unexpectedly, there were insignificant changes in the CT values of the analyzed oil samples independent of the RF heating temperature of tree peony; the value ranged from 0.16–0.24. Conversely, the CT amounts of TPSO were found to be insignificantly different between MW (lowest 0.16) and RT (highest 0.25). P-AV was the measurement of the secondary oxidation product of aldehydes and ketones. The P-AV ranged from 1.19 to 1.49, which showed significant differences ($p < 0.05$) between the three different pretreatments. P-AV values were 1.26, 1.49, and 1.43 after different pretreatment with RF, MW, and RT respectively. MW and RT had the highest values, caused by the oxidative degradation of linoleic acid (abstraction C11 and C9), the Maillard reaction, Strecker degradation, and oxidation processes, which result in the formation of heterocycles, ketones, and aldehydes, and decompose the hydroperoxides, resulting in the formation of aldehydes and ketones [50]. The results are consistent with the previously determined volatile compounds. In addition, the IP value of untreated TPSO was found to be 1.07 h, mainly caused by the high content of unsaturated fatty acids. the value significantly increased to 4.33 h, 4.53 h and 3.11 h after RF, MW and RT treatments at 140 °C, respectively.

Table 4. Correlation between main bioactive components and antioxidant capacities of TPSO.

	β-Sitosterol	Fucosterol	Squalene	FRAP	DPPH	TPC
γ-tocopherol	0.982 **	0.967 **	0.895 *	0.900 *	0.965 **	0.846
β-sitosterol		0.944 *	0.915 *	0.960 **	0.996 **	0.929 *
Fucosterol			0.939 *	0.895 *	0.925 *	0.819
Squalene				0.932 *	0.902 *	0.864
FRAP					0.974 **	0.986 **
DPPH						0.955 *

** Significantly correlated at the 0.01 level (two-sided). * Significantly correlated at the 0.05 level (two-sided).

4. Conclusions

In this study, the effects of RF pretreatment at three different temperatures on the quality in TPSO and its comparison with MW and RT pretreatments at 140 °C were investigated. The results demonstrate that the RF pretreatment could significantly improve TPSO qualities, since it enhances oil yield, increases bioactive compounds and it improves flavor and the oxidation resistance. TPSO was found to have improved qualities when the seed was treated by RF at 140 °C. RF can be applied to the processing of TPS, effectively destroying its internal structure and increasing the oil yield (15.23%). It significantly affected the TPC and antioxidant (DPPH and FRAP) capacities of TPSO ($p < 0.05$), but the flavor was not as strong as the oil produced by MW and RT pretreatments. These findings may serve as a reference for improving the application of RF technology to oilseeds. In addition, RF, MW, and RT treatments at 140 °C significantly affected the characteristic compositions compared to untreated TPSO (the phytosterol increased by 331.06, 341.35 and 272.37 mg/kg and tocopherol increased by 51.63, 35.40, and 18.43 mg/kg, respectively). RF pretreatment could more effectively enhance the bioactive compounds of the oil compared with RT, thereby increasing their nutritional value. By comparing all three pretreatments, we found that the strongest antioxidant activity was produced by MW. The primary oxidation POV and CD were enhanced further by RF; the secondary oxidation products of the P-AV and CT showed no significant changes ($p < 0.05$). In conclusion, the three preheating treatments not only induce the rapid extraction of edible oil but also improve the quality of TPSO. Deacidification and the mechanism of producing volatile compounds after different pretreatments should be further considered for TPSO production. These results can provide guidance for the industrial applications of TPSO.

Supplementary Materials: The following are available online at https://www.mdpi.com/article/10.3390/foods10123062/s1, Table S1: Oil yields and moisture contents of radio frequency (RF) pretreated tree peony seeds samples. Table S2: Odorants and odor descriptors determined by GC–MS for different pretreatments.

Author Contributions: Z.W. carried out the experimental plan, data interpretation, and manuscript writing; C.Z. performed review, editing, and supervision; C.L. was involved in the data interpretation and manuscript redaction, in addition to being responsible for the project. F.H. was involved in validation, investigation, and project administration. Y.H. and W.W. were involved in validation, investigation and project administration. All authors have read and agreed to the published version of the manuscript.

Funding: This work was financially supported by the National Key Research and Development Project of China (No. 2018YFD0401104). Wuhan Scientific and Technical Payoffs Transformation Project (2019030703011505). The Agricultural Science and Technology Innovation Project of Chinese Academy of Agricultural Sciences (CAAS-ASTIP-2016-OCRI).

Institutional Review Board Statement: The conditions for the sensory descriptive analysis were in accordance with the guidelines in ISO 13299:2003.

Informed Consent Statement: Informed consent was obtained from all subjects involved in the study.

Conflicts of Interest: The authors declare no competing financial interest.

References

1. Yu, S.; Du, S.; Yuan, J.; Hu, Y. Fatty acid profile in the seeds and seed tissues of *Paeonia* L. species as new oil plant resources. *Sci. Rep.* **2016**, *6*, 26944. [CrossRef]
2. Gao, L.L.; Li, Y.Q.; Wang, Z.S.; Sun, G.J.; Qi, X.M.; Mo, H.Z. Physicochemical characteristics and functionality of tree peony (*Paeonia suffruticosa* Andr.) seed protein. *Food Chem.* **2018**, *240*, 980–988. [CrossRef]
3. Wang, X.; Li, C.; Contreras, M.D.M.; Verardo, V.; Gomez-Caravaca, A.M.; Xing, C. Integrated profiling of fatty acids, sterols and phenolic compounds in tree and herbaceous peony seed oils: Marker Screening for New Resources of Vegetable Oil. *Foods* **2020**, *9*, 770. [CrossRef]
4. Ning, C.; Jiang, Y.; Meng, J.; Zhou, C.; Tao, J. Herbaceous peony seed oil: A rich source of unsaturated fatty acids and γ-tocopherol. *Eur. J. Lipid Sci. Technol.* **2015**, *117*, 532–542. [CrossRef]
5. Chang, M.; Wang, Z.; Zhang, T.; Wang, T.; Liu, R.; Wang, Y.; Jin, Q.; Wang, X. Characterization of fatty acids, triacylglycerols, phytosterols and tocopherols in peony seed oil from five different major areas in China. *Food Res. Int.* **2020**, *137*, 109416. [CrossRef]
6. Su, J.; Ma, C.; Liu, C.; Gao, C.; Nie, R.; Wang, H. Hypolipidemic activity of peony seed oil rich in alpha-Linolenic, is mediated through inhibition of lipogenesis and upregulation of fatty acid beta-oxidation. *J. Food Sci.* **2016**, *81*, H1001–H1009. [CrossRef] [PubMed]
7. Su, J.; Wang, H.; Ma, C.; Lou, Z.; Liu, C.; Tanver Rahman, M.; Gao, C.; Nie, R. Anti-diabetic activity of peony seed oil, a new resource food in STZ-induced diabetic mice. *Food Funct.* **2015**, *6*, 2930–2938. [CrossRef]
8. Wei, X.B.; Xue, J.Q.; Wang, S.L.; Xue, Y.Q.; Lin, H.; Shao, X.F.; Xu, D.H.; Zhang, X.X. Fatty acid analysis in the seeds of 50 *Paeonia ostii* individuals from the same population. *J. Integr. Agric.* **2018**, *17*, 1758–1767. [CrossRef]
9. Chew, S.C. Cold-pressed rapeseed (*Brassica napus*) oil: Chemistry and functionality. *Food Res. Int.* **2020**, *131*, 108997. [CrossRef]
10. Szydłowska-Czerniak, A.; Tymczewska, A.; Momot, M.; Włodarczyk, K. Optimization of the microwave treatment of linseed for cold-pressing linseed oil—Changes in its chemical and sensory qualities. *LWT Food Sci. Technol.* **2020**, *126*, 109317. [CrossRef]
11. Li, W.G.; Sun, X.L.; Zu, Y.G.; Zhao, X.H. Optimization peony seed oil extraction process at suitable temperature and physicochemical property analysis. *Bull. Bot. Res.* **2020**, *40*, 73–78. [CrossRef]
12. Prakesh, A.; Dave, V.; Sur, S.; Sharma, P. Vivid techniques of pretreatment showing promising results in biofuel production and food processing. *J. Food Process. Eng.* **2020**, *44*, e13580. [CrossRef]
13. Jiao, Y.; Tang, J.; Wang, Y.; Koral, T.L. Radio-frequency applications for food processing and safety. *Annu. Rev. Food Sci. Technol.* **2018**, *9*, 105–127. [CrossRef]
14. Ghafoor, K.; Ahmed, I.A.M.; Ozcan, M.M.; Al-Juhaimi, F.Y.; Babiker, E.E.; Azmi, I.U. An evaluation of bioactive compounds, fatty acid composition and oil quality of chia (*Salvia hispanica* L.) seed roasted at different temperatures. *Food Chem.* **2020**, *333*, 127531. [CrossRef] [PubMed]
15. Jin, F.; Xu, J.; Liu, X.R.; Regenstein, J.M.; Wang, F.J. Roasted tree peony (*Paeonia ostii*) seed oil: Benzoic acid levels and physico-chemical characteristics. *Int. J. Food Prop.* **2019**, *22*, 499–510. [CrossRef]
16. Azadmard-Damirchi, S.; Habibi-Nodeh, F.; Hesari, J.; Nemati, M.; Achachlouei, B.F. Effect of pretreatment with microwaves on oxidative stability and nutraceuticals content of oil from rapeseed. *Food Chem.* **2010**, *121*, 1211–1215. [CrossRef]

17. Ye, M.; Zhou, H.; Hao, J.; Chen, T.; He, Z.; Wu, F.; Liu, X. Microwave pretreatment on microstructure, characteristic compounds and oxidative stability of Camellia seeds. *Ind. Crop. Prod.* **2021**, *161*, 113193. [CrossRef]

18. Al Juhaimi, F.; Musa Ozcan, M.; Ghafoor, K.; Babiker, E.E. The effect of microwave roasting on bioactive compounds, antioxidant activity and fatty acid composition of apricot kernel and oils. *Food Chem.* **2018**, *243*, 414–419. [CrossRef]

19. Zhou, X.; Li, R.; Lyng, J.G.; Wang, S.J. Dielectric properties of kiwifruit associated with a combined radio frequency vacuum and osmotic drying. *J. Food Eng.* **2018**, *239*, 72–82. [CrossRef]

20. Zheng, J.; Li, H.; Wang, D.; Li, R.; Wang, S.; Ling, B. Radio frequency assisted extraction of pectin from apple pomace: Process optimization and comparison with microwave and conventional methods. *Food Hydrocoll.* **2021**, *121*, 107031. [CrossRef]

21. Lan, R.; Qu, Y.; Ramaswamy, H.S.; Wang, S. Radio frequency reheating behavior in a heterogeneous food: A case study of pizza. *Innov. Food Sci. Emerg. Technol.* **2020**, *65*, 102478. [CrossRef]

22. Liu, P.; Zhang, L.N.; Wang, X.S.; Gao, J.-Y.; Yi, J.P.; Deng, R.X. Characterization of *Paeonia ostii* seed and oil sourced from different cultivation areas in China. *Ind. Crop. Prod.* **2019**, *133*, 63–71. [CrossRef]

23. Li, S.S.; Yuan, R.Y.; Chen, L.G.; Wang, L.S.; Hao, X.H.; Wang, L.J.; Zheng, X.C.; Du, H. Systematic qualitative and quantitative assessment of fatty acids in the seeds of 60 tree peony (*Paeonia* section *Moutan* DC.) cultivars by GC-MS. *Food Chem.* **2015**, *173*, 133–140. [CrossRef] [PubMed]

24. Cong, Y.; Zheng, M.; Huang, F.; Liu, C.; Zheng, C. Sinapic acid derivatives in microwave-pretreated rapeseeds and minor components in oils. *J. Food Compos. Anal.* **2020**, *87*, 103394. [CrossRef]

25. National Health Commission. *National Standard of the People's Republic of China*; National Health Commission: Beijing, China, 2016.

26. Cong, Y.; Cheong, L.Z.; Huang, F.; Zheng, C.; Wan, C.; Zheng, M. Effects of microwave irradiation on the distribution of sinapic acid and its derivatives in rapeseed and the antioxidant evaluation. *LWT Food Sci. Technol.* **2019**, *108*, 310–318. [CrossRef]

27. MPOB. *Determination of Carotene Content*; Method No. p2.6: 2004; Malaysian Palm Oil Board: Kuala Lumpur, Malaysia, 2005.

28. AOCS Official Method Cd18-90. *p-Anisidine Value*; The American Oil Chemists' Society: Urbana, IL, USA, 2011.

29. Xiao, J.; Deng, Q.; Yang, Y.; Xiang, X.; Zhou, X.; Tan, C.; Zhou, Q. Unraveling of the aroma-active compounds in virgin camellia oil (*Camellia oleifera* Abel) using gas chromatography–mass spectrometry-olfactometry, aroma recombination, and omission studies. *J. Agric. Food Chem.* **2021**, *69*, 9043–9055. [CrossRef]

30. ISO-3656:2002. *Animal and Vegetable Fats and Oils. Determination of Ultraviolet Absorbance Expressed as Specific UV Extinction*; ISO: Geneva, Switzerland, 2002.

31. AOCS Official Method Ce 8-89. *Determination of Tocopherols and Tocotrienols in Vegetable Oils and Fats by HPLC*; The American Oil Chemists' Society: Urbana, IL, USA, 2009.

32. Amini Khoozani, A.; Birch, J.; Bekhit, A.E.D.A. Textural properties and characteristics of whole green banana flour produced by air-oven and freeze-drying processing. *J. Food Meas. Charact.* **2020**, *14*, 1533–1542. [CrossRef]

33. Taheri, S.; Brodie, G. Gupta, DFluidisation of lentil seeds during microwave drying and disinfection could prevent detrimental impacts on their chemical and biochemical characteristics. *LWT Food Sci. Technol.* **2020**, *129*, 109534. [CrossRef]

34. Hatamian, M.; Noshad, M.; Abdanan-Mehdizadeh, S.; Barzegar, H. Effect of roasting treatment on functional and antioxidant properties of chia seed flours. *NFS J.* **2020**, *21*, 1–8. [CrossRef]

35. Zhong, Y.; Wang, Z.; Zhao, Y. Impact of radio frequency, microwaving, and high hydrostatic pressure at elevated temperature on the nutritional and antinutritional components in black soybeans. *J. Food Sci.* **2015**, *80*, C2732–C2739. [CrossRef]

36. Wroniak, M.; Rêkas, A.; Siger, A.; Janowicz, M. Microwave pretreatment effects on the changes in seeds microstructure, chemical composition and oxidative stability of rapeseed oil. *LWT Food Sci. Technol.* **2016**, *68*, 634–641. [CrossRef]

37. Hu, B.; Li, C.; Qin, W.; Zhang, Z.; Liu, Y.; Zhang, Q.; Liu, A.; Jia, R.; Yin, Z.; Han, X.; et al. A method for extracting oil from tea (*Camelia sinensis*) seed by microwave in combination with ultrasonic and evaluation of its quality. *Ind. Crop. Prod.* **2019**, *131*, 234–242. [CrossRef]

38. Ling, B.; Yang, X.; Li, R.; Wang, S. Physicochemical properties, volatile compounds, and oxidative stability of cold pressed kernel oils from raw and roasted pistachio (*Pistacia vera* L. Var Kerman). *Eur. J. Lipid Sci. Technol.* **2016**, *118*, 1368–1379. [CrossRef]

39. Suri, K.; Singh, B.; Kaur, A. Impact of microwave roasting on physicochemical properties, maillard reaction products, antioxidant activity and oxidative stability of nigella seed (*Nigella sativa* L.) oil. *Food Chem.* **2021**, *368*, 130777. [CrossRef] [PubMed]

40. Özcan, M.M.; Atalay, C. Determination of seed and oil properties of some poppy (*Papaver somniferum* L.) varieties. *Grasas Y Aceites* **2006**, *57*, 169–174. [CrossRef]

41. Zheng, L.Y.; Jin, J.; Shi, L.K.; Huang, J.H.; Chang, M.; Wang, X.G.; Jin, Q.Z. Gamma tocopherol, its dimmers, and quinones: Past and future trends. *Crit. Rev. Food Sci. Nutr.* **2020**, *60*, 3916–3930. [CrossRef]

42. Thompson, M.D.; Cooney, R.V. The potential physiological role of gammatocopherol in human health: A qualitative review. *Nutr. Cancer -Int. J.* **2019**, *72*, 808–825. [CrossRef]

43. Yu, P.; Yang, Y.; Sun, J.; Jia, X.; Zheng, C.; Zhou, Q.; Huang, F. Identification of volatile sulfur-containing compounds and the precursor of dimethyl sulfide in cold-pressed rapeseed oil by GC-SCD and UPLC-MS/MS. *Food Chem.* **2022**, *367*, 130741. [CrossRef]

44. Xiao, L.; Lee, J.; Zhang, G.; Ebeler, S.E.; Wickramasinghe, N.; Seiber, J.; Mitchell, A.E. HS-SPME GC/MS Characterization of Volatiles in Raw and Dry-Roasted Almonds (*Prunus dulcis*). *Food Chem.* **2014**, *151*, 31–39. [CrossRef]

45. Kim, H.K.; Tak, J.H.; Ahn, Y.J. Acaricidal activity of *Paeonia suffruticosa* root bark-derived compounds against *Dermatophagoides farinae* and *Dermatophagoides pteronyssinus* (Acari: Pyroglyphidae). *J. Agric. Food Chem.* **2004**, *52*, 7857–7861. [CrossRef]

46. Zhang, W.Z.; Wang, R.; Yuan, Y.H.; Yang, T.K.; Liu, S.Q. Changes in Volatiles of Palm Kernel Oil before and after Kernel Roasting. *LWT Food Sci. Technol.* **2016**, *73*, 432–441. [CrossRef]
47. Chinma, C.E.; Adedeji, O.E.; Etim, I.I.; Aniaka, G.I.; Mathew, E.O.; Ekeh, U.B.; Anumba, N.L. Physicochemical, nutritional, and sensory properties of chips produced from germinated African yam bean (*Sphenostylis stenocarpa*). *LWT Food Sci. Technol.* **2021**, *136*, 110330. [CrossRef]
48. Diniyah, N.; Alam, M.B.; Lee, S. Antioxidant potential of non-oil seed legumes of Indonesian's ethnobotanical extracts. *Arab. J. Chem.* **2020**, *13*, 5208–5217. [CrossRef]
49. Ghafoor, K.; Özcan, M.M.; Al-Juhaimi, F.; Babiker, E.E.; Fadimu, G.J. Changes in quality, bioactive compounds, fatty acids, tocopherols, and phenolic composition in oven- and microwave-roasted poppy seeds and oil. *LWT Food Sci. Technol.* **2019**, *99*, 490–496. [CrossRef]
50. Pan, F.; Wen, B.; Wang, X.; Ma, X.; Zhao, J.; Liu, C.; Xu, Y.; Dang, W. Effect of the chemical refining process on perilla seed oil composition and oxidative stability. *J. Food Process. Preserv.* **2019**, *43*, e14094. [CrossRef]

 *foods*

Article

Controlled Release of Flavor Substances from Sesame-Oil-Based Oleogels Prepared Using Biological Waxes or Monoglycerides

Min Pang [1,2], Lulu Cao [1,2], Shengmei Kang [1,2], Shaotong Jiang [1,2] and Lili Cao [1,2,*]

1   School of Food Science and Bioengineering, Hefei University of Technology, Hefei 230009, China; pangmin@hfut.edu.cn (M.P.); 2018111236@mail.hfut.edu.cn (L.C.); 2020171424@mail.hfut.edu.cn (S.K.); jiangshaotong@163.com (S.J.)
2   Key Laboratory for Agricultural Products Processing of Anhui Province, Hefei 230009, China
*   Correspondence: lilycao504@hfut.edu.cn

Abstract: The flavor substances in sesame oil (SO) are volatile and unstable, which causes a decrease in the flavor characteristics and quality of SO during storage. In this study, the effect of gelation on the release of flavor substances in SO was investigated by preparing biological waxes and monoglycerides oleogels. The results showed that the release of flavor substances in SO in an open environment is in accordance with the Weibull equation kinetics. The oleogels were found to retard the release of volatiles with high saturated vapor pressures and low hydrophobic constants in SO. The release rate constant k value of 2-methylpyazine in BW oleogel is 0.0022, showing the best retention effect. In contrast, the addition of gelling agents had no significant retention effect on the release of volatiles with low saturated vapor pressures or high hydrophobic constants in SO, and even promoted the release of these compounds to some extent. This may be due to the hydrophilic structural domains formed by the self-assembly of gelling agents, which reduces the hydrophobicity of SO. This work provides a novel approach for retaining volatile compounds in flavored vegetable oils. As a new type of flavor delivery system, oleogels can realize the controlled release of volatile compounds.

Keywords: sesame oil; gelation; oleogels; controlled volatile release

Citation: Pang, M.; Cao, L.; Kang, S.; Jiang, S.; Cao, L. Controlled Release of Flavor Substances from Sesame-Oil-Based Oleogels Prepared Using Biological Waxes or Monoglycerides. *Foods* **2021**, *10*, 1828.

https://doi.org/10.3390/foods10081828

Academic Editors: Qiang Wang and Aimin Shi

Received: 16 July 2021
Accepted: 3 August 2021
Published: 7 August 2021

Publisher's Note: MDPI stays neutral with regard to jurisdictional claims in published maps and institutional affiliations.

Copyright: © 2021 by the authors. Licensee MDPI, Basel, Switzerland. This article is an open access article distributed under the terms and conditions of the Creative Commons Attribution (CC BY) license (https://creativecommons.org/licenses/by/4.0/).

1. Introduction

Sesame (*Sesamum indicum* L.) oil (SO), a traditional vegetable oil used for seasoning in China, has a strong and unique flavor and is rich in nutrients. The flavor substances in SO are formed by a series of Maillard reactions, oxidation reactions, and caramelization reactions during processing, which produce furans, pyrazines, aldehydes, alcohols, ketones, and pyrroles. In addition, SO can have an unsaturated fatty acid content of more than 80% [1] and is rich in phytosterols, tocopherols, lignans, vitamins, and trace elements such as iron, zinc, and copper [2,3]. Owing to its flavor and nutritional value, SO has become widely used for seasoning in China and is popular with consumers. However, the flavor substances in SO undergo osmotic diffusion during processing and storage owing to their volatility and poor stability, which leads to flavor loss and quality degradation. The diffusion release process of small flavor molecules follows the zero-order release kinetic equation, first-order release kinetic equation, Higuchi equation, or Weibull equation [4–7]. Under the action of a concentration gradient, small flavor molecules diffuse through the vegetable oil matrix and volatilize on the outer surface. Because SO has a strong flavor, the loss of flavor substances is of particular significance. Therefore, research on effective SO flavor preservation technologies is important to enhance the nutritional function of SO and promote the modernization of traditional flavored oils.

In recent years, gel formation for vegetable oil solidification and corresponding application technologies has become a research hotspot [8]. Gels of vegetable oils are formed by melt-cooling and dispersing a gelling agent in vegetable oils. Subsequently, the system is gelled by forming a thermally reversible three-dimensional network structure through non-covalent physical bonds, which prevent the flow of vegetable oils [9]. Oleogels are as strong and malleable as solid fats and their use can radically reduce the content of trans and saturated fats in food. Owing to these advantages, oleogels are considered a healthy fat type that can replace traditional solid fats in various applications [10,11]. The n-alkanes are important for the gelation of vegetable oils. Studies conducted on vegetable oils found that the n-alkane composition varied depending on the botanical origin [12]. Sesame oil [13], sunflower oil [14], olive oil [15,16], palm oil [17], peanut oil [14] and linum seed oil [18] all have specific patterns of n-alkane carbon chain length composition and content. For instance, the carbon atom chain of n-alkanes in olive oil [16] ranges from 24 to 260 mg/kg, and the carbon atom chain of n-alkanes in sesame oil [13] ranges from 22 to 82 mg/kg. Currently, oleogels are typically prepared using small-molecule gelling agents such as γ-oryzanol, sterols, monoglycerides, fatty acids, and natural waxes. Monoglycerides (GMS) refers to a series of surfactants produced by interesterification of fats or oils with glycerol. GMS is one of the commonly used gelling agents in food production, which can form a crystalline network in oils and then form elastic gels [19]. Further studies on the controlled release of volatile substances by grease gels have revealed that wax-based oleogels can effectively control and slow the release of flavor substances, thus providing a promising and adjustable delivery strategy for flavor compounds [20,21]. However, research on oleogels has mainly focused on the gelation mechanism, physicochemical properties, nutritional function, and applications. In contrast, there is insufficient knowledge about the effect of gel structures and intermolecular interaction mechanisms on the release of small flavor molecules. In particular, the effect of gelation on the mechanism of controlled flavor substance release from SO has not been investigated.

In this study, sesame-oil-based oleogels were prepared using biological waxes and monoglycerides as gelling agents, and the controlled release of volatile flavor substances was investigated after gelation. The physical properties, microstructures, thermodynamic properties, and crystal morphologies of the oleogels were characterized using texture analysis, polarized light microscopy, differential scanning calorimetry (DSC), and X-ray diffraction (XRD). The release kinetics of the characteristic flavor substances in SO was studied in the oleogels to clarify the effect of SO self-assembly behavior on the controlled release of small flavor molecules. These results can aid in establishing controllable flavor delivery systems and technologies for maintaining the nutritional value and flavor of SO.

2. Materials and Methods

2.1. Materials and Chemicals

Candelilla (*Euphorbia antisyphillitica*) wax (CLW), rice bran (*Oryza sativa* L.) wax (RBW), carnauba (*Copernicia prunifera* (Mill.) H.E. Moore) wax (CRW) and beeswax (*Apis mellifera*) (BW) were purchased from Changge City, Yi Heng Jian apiculture, Ltd. (Henan, China). Glycerol monostearate (GMS) was purchased from Shanghaiyuanye Bio-Technology Co., Ltd. (Shanghai, China). Standards of the flavor compounds were obtained from Sigma Chemical Co., Ltd. (Shanghai, China). All organic reagents used in the experiments were of analytical grade and obtained from Sinopharm Holdings Ltd. (Shanghai, China). Sesame oil was obtained from a local market (Hefei, China).

2.2. Preparation of Oleogels

The gelling agents (CLW, RBW, CRW, BW or GMS) and SO were accurately weighed and placed in a clean, dry glass container, which was then covered with cling film. The samples were heated with stirring (magnetic stirrer, 250 rpm) for 15 min (90 °C) to completely dissolve the gelling agents, after which time, the mixtures were deposited onto a receptacle suitable for analysis. Then, the mixtures were cooled to room temperature (~25 °C). All characterizations were performed 24 h after gel preparation. Testing was carried out on triplicate sample sets and all gel concentrations were reported as percentage (wt.%). wt.% (weight percent) = [g solute/(g solute + g solvent)] × 100, so 5 wt.% CLW in SO should be: 5 g CLW in 95 g SO.

2.3. Determination of Critical Concentrations

Sesame oil and gelling agent were heated in a water bath until the gelling agent was completely dissolved (90 °C, 15 min), and poured into 10 mL glass vials (20 mm diameter × 50 mm length). The test was started with 1% (wt.) of the additive and subsequently increased the dosage to 10% in 1% increments. This series of samples was stored at room temperature (~25 °C) for 24 h, and the sample was inverted to observe whether there was gravity flow. In this way, the minimum gelling agent concentration without flowing was the critical concentration. The experiment was repeated three times for each test.

2.4. Microscopy

The crystal morphology of the oleogels were observed using a MP 30 polarized light microscope (Mshot, Guangzhou, China). 20 μL of the molten sample was gently spread on a preheated slide and covered with a coverslip. The samples were stored at room temperature (~25 °C) for 24 h and the microstructure of all the samples was observed and the crystalline morphology of the samples was photographed.

2.5. Texture Analysis

The firmness and stickiness of gel samples were measured using a texture analyzer TA-XT plus (Stable Microsystems, Surrey, UK) equipped with a 12.7 mm cylindrical probe (P/0.5). All samples were measured after 24 h of preparation at room temperature (~25 °C). Texture analyses were performed by inserting the probe into the sample at a speed of 1 mm/s to a maximum penetration depth of 10 mm and pull the probe out at a speed of 10 mm/s. The firmness and stickiness were calculated using Texture Exponent v.6.1.16.0 software (Stable Microsystems).

2.6. Thermal Properties

To investigate the melting and crystallization behaviors of the samples, DSC thermograms were obtained using a Q2000 differential scanning calorimeter (TA Instruments, New Castle, DE, USA). Samples with a mass of ca. 10–15 mg were sealed in aluminum discs and heated from room temperature (~25 °C) to 100 °C at a rate of 20 °C/min, held at 100 °C for 5 min to remove crystallization memory, then cooled to 0 °C at a rate of 10 °C/min and held for 5 min to fully crystallize, and finally heated to 100 °C at a rate of 10 °C/min. Meanwhile, empty aluminum discs were used as controls in the experiment. The onset crystallization temperature (T_g, °C), peak crystallization temperature (T_c, °C) and crystalline enthalpy (ΔH_c, J/g) during the cooling phase and peak melting temperature (T_m, °C) and melting enthalpy (ΔH_m, J/g) during the heating phase were calculated using TA Universal Analysis software (TA Instrument, USA).

2.7. XRD Measurements

An X-ray diffractometer (PANalytical B.V., Almelo, The Netherlands) was used to determine the crystal structure of samples using a copper X-ray tube (λ = 1.54 Å) with an operating voltage of 40 kV and a current of 40 mA as the X-ray source. Samples were prepared on a slide, and the sample surface was smoothly treated by rotating the slide around an axis perpendicular to the plate surface. Diffraction patterns were measured by scintillation detector with a scanning speed of 4°/min and a scanning range of 5.0°–50° (2θ) at room temperature (~25 °C). The X-ray diffraction patterns of the samples were analyzed using MDI Jade 6.0 software (Materials Data Ltd., Livermore, CA, USA).

2.8. Determination of Cumulative Release Rate of Flavor Compounds

Sesame oil (5 g) or an oleogel sample with 8 wt.% gelling agent (5.435 g) was accurately weighed and placed in a 20 mL headspace vial (to equalize the SO content). The vial then was left open and stored at 25 °C for 60 d. The flavor compounds in the sample were analyzed by GC-MS after 0, 1, 3, 4, 5, 7, 9, 11, 13, 16, 19, 21, 23, 27, 32, 37, 42, 47, 54, and 60 d. Before analysis, the lid was sealed (silicone/polytetrafluoroethylene seal, La-Pha-Pack GmbH, Germany) and the sample was equilibrated in a constant temperature water bath at 60 °C with magnetic stirring for 20 min. An activated SPME fiber (Supelco, Bellefonte, PA, USA) was inserted into the headspace vial. After adsorption for 50 min, the SPME fiber was inserted into the GC inlet, where thermal desorption at 250 °C was allowed to proceed for 6 min before data collection. Three parallel samples were prepared for each time point, and each sample was tested only once.

GC-MS analysis was performed using an Agilent 6890 GC-5975I MS instrument (Agilent, Santa Clara, CA, USA) equipped with a DB-5MS capillary column (60 m length × 1 mm inner diameter, 0.32 μm film thickness). Analyses were carried out in splitless mode with helium as the carrier gas at a flow rate of 1.7 mL/min. The oven temperature was initially held at 50 °C for 2 min, increased to 220 °C at a rate of 4 °C/min, and then held at 220 °C for 10 min. The mass spectrometry interface and ion source temperatures were both 250 °C. Ionization was performed in electron impact mode at 70 eV. Chromatograms were collected in full scan mode with a mass scan range of m/z 30–550.

The cumulative release rate, Q, was used to describe the dynamic release process of the flavor compounds. The Q values of 28 flavor compounds in SO were calculated using the following equation: $Q = (C_0 - C_t)/C_0 \times 100\%$, where C_0 is the initial concentration of the flavor compound and C_t is the concentration of the flavor compound on day t (t, the storage time). Volatile compounds were qualitatively analyzed by standard compounds, mass spectra in the database (NIST18. LIB) and retention index. The external standard method was used for the quantitative analysis of flavor compounds. Using a series of working standard solutions, the peak area corresponding to each standard in the mixture was determined at varying concentrations, and standard curves were constructed. The concentrations of the flavor compounds (μg/kg) in the SO or oleogel samples were then determined from the standard curve using the corresponding peak areas, which were the average values from three measurements.

2.9. Data Analysis

The analysis of variance (ANOVA) and significant differences were calculated by using SPSS 18.0 statistical software (SPSS Inc, Chicago, IL, USA). Sample preparation and analyses were performed in triplicate, and the results were expressed as mean ± standard deviation (x ± SD). Data analysis and plotting were performed by using Origin 9.0 (OriginLab, Northampton, MA, UK) and MATLAB R2018a (MathWorks, Neddick, MA, USA.

3. Results and Discussion

3.1. Determination of Critical Concentration

As shown in Figure 1, CLW had the best gelation effect among the investigated samples, forming a gel with SO at an addition level of 2 wt.%. Table S1 indicates the minimum quantity (wt.%) required for SO gelation and composition of gelling agents. As previously reported [22–24], CLW predominantly consists of 49–50% n-alkanes (C29–C33), 20–29% esters of acids and alcohols with even-numbered carbon chains (C28–34), 12–14% resins (mainly triterpenoid esters), and 7–9% free acids. Abdallah and Weiss [25] showed that n-alkanes are the structurally simplest organic gelling agents and that the stability of a gelled oil increases as the n-alkane chain length increases. Morales-Rueda, Dibildox-Alvarado, Charó-Alonso, Weiss and Toro-Vazquez [23] investigated the thermal and oil-holding properties of the main component of CLW, n-tridodecane (C32), when added to safflower oil to prepare oleogels. They found that C32 oleogels have a better self-assembly ability than CLW oleogels, which indicates that the n-alkanes in CLW have a better gelation ability than the minor components. Thus, the high content of C29–C33 n-alkanes in CLW (49–50%) may be responsible for the better gelling ability of CLW than that of other waxes. BW is a mixture of odd-numbered hydrocarbons (mainly heptacosane), even-numbered hydrocarbons, odd-numbered monounsaturated hydrocarbons, fatty acid esters combined with long-chain alcohols, and free wax acids [26]. BW was also shown to be a good gelling agent for SO, with the formation of a stable gel observed at a concentration of 3 wt.%. Esters derived from long-chain saturated fatty acids (C24 and C22) and long-chain saturated fatty alcohols (C28–C34) are the major components in RBW [27]. In contrast to the results reported by Hwang, Kim, Singh, Winkler-Moser and Liu [24], which showed good gelation of soybean oil with RBW, the critical concentration of RBW for SO gelation was 5 wt.%. This difference in behavior may be due to variations in the wax composition with season, region, and purity, indicating that knowledge of the exact composition is important for understanding the gelation properties of a wax. The main components of CRW are 84–85% aliphatic and aromatic esters (including 40% aliphatic esters, 13% ω-hydroxy esters, and 8% cinnamic aliphatic diesters) [28]. The critical concentration of CRW was 5 wt.% owing to the weak gelation ability of the mixture of aliphatic esters and p-hydroxycinnamic aliphatic diesters. GMS, consisting of 84.5% monostearin and 12.2% monopalmitin, is one of the most commonly used gelling agents. The critical concentration of GMS in SO was 5 wt.%. Thus, the gelation ability of GMS was similar to that of RBW and CRW but weaker than that of CLW and BW.

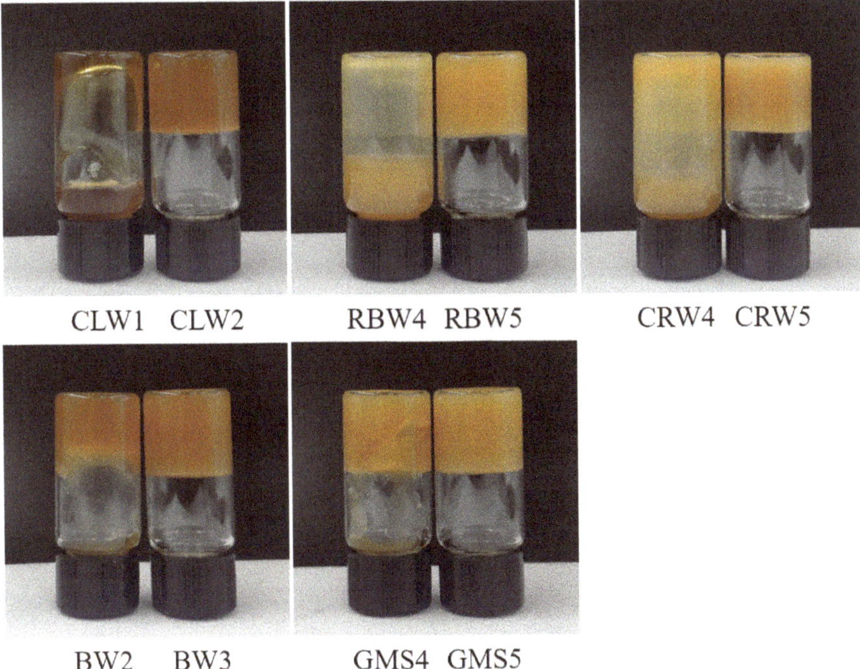

Figure 1. Determination of critical concentration of different types of oleogel (wt.%). CLW, candelilla wax; RBW, rice bran wax; CRW, carnauba wax; BW, beeswax; GMS, glycerol monostearate.

3.2. Microscopic Analysis

Figure 2 shows the polarized light micrographs of the oleogels formed with different gelling agents at critical and 8 wt.% concentrations. In the polarized light field, crystals are birefringent, whereas the liquid oil phase is optically isotropic and therefore black. As shown in Figure 2, the microstructures of the oleogels formed by different gelling agents vary considerably. The crystals in the CLW oleogel at the critical concentration were short rod-like structures of approximately 4 µm in length, which were uniformly scattered in the SO. With an increase in the CLW concentration, longer crystals (20–30 µm) in the shape of slender needles were observed. These long needle-like structures can form a good crystal matrix and interlock well at the intercrystalline interface to form organogels. The oleogel with 5 wt.% RBW contained crescent-shaped crystalline units, which were relatively sparsely distributed, and a tight gel network structure was not formed. When the amount of added RBW was 8 wt.%, flocculent crystallization occurred and the crystal distribution density increased. Bundled crystals with lengths of approximately 20 µm and a regular distribution were observed in the CRW oleogel. However, the crystals tended to be arranged in the same direction, which may be unfavorable for the formation of a three-dimensional network structure. A small number of fine needle-like crystals were observed at a BW concentration of 3 wt.%. In contrast, when the concentration of BW was 8 wt.%, the fine needle-like crystals formed clusters, and an interwoven three-dimensional network structure could be clearly observed, suggesting that this specific microstructure could provide control over the diffusion of volatile flavor substances. Additionally, at 8 wt.% gelator, the hazy appearance of RBW, CRW, and BW may be due to the formation of spherulitic-type structures made of assemblies of needles or platelets which are out of the focal plane [29,30]. The crystal structure of the GMS oleogel was feather-like, and the crystal length tended to increase with an increase in the amount of gelling agent. Compared with wax-based oleogels, the distribution of crystals in the GMS oleogels was sparser, resulting in numerous blank areas and the formation of a less dense network structure, which is consistent with the texture analysis results. It indicates that GMS oleogels were weaker than the plant wax-based systems at equivalent concentrations. The different microstructures of the various oleogels may have different effects on controlling the diffusion of volatile compounds.

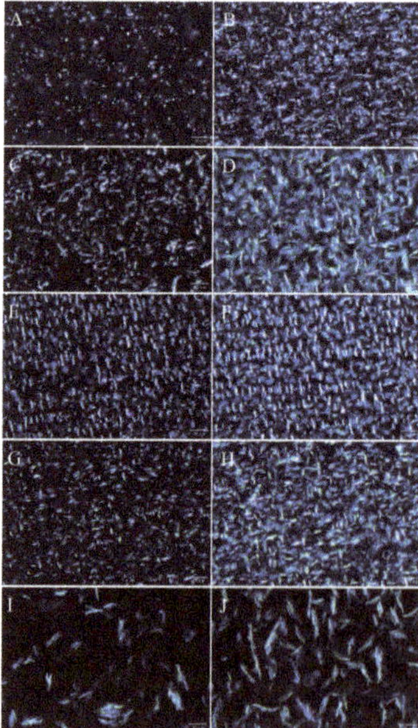

Figure 2. Polarized light microscopy images of CLW oleogels at critical (**A**) and 8 wt.% concentrations (**B**). Polarized light microscopy images of RBW oleogels at critical (**C**) and 8 wt.% concentrations (**D**). Polarized light microscopy images of CRW oleogels at critical (**E**) and 8 wt.% concentrations (**F**). Polarized light microscopy images of BW oleogels at critical (**G**) and 8 wt.% concentrations (**H**). Polarized light microscopy images of GMS oleogels at critical (**I**) and 8 wt.% concentrations (**J**). All images were taken at 40× magnification at 25 °C. Scale bar = 30 µm.

3.3. Texture Analysis

In order to evaluate the feasibility of replacing solid fats with oleogels from various aspects, the texture parameters and thermodynamic properties of different oleogels were analyzed. The firmness depends on the density of a gel network structure, whereas the stickiness indicates its cohesiveness. A high density and cohesiveness provide a gel with structural integrity and resistance to external forces. As shown in Figure 3, the firmness of each wax-based oleogel sample was close to zero at the critical concentration, although there was no gravitational flow when the samples were inverted. At higher concentrations, the firmness and stickiness of each oleogel increased because the total crystalline solids increased, and the gel network structure became tighter (Figure 2). At a gelling agent concentration of 8 wt.%, the firmness of CLW and BW were 1051.71 and 1188.53 g, respectively, whereas those of RBW, CRW, and GMS were 457.39, 485.28, and 285.68 g, respectively. CLW and BW showed significantly higher firmness than the other waxes or GMS, which may be due to their lower critical concentrations and the formation of more crystalline solids at the same concentration. The formation of closed spaces within the oleogel is the basis for flavor retention. The harder the oleogel is, the stronger the oil binding property is. Therefore, the oil mobility within the crystalline network is limited, which may have a greater effect on the migration of flavor molecules.

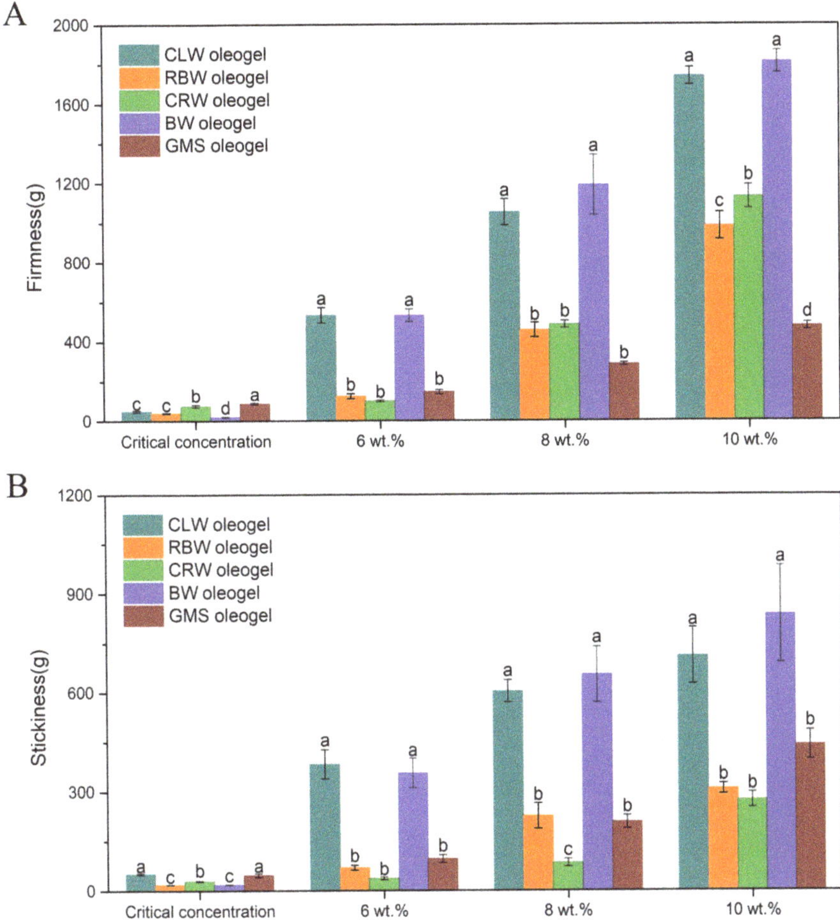

Figure 3. Firmness (**A**) and stickiness (**B**) of the oleogels at room temperature (~25 °C). Sample preparation and analyses were performed in triplicate, and the results were expressed as mean with error bar. Different letters represent statistical differences between values within each panel ($p < 0.05$).

3.4. Thermal Properties

Figure 4 shows the melting and crystallization behaviors of the oleogels formed with different gelling agents at critical and 8 wt.% concentrations as well as the thermal behavior of each gelling agent alone for comparison. Waxes are complex mixtures of multiple compounds with both major and minor components. The oleogels were obtained by diluting the gelling agents in SO, which caused the exothermic and endothermic peaks of the oleogel to be broadened and shifted to lower temperatures as compared to those of the gelling agent alone. In addition, the melting and crystallization behavior of the oleogel was determined by the main components of the gelling agent because of the dilution effect. As shown in Figure 4A,B, the CLW oleogel exhibited no significant exothermic or endothermic peaks at the critical concentration of 2 wt.%, which may be due to the low gelling agent concentration, less crystallization in the oleogel, or a nucleation rate too slow to be detected. Broad endothermic and exothermic peaks were observed at a CLW concentration of 8 wt.%, which are mainly related to the *n*-alkanes in CLW.

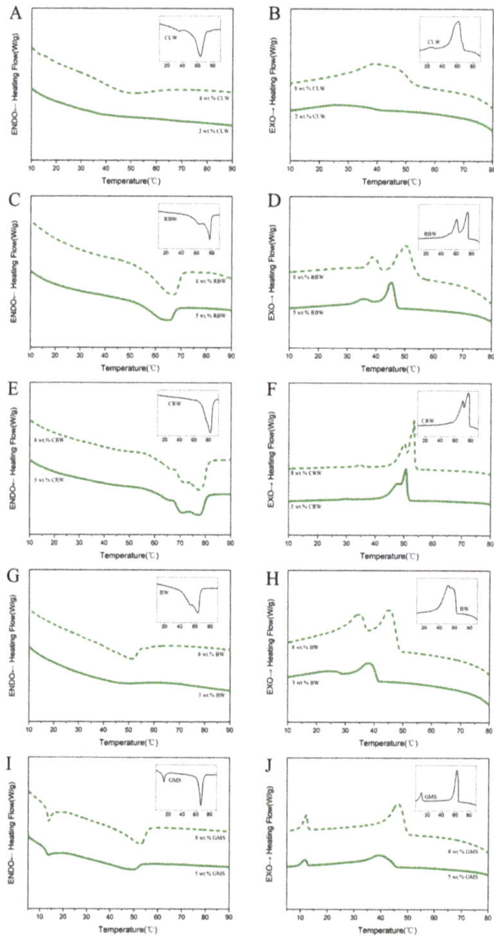

Figure 4. DSC thermograms during heating (**A**) and cooling (**B**) of CLW oleogels. DSC thermograms during heating (**C**) and cooling (**D**) of RBW oleogels. DSC thermograms during heating (**E**) and cooling (**F**) of CRW oleogels. DSC thermograms during heating (**G**) and cooling (**H**) of BW oleogels. DSC thermograms during heating (**I**) and cooling (**J**) of GMS oleogels.

As shown in Figure 4D, two exothermic peaks were observed in the crystallization curve of the RBW oleogel, as in the case of RBW alone. For the 5 and 8 wt.% RBW oleogels, the first exothermic peak was located at 44.51 ± 0.56 °C and 49.38 ± 1.00 °C, respectively, whereas the second exothermic peak was located at 34.54 ± 0.19 °C and 38.28 ± 0.75 °C, respectively. The crystallization peaks gradually shifted toward higher temperatures as more RBW was added. In contrast, only one endothermic peak was observed during the melting of the RBW oleogel (Figure 4C), which may be because the constituents of RBW have similar melting temperatures. As shown in Figure 4E,F, the CRW oleogels exhibited complex melting and crystallization behaviors with multiple endothermic and exothermic peaks, indicating the presence of a series of different solids in CRW. In addition, the melting peaks of the CRW oleogels varied between 69.87 ± 0.35 °C and 77.72 ± 0.38 °C, which indicates that CRW has a higher melting temperature. Figure 4G,H show that the melting and crystallization behaviors of the BW oleogels with different concentrations of the gelling agent are similar. Two major broad exothermic peaks were observed, which are related to the major components (plant waxes) and minor components (hydrocarbons) in BW [31]. The onset gelation temperature (T_g), peak temperature of crystallization (T_c), and enthalpy change of crystallization (ΔH_c) of the oleogel during the cooling phase increased with increasing BW content. As shown in Figure 4I,J, two endothermic and exothermic peaks were observed for the GMS oleogels, which are related to monopalmitin and monostearin in GMS. The peak temperatures (T_c, T_m) in the low-temperature region did not increase significantly when the gelling agent content was increased. In contrast, similar to the trend observed for the melting behavior, the endothermic peak in the high-temperature region shifted toward a higher temperature as the concentration of the gelling agent increased.

The enthalpy change reflects that the free energy is required to cause the state change of the material, with a larger enthalpy change indicating more spontaneous oleogel formation. As shown in Table 1, the type and amount of gelling agent influence the enthalpy change of crystallization (ΔH_c) and the enthalpy change of melting (ΔH_m) of the oleogels. In general, the ΔH_c and ΔH_m values tend to increase with an increase in the gelling agent concentration, which indicates that the arrangement of gelant molecules in the system is more organized and the spontaneity of oleogel formation increases. Moreover, for each oleogel, the ΔH_c value was higher than the ΔH_m value, and this hysteresis can be explained by the heat of dissolution of the gelling agent during heating. As reported by Abdallah et al. [32], the increase in temperature leads to network fracture, and the dissolution of solid materials as exothermic materials may inhibit the visualization of heat-absorbing melting events. The results of DSC showed that the melting points of CLW, BW and GMS oleogels were relatively close to butter (32–35 °C), which proved the feasibility of partial substitution of butter for preparing low-trans and low-saturated fatty acid foods. In contrast, the melting temperatures of RBW and CRW oleogels were too high, which might cause undesirable sensory experiences during consumption.

Table 1. Thermodynamic parameters of oleogels formed by different types of gelling agent.

Sample	Crystallization Process				Melting Process			
	T_g (°C)	T_{c1} (°C)	T_{c2} (°C)	ΔH_c (J/g)	T_e (°C)	T_{m1} (°C)	T_{m2} (°C)	ΔH_m (J/g)
2 wt.% CLW	ND [i]	ND [h]	ND [g]	ND [i]	ND [i]	ND [i]	ND [d]	ND [i]
8 wt.% CLW	52.73 ± 0.09 [b]	41.34 ± 0.10 [f]	ND [g]	7.68 ± 0.04 [e]	58.69 ± 0.09 [e]	46.89 ± 0.18 [f]	ND [d]	7.23 ± 0.12 [d]
5 wt.% RBW	47.54 ± 0.25 [f]	44.51 ± 0.56 [e]	34.54 ± 0.19 [d]	8.35 ± 0.18 [d]	70.04 ± 0.72 [d]	63.28 ± 0.38 [d]	ND [d]	7.90 ± 0.11 [c]
8 wt.% RBW	53.52 ± 0.31 [a]	49.38 ± 1.00 [c]	38.28 ± 0.75 [c]	12.34 ± 0.02 [a]	71.26 ± 0.06 [c]	65.95 ± 0.10 [c]	ND [d]	12.11 ± 0.12 [a]
5 wt.% CRW	51.60 ± 0.23 [c]	50.95 ± 0.27 [b]	47.94 ± 0.09 [b]	7.02 ± 0.03 [f]	79.97 ± 0.14 [b]	69.87 ± 0.35 [b]	76.80 ± 0.08 [a]	3.90 ± 0.05 [g]
8 wt.% CRW	53.71 ± 0.02 [a]	53.30 ± 0.04 [a]	49.81 ± 0.17 [a]	11.41 ± 0.03 [b]	81.26 ± 0.55 [a]	71.21 ± 0.09 [a]	77.72 ± 0.38 [a]	6.12 ± 0.11 [e]
3 wt.% BW	41.31 ± 0.03 [h]	38.66 ± 0.07 [g]	23.28 ± 0.02 [e]	1.94 ± 0.04 [h]	53.00 ± 0.14 [h]	45.01 ± 0.01 [g]	ND [d]	1.76 ± 0.02 [h]
8 wt.% BW	48.45 ± 0.01 [e]	45.59 ± 0.19 [de]	34.13 ± 0.11 [d]	7.02 ± 0.01 [f]	54.68 ± 0.12 [g]	49.76 ± 0.42 [e]	ND [d]	4.98 ± 0.04 [f]
5 wt.% GMS	44.58 ± 0.03 [g]	40.20 ± 0.02 [f]	11.34 ± 0.08 [f]	4.33 ± 0.03 [g]	52.95 ± 0.05 [h]	12.84 ± 0.06 [h]	47.18 ± 0.11 [c]	4.11 ± 0.04 [g]
8 wt.% GMS	49.52 ± 0.04 [d]	46.67 ± 0.06 [d]	11.92 ± 0.10 [f]	9.86 ± 0.07 [c]	56.59 ± 0.10 [f]	13.03 ± 0.08 [h]	51.58 ± 0.25 [b]	9.74 ± 0.05 [b]

Each sample was analyzed in triplicate, and the results were expressed as mean ± standard deviation (x ± SD). Values in the same column with different superscript letters are significantly different at $p < 0.05$. T_g, onset gelation temperature; T_{c1}, first peak temperature of crystallization; T_{c2}, second peak temperature of crystallization; ΔH_c, enthalpy change of crystallization; T_e, melting complete temperature; T_{m1}, first peak temperature in the heating phase; T_{m2}, second peak temperature in the heating phase; ΔH_m, enthalpy change of melting. ND = not detectable.

3.5. XRD Analysis

Polymorphism refers to the phenomenon in which a molecule (or atom) forms two or more different packing modes and crystal structures under different conditions. Polymorphism mainly depends on the arrangement of the molecules in the crystal lattice. Changes in the hydrocarbon chain stacking pattern and tilt angles in lipids can cause differences in crystal forms, each of which has a different short spacing (the distance between parallel acyl groups on triglycerides), which can be analyzed using XRD [33]. Figure 5 shows the XRD patterns of the oleogels formed with different gelling agents at critical and 8 wt.% concentrations, as well as those of the gelling agents alone for comparison. All the wax-based oleogels showed XRD patterns similar to that of the corresponding waxes, with two strong wide-angle diffraction peaks appearing at 4.1 and 3.7 Å, which indicates the orthorhombic perpendicular ($O_\perp$) subcell packing (β' morphology) [34]. Among the three different polymorphs of triglycerides (α, β, and β'), the β' crystal form has the strongest plasticity, showing better spreadability and mouthfeel. The main diffraction peaks of GMS appeared at d = 15.7, 11.8, 7.9, 4.5, 4.3, and 3.9 Å, whereas those of the GMS oleogel were observed at d = 16.6, 12.6, and 4.1 Å, which correspond to the characteristic diffraction peaks of the α morphology with hexagonal symmetry. The peaks (T_c, T_m) observed in the low-temperature region during crystallization and cooling of the GMS oleogels (Figure 4I,J) are consistent with the results of Doan et al. [35]. The orientational order of the 'zig-zag' planes disappears, and the molecules can rotate around their long axis to form a hexagonally symmetric structure when the orthorhombic gelling agent undergoes a phase transition at a temperature close to its melting point. In addition, for different gelling agents, the intensity of the short-spacing peaks increased with an increase in the gelling agent concentration, indicating that the stacking pattern and polycrystalline structure of the oleogel molecules depend on the gelling agent concentration.

3.6. Release Kinetics of Flavor Compounds

The contents of 28 of the main flavor substances in SO were analyzed before and after storage (Table 2). Based on the retention effects in oleogels, these compounds can be divided into two categories. The first category includes 18 flavor substances with high saturated vapor pressures and low hydrophobic constants, which are extremely volatile in SO. After 60 d of open storage, 2-pentanone, 2-methylbutanal, pyrazine, 2-methylthiazole, 4-methylthiazole, and (5-methyl-2-furyl)methanol were depleted in SO, and the content of the other flavor substances also decreased significantly, resulting in a decline in SO quality and flavor characteristics. This behavior is consistent with the results of Chen et al. [36], who found that for an active ingredient with high vapor pressure, the diffusion rate often depends on the volatility of the substance, which is usually high. However, the addition of gelling agents slowed the release of these compounds in SO to varying degrees. This effect is due to the rigid crystal network, which traps the volatile substances within the internal phase and acts as a physical barrier. The CLW and BW oleogels were more effective in retaining such substances, which may be related to their higher textural parameters. The firmness of the CLW and BW oleogels was significantly higher than that of the oleogels with other waxes and monoglycerides (Figure 3A). These oleogels formed interwoven three-dimensional network structures (Figure 2B,H) inside, which can impede the mass transfer of flavor substances across the oil to the headspace, thereby inhibiting the loss of flavor substances. Yin et al. [37] showed that the sensory attributes "sweet", "toasted", "nutty", "persistent", "high-intense flavor" and "cooked sesame seed flavor" were drivers of liking. Pyrazines contributed to improve roasted flavors and increase consumer acceptance of sesame oil. Therefore, the retention of such compounds is of relevance.

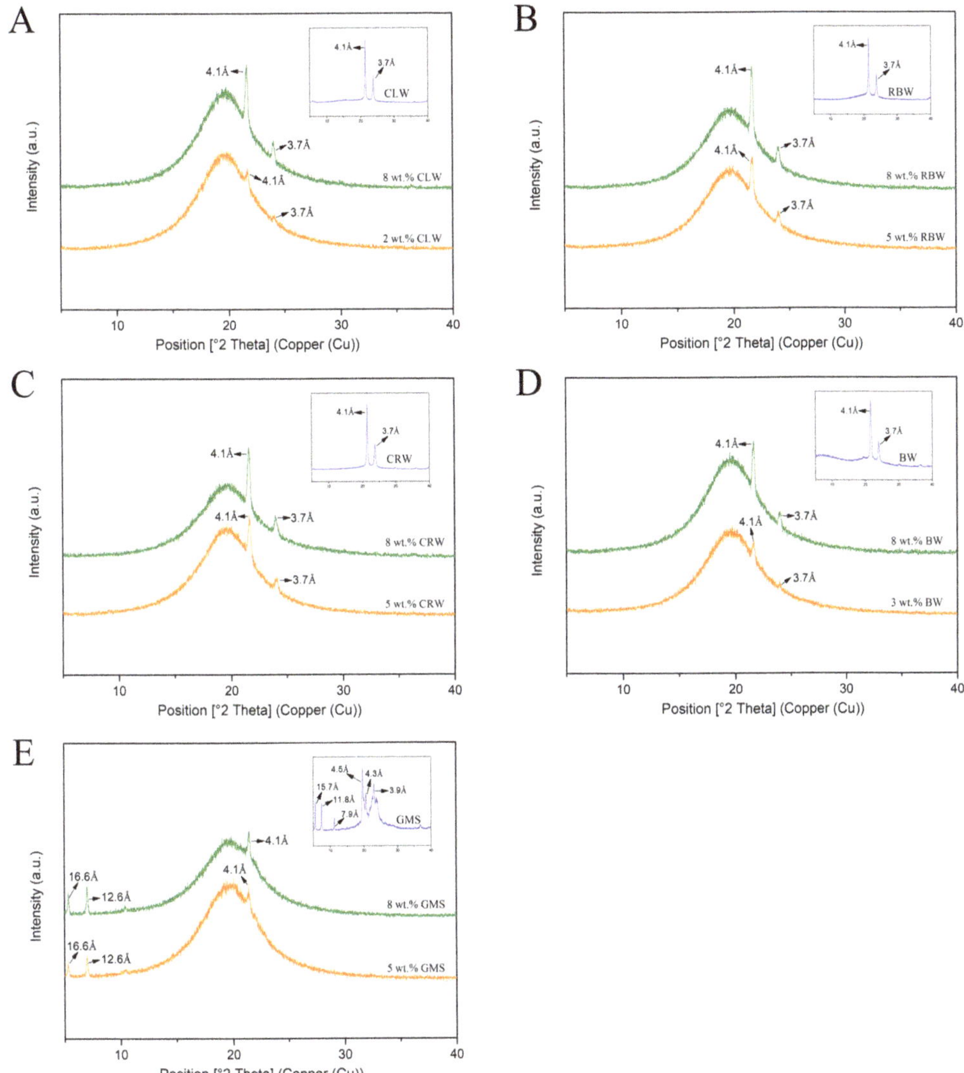

Figure 5. The X-ray diffractograms of oleogels formed by different gel factors at critical and 8 wt.% concentrations, (**A**) CLW oleogels; (**B**) RBW oleogels; (**C**) CRW oleogels; (**D**) BW oleogels; (**E**) GMS oleogels.

The second category mainly includes compounds with low saturated vapor pressures or hydrophobic compounds. The changes in the contents of these compounds before and after storage were relatively small, and the retention effect in SO was better. Unlike the first category of compounds, the addition of a gelling agent promoted the release of compounds of the second category to a certain extent. This behavior may be due to the self-assembly of the gelling agent in SO to form hydrophilic structural domains [38], which can reduce the hydrophobicity of SO and increase the release of lipophilic volatiles.

To further investigate the effect of gelation on the controlled release behavior of the flavor substances in SO, the cumulative release rates of the flavor substances in different oleogels and SO were fitted using the zero-order release, first-order release, Higuchi model,

Kormeyer–Peppas model, and Weibull equations (see Supplementary Materials Table S2). The model with the best fitting effect (highest correlation coefficient, R^2) was the Weibull equation ($R^2 > 0.97$), and the flavor release kinetics were analyzed using this model. During the open storage process, the flavor substances in the oil phase were released owing to the effect of internal and external concentration differences. Because the concentrations of the flavor substances in the samples were higher during the initial stage, the release rate was faster. At longer storage times, the concentrations of the flavor substances in the samples decreased, which decreased the driving force for release and consequently the release rate. Therefore, the dynamic law obtained from the Weibull equation is as follows: $Q = 1 - \exp[-(kt)^n]$ (Table S2).

For each category of compounds with different retention effects in the oleogel, a representative example was selected to study the change in the cumulative release rate during storage for 60 d. The results showed that the release of the flavor compounds in the open environment followed the kinetic law of the Weibull equation and a high degree of fit was obtained. The first category of compounds was exemplified by 2-methylpyrazine, which is highly volatile in SO because of its high saturated vapor pressure. As shown in Figure 6A the cumulative release rate of 2-methylpyrazine reached 96.08% after 60 d of storage, implying that it is quickly depleted from SO. The addition of the gelling agent significantly ($p < 0.05$) reduced the release rate and quantity of 2-methylpyrazine. The CLW and BW oleogels had the best retention effects on 2-methylpyrazine, with cumulative release rates of 39.09% and 26.76%, respectively, at the end of the storage. The RBW, CRW, and GMS oleogels exhibited smaller retention effects, with cumulative release rates of 68.12%, 61.13%, and 64.33%, respectively. The release rate constant, k, indicates the rapidity of release, with a larger the k value corresponding to faster release. As shown in Table 3, the largest k value for 2-methylpyrazine was observed in SO (0.0445). In contrast, the k values for the RBW, CRW, and GMS oleogels were 0.0207, 0.0162, and 0.0173, respectively. The k values for the CLW and BW oleogels were 0.0047 and 0.0022, respectively, indicating a significant decrease in the release rate, which is consistent with the cumulative release rates of 2-methylpyrazine after storage for 60 d. The n value in the Weibull equation corresponds to the release mechanism. The n values for 2-methylpyrazine in the oleogels ranged from 0.5402 to 0.6572, indicating a release mechanism between restricted diffusion and a first-order reaction. However, the n value for 2-methylpyrazine in SO was 1.1422, indicating controlled release with an initial induction time.

Table 2. Changes in the content of the main volatile compounds in samples before and after storage (µg/kg, mean ± SD).

Volatiles	Aroma Description	Log p	Vapor Pressure (mmHg at 25 °C)	Initial Concentration (µg/kg)	Post-Storage Concentrations (µg/kg)					
					CLW	RBW	CRW	BW	GMS	SO
Significant retention effect										
2-Pentanon	ethereal, fruity	0.9	38.6 ± 0.2	1920 ± 114	1985 ± 46 [a]	419 ± 46 [c]	345 ± 19 [c]	2013 ± 76 [a]	606 ± 37 [b]	Nd [d]
2-Methylbutanal	flowers, grass	1.25	49.3 ± 0.2	2151 ± 136	1907 ± 54 [b]	662 ± 8 [c]	379 ± 40 [d]	2142 ± 58 [a]	427 ± 79 [d]	Nd [e]
Pyrazine	pungent, sweet, corn, nutty	−0.28	19.7 ± 0.2	4891 ± 367	2787 ± 140 [b]	1201 ± 56 [d]	1583 ± 86 [c]	3291 ± 180 [a]	1376 ± 85 [cd]	Nd [e]
Hexanal	fatty-green, grassy	1.97	10.9 ± 0.2	2171 ± 126	2144 ± 59 [a]	1703 ± 15 [b]	1779 ± 39 [b]	2184 ± 36 [a]	1702 ± 143 [b]	1127 ± 68 [c]
2-Methylthiazole	nutty, green	1.10	12.9 ± 0.2	572 ± 19	451 ± 29 [b]	Nd [d]	253 ± 20 [c]	569 ± 23 [a]	289 ± 19 [c]	Nd [d]
4-Methylthiazole	nutty, green	0.90	10.0 ± 0.2	3242 ± 241	2442 ± 137 [a]	680 ± 89 [b]	883 ± 40 [b]	2450 ± 22 [a]	930 ± 162 [b]	Nd [c]
2-Methylpyrazine	nutty, cocoa-like	0.18	9.7 ± 0.2	43,839 ± 1291	26,702 ± 532 [b]	13,976 ± 243 [d]	17,040 ± 445 [c]	32,108 ± 611 [a]	15,637 ± 376 [c]	1718 ± 78 [e]
Furfural	almond-like	0.73	2.2 ± 0.3	13,949 ± 67	7431 ± 25 [b]	2148 ± 219 [d]	1722 ± 28 [de]	9332 ± 214 [a]	4348 ± 240 [c]	1356 ± 69 [e]
2-Furanmethanol	faint, burning odor	0.20	1.0 ± 0.3	15,654 ± 650	6936 ± 316 [b]	3718 ± 81 [c]	2583 ± 252 [d]	9267 ± 282 [a]	3957 ± 92 [c]	1012 ± 47 [e]
3-Methylpyridine	green	1.19	6.7 ± 0.3	3109 ± 352	1616 ± 43 [cd]	2107 ± 66 [ab]	1839 ± 55 [bc]	2322 ± 115 [a]	1329 ± 130 [d]	1675 ± 65 [c]
2,4-Dimethylthiazole	meat, cocoa-like	1.56	4.3 ± 0.3	1489 ± 47	947 ± 34 [b]	628 ± 8 [d]	721 ± 18 [cd]	1354 ± 70 [a]	790 ± 12 [c]	275 ± 9 [e]
1-(2-Furyl)ethanone	coffee-like	0.52	0.8 ± 0.3	2131 ± 207	1279 ± 22 [b]	824 ± 80 [de]	1035 ± 32 [cd]	1746 ± 77 [a]	1187 ± 112 [bc]	737 ± 15 [e]
2,5-Dimethylpyrazine	nutty, coffee-like	0.64	4.0 ± 0.2	37,933 ± 2490	24772 ± 243 [b]	16,760 ± 434 [d]	21,229 ± 1096 [c]	33103 ± 797 [a]	21,964 ± 1712 [bc]	12,523 ± 541 [e]
2-Ethylpyrazine	nutty, peanut butter	0.71	4.0 ± 0.3	6609 ± 474	3791 ± 163 [b]	2141 ± 33 [d]	2762 ± 58 [c]	4650 ± 98 [a]	2898 ± 350 [c]	1566 ± 85 [e]
2,3-Dimethylpyrazine	nutty, cocoa-like	0.64	3.4 ± 0.3	9595 ± 544	5026 ± 123 [b]	3528 ± 192 [de]	4205 ± 49 [cd]	6302 ± 51 [a]	4656 ± 394 [bc]	3207 ± 229 [e]
2-Vinylpyrazine	coffee-like	0.58	3.3 ± 0.3	2822 ± 178	1383 ± 53 [b]	960 ± 14 [cd]	1118 ± 88 [c]	1902 ± 37 [a]	1332 ± 32 [b]	914 ± 22 [d]
(5-Methyl-2-furyl)methanol	sweet caramel-like	0.66	0.6 ± 0.4	3488 ± 129	564 ± 33 [b]	Nd [c]	Nd [c]	912 ± 68 [a]	Nd [c]	Nd [c]
5-Methyl-2-furaldehyde	spicy-sweet, slightly caramellic	1.19	0.6 ± 0.3	19,949 ± 1301	9361 ± 482 [d]	13,288 ± 311 [c]	15,386 ± 224 [b]	17,974 ± 593 [a]	17,995 ± 264 [a]	14383 ± 434 [bc]
No significant retention effect										
2-Ethyl-6-methylpyrazine	roasted baked potato	1.17	2.0 ± 0.3	11,203 ± 492	5869 ± 71 [bc]	5209 ± 148 [c]	6547 ± 162 [b]	8711 ± 161 [a]	8034 ± 253 [a]	8424 ± 334 [a]
2,3,5-Trimethylpyrazine	roasted nut, baked potato	1.10	1.7 ± 0.3	26,542 ± 1612	15,006 ± 169 [c]	12,903 ± 275 [d]	17,151 ± 109 [b]	21,770 ± 555 [a]	20,495 ± 870 [a]	21,616 ± 754 [a]
1H-Pyrrole-2-carbaldehyde	grassy	0.64	0.1 ± 0.4	18,461 ± 877	9879 ± 34 [d]	10,000 ± 567 [d]	12,381 ± 459 [c]	14,916 ± 454 [b]	14,085 ± 219 [b]	17815 ± 667 [a]
2-Acetylpyrazine	nutty, popcorn, breadcrust	0.16	0.2 ± 0.4	16,477 ± 398	8040 ± 116 [d]	7261 ± 240 [d]	9427 ± 119 [c]	11,794 ± 405 [b]	12,424 ± 231 [b]	13,373 ± 287 [a]
1-(2-Pyridinyl)ethanone	tobacco, heavy-oily-fatty	0.87	0.5 ± 0.4	11,234 ± 701	5926 ± 55 [d]	5122 ± 152 [e]	8629 ± 209 [c]	9433 ± 337 [b]	9076 ± 188 [bc]	11039 ± 148 [a]
3-Ethyl-2,5-dimethylpyrazine	nutty, coffee-like	1.63	1.2 ± 0.3	23,354 ± 1357	10,051 ± 435 [d]	9395 ± 83 [d]	12,325 ± 234 [c]	14,329 ± 478 [b]	15,420 ± 349 [b]	18,050 ± 236 [a]
2-Methoxyphenol	smoke, somewhat medicinal	1.19	0.2 ± 0.4	38,612 ± 1786	21,634 ± 691 [d]	22,497 ± 142 [d]	27,067 ± 697 [c]	31,870 ± 1387 [b]	34,329 ± 251 [b]	37,909 ± 605 [a]
Nonanal	fruity	3.56	0.5 ± 0.4	11,488 ± 504	5147 ± 186 [d]	6751 ± 169 [c]	6293 ± 157 [c]	6731 ± 212 [c]	7425 ± 140 [b]	8090 ± 167 [a]
2-Acetyl-6-methylpyrazine	nutty	0.62	0.1 ± 0.5	26,345 ± 1182	10,545 ± 72 [d]	13,628 ± 108 [c]	13,967 ± 173 [c]	15,098 ± 207 [b]	15,554 ± 571 [b]	18,861 ± 510 [a]
2-Methoxy-4-vinylphenol	spicy, roasted peanut	1.93	0.0 ± 0.5	43,737 ± 1419	16,196 ± 162 [d]	20,740 ± 518 [bc]	23,432 ± 393 [b]	20,378 ± 284 [c]	21,890 ± 799 [bc]	27,162 ± 1896 [a]

Values in the same row with different superscript letters are significantly different at $p < 0.05$. Nd = not detectable.

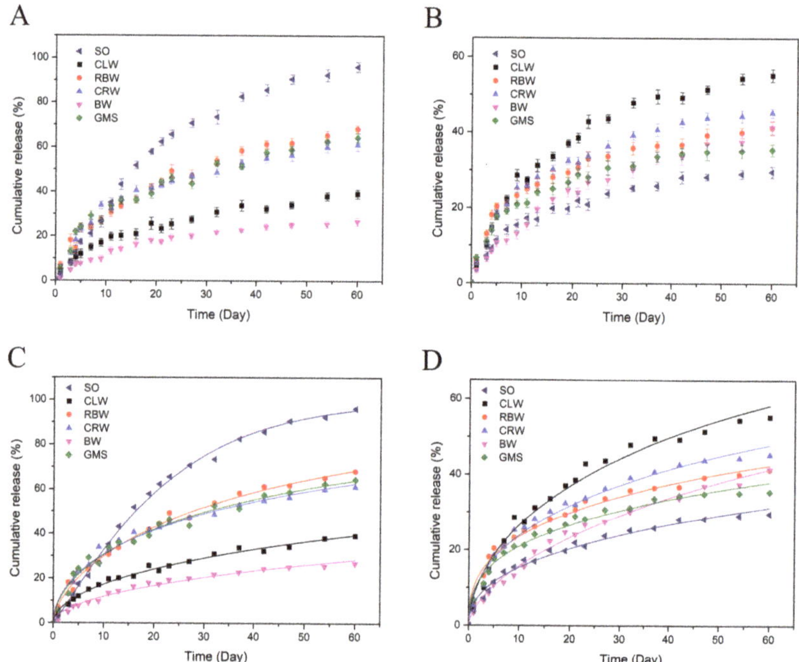

Figure 6. Dynamic release curves of 2-methylpyrazine (**A**) and nonanal (**B**). Weibull model fitting release curves of 2-methylpyrazine (**C**) and nonanal (**D**).

The second category of compounds was exemplified by nonanal. As shown in Figure 6B, after 60 d of open storage, nonanal had the best retention effect in SO, with a cumulative release rate of only 29.58% and a k value of 0.0019. For the GMS oleogel, the cumulative release rate was 35.37% with a k value of 0.0025. When the release process was finished, there was no significant difference in the nonanal content in the RBW, CRW, and BW oleogels, and the cumulative release rate was 41.23–45.22%. Unlike the retention effect observed for the first type of compound, the addition of CLW promoted the release of nonanal. It is speculated that the main *n*-alkane components of CLW may compete with the binding sites for the hydrophobic flavor substances, thus decreasing the stability. In addition, the n values for nonanal in SO and the oleogels were 0.3943–0.6352, which corresponds to the release mechanism between restricted diffusion and a first-order reaction. In general, the addition of gelling agents promoted the release of the second type of compounds to varying degrees, but due to their low saturated vapor pressures, losses after storage is relatively low and has little effect on the overall flavor. Overall, the BW oleogel exhibited a good retention effect on flavor molecules with different physicochemical properties. Thus, this gelling agent can be used to establish a controllable flavor delivery system and realize SO flavor stabilization and retention technologies.

Table 3. Fitting results of Weibull equations of 2-methylpyrazine and nonanal in different samples.

Sample	2-Methylpyrazine			Nonanal		
	k	n	R^2	k	n	R^2
SO	0.0445	1.1422	0.9966	0.0019	0.4518	0.9799
CLW oleogel	0.0047	0.5424	0.9877	0.0132	0.5791	0.9804
RBW oleogel	0.0207	0.6572	0.9898	0.0038	0.4019	0.9770
CRW oleogel	0.0162	0.5483	0.9820	0.0069	0.4840	0.9799
BW oleogel	0.0022	0.5402	0.9763	0.0062	0.6352	0.9897
GMS oleogel	0.0173	0.5759	0.9795	0.0025	0.3943	0.9750

k, the release rate constant; n, the release mechanism constant; R^2, coefficient of determination

4. Conclusions

In this study, sesame-oil-based oleogels were prepared with biological waxes and monoglycerides as gelling agents to form controllable flavor delivery systems. The obtained results showed clearly that sesame oil gelation enables controlled release of flavor substances, and that this retention is more significant for volatiles with high saturated vapor pressures and low hydrophobic constants. This is due to the rigid crystal network formed by gelation, which traps the volatiles within the internal phase and acts as a physical barrier. The release of flavor substances in SO during storage can be described by the Weibull equation kinetic model. BW oleogel is the most effective gel system for the controlled release of flavor substances, with good retention of flavor substances with different physicochemical properties. This is related to the high firmness of the BW oleogel and the interwoven three-dimensional network structure inside. DSC and XRD results showed that the wax-based oleogels had good plasticity and suitable melting temperatures, which confirmed the feasibility of these systems for replacing solid fats. We hope that these findings will assist in the design and construction of gel systems for vegetable oils with different flavor profiles to achieve the controlled release of specific flavors.

Supplementary Materials: The following are available online at https://www.mdpi.com/article/10.3390/foods10081828/s1, Table S1: Minimum quantity (wt.%) required for SO gelation and composition of gelling agents, Table S2: Dynamics equation.

Author Contributions: Conceptualization, M.P. and L.C. (Lili Cao); Data curation, M.P. and S.K.; Funding acquisition, L.C. (Lili Cao); Investigation, L.C. (Lulu Cao) and S.K.; Methodology, L.C. (Lulu Cao) and L.C. (Lili Cao); Project administration, S.J.; Resources, S.J.; Supervision, M.P., S.J. and L.C. (Lili Cao); Validation, S.J.; Writing—original draft, L.C. (Lulu Cao); Writing—review and editing, M.P. All authors have read and agreed to the published version of the manuscript.

Funding: This research was funded by the National Natural Science Foundation of China (No. 31801504).

Institutional Review Board Statement: Not applicable.

Informed Consent Statement: Not applicable.

Data Availability Statement: All data included in this study are available upon request by contact with the corresponding author.

Conflicts of Interest: The authors declare no conflict of interest.

References

1. Wan, Y.; Li, H.; Fu, G.; Chen, X.; Chen, F.; Xie, M. The relationship of antioxidant components and antioxidant activity of sesame seed oil. *J. Sci. Food Agric.* **2015**, *95*, 2571–2578. [CrossRef] [PubMed]
2. Konsoula, Z.; Liakopoulou-Kyriakides, M. Effect of endogenous antioxidants of sesame seeds and sesame oil to the thermal stability of edible vegetable oils. *LWT Food Sci. Technol.* **2010**, *43*, 1379–1386. [CrossRef]
3. Kumar, C.M.; Singh, S.A. Bioactive lignans from sesame (*Sesamum indicum* L.): Evaluation of their antioxidant and antibacterial effects for food applications. *J. Food Sci. Technol.* **2015**, *52*, 2934–2941. [CrossRef] [PubMed]
4. Varelas, C.G.; Dixon, D.G.; Steiner, C.A. Zero-order release from biphasic polymer hydrogels. *J. Control. Release* **1995**, *34*, 185–192. [CrossRef]

5.   Gibaldi, M.; Feldman, S. Establishment of sink conditions in dissolution rate determinations. Theoretical considerations and application to nondisintegrating dosage forms. *J. Pharm. Sci.* **1967**, *56*, 1238–1242. [CrossRef]
6.   Higuchi, T. Rate of Release of Medicaments from Ointment Bases Containing Drugs in Suspension. *J. Pharm. Sci.* **1961**, *50*, 874–875. [CrossRef]
7.   Neoh, T.-L.; Yoshii, H.; Furuta, T. Encapsulation and Release Characteristics of Carbon Dioxide in α-Cyclodextrin. *J. Incl. Phenom. Macrocycl. Chem.* **2006**, *56*, 125–133. [CrossRef]
8.   Puşcaş, A.; Mureşan, V.; Socaciu, C.; Muste, S. Oleogels in Food: A Review of Current and Potential Applications. *Foods* **2020**, *9*, 70. [CrossRef] [PubMed]
9.   Patel, A.R. A colloidal gel perspective for understanding oleogelation. *Curr. Opin. Food Sci.* **2017**, *15*, 1–7. [CrossRef]
10. Meng, Z.; Qi, K.; Guo, Y.; Wang, Y.; Liu, Y. Macro-micro structure characterization and molecular properties of emulsion-templated polysaccharide oleogels. *Food Hydrocoll.* **2018**, *77*, 17–29. [CrossRef]
11. Chopin-Doroteo, M.; Morales-Rueda, J.A.; Dibildox-Alvarado, E.; Charó-Alonso, M.A.; de la Peña-Gil, A.; Toro-Vazquez, J.F. The Effect of Shearing in the Thermo-mechanical Properties of Candelilla Wax and Candelilla Wax–Tripalmitin Organogels. *Food Biophys.* **2011**, *6*, 359–376. [CrossRef]
12. Troya, F.; Lerma-García, M.J.; Herrero-Martínez, J.M.; Simó-Alfonso, E.F. Classification of vegetable oils according to their botanical origin using n-alkane profiles established by GC–MS. *Food Chem.* **2015**, *167*, 36–39. [CrossRef] [PubMed]
13. McGill, A.S.; Moffat, C.F.; Mackie, P.R.; Cruickshank, P. The composition and concentration of n-alkanes in retail samples of edible oils. *J. Sci. Food Agric.* **1993**, *61*, 357–362. [CrossRef]
14. Moreda, W.; Pérez-Camino, M.C.; Cert, A. Gas and liquid chromatography of hydrocarbons in edible vegetable oils. *J. Chromatogr. A* **2001**, *936*, 159–171. [CrossRef]
15. Giuffrè, A.M. n-Alkanes and n-Alkenes in Virgin Olive Oil from Calabria (South Italy): The Effects of Cultivar and Harvest Date. *Foods* **2021**, *10*, 290. [CrossRef]
16. Giuffrè, A.M. The effect of cultivar and harvest season on the n-alkane and the n-alkene composition of virgin olive oil. *Eur. Food Res. Technol.* **2021**, *247*, 25–36. [CrossRef]
17. Goh, S.H.; Gee, P.T. Noncarotenoid hydrocarbons in palm oil and palm fatty acid distillate. *J. Am. Oil Chem. Soc.* **1986**, *63*, 226–230. [CrossRef]
18. Herchi, W.; Saoussem, H.; Rochut, S.; Boukhchina, S.; Kallel, H.; Pepe, C. Characterization and Quantification of the Aliphatic Hydrocarbon Fraction during Linseed Development (*Linum usitatissimum* L.). *J. Agric. Food Chem.* **2009**, *57*, 5832–5836. [CrossRef]
19. Chen, C.H.; Terentjev, E.M. Aging and Metastability of Monoglycerides in Hydrophobic Solutions. *Langmuir* **2009**, *25*, 6717–6724. [CrossRef] [PubMed]
20. Yılmaz, E.; Öğütcü, M.; Yüceer, Y.K. Physical Properties, Volatiles Compositions and Sensory Descriptions of the Aromatized Hazelnut Oil-Wax Organogels. *J. Food Sci.* **2015**, *80*, S2035–S2044. [CrossRef]
21. Öğütcü, M.; Yılmaz, E.; Güneşer, O. Influence of Storage on Physicochemical and Volatile Features of Enriched and Aromatized Wax Organogels. *J. Am. Oil Chem. Soc.* **2015**, *92*, 1429–1443. [CrossRef]
22. Toro-Vazquez, J.F.; Morales-Rueda, J.A.; Dibildox-Alvarado, E.; Charó-Alonso, M.; Alonzo-Macias, M.; González-Chávez, M.M. Thermal and Textural Properties of Organogels Developed by Candelilla Wax in Safflower Oil. *J. Am. Oil Chem. Soc.* **2007**, *84*, 989–1000. [CrossRef]
23. Morales-Rueda, J.A.; Dibildox-Alvarado, E.; Charó-Alonso, M.A.; Weiss, R.G.; Toro-Vazquez, J.F. Thermo-mechanical properties of candelilla wax and dotriacontane organogels in safflower oil. *Eur. J. Lipid Sci. Technol.* **2009**, *111*, 207–215. [CrossRef]
24. Hwang, H.-S.; Kim, S.; Singh, M.; Winkler-Moser, J.K.; Liu, S.X. Organogel Formation of Soybean Oil with Waxes. *J. Am. Oil Chem. Soc.* **2012**, *89*, 639–647. [CrossRef]
25. Abdallah, D.J.; Weiss, R.G. n-Alkanes Gel n-Alkanes (and Many Other Organic Liquids). *Langmuir* **2000**, *16*, 352–355. [CrossRef]
26. Maia, M.; Nunes, F.M. Authentication of beeswax (*Apis mellifera*) by high-temperature gas chromatography and chemometric analysis. *Food Chem.* **2013**, *136*, 961–968. [CrossRef] [PubMed]
27. Wijarnprecha, K.; Aryusuk, K.; Santiwattana, P.; Sonwai, S.; Rousseau, D. Structure and rheology of oleogels made from rice bran wax and rice bran oil. *Food Res. Int.* **2018**, *112*, 199–208. [CrossRef]
28. Dassanayake, L.S.K.; Kodali, D.R.; Ueno, S.; Sato, K. Physical Properties of Rice Bran Wax in Bulk and Organogels. *J. Am. Oil Chem. Soc.* **2009**, *86*, 1163. [CrossRef]
29. Doan, C.D.; Tavernier, I.; Okuro, P.K.; Dewettinck, K. Internal and external factors affecting the crystallization, gelation and applicability of wax-based oleogels in food industry. *Innov. Food Sci. Emerg. Technol.* **2018**, *45*, 42–52. [CrossRef]
30. Blake, A.I.; Marangoni, A.G. Plant wax crystals display platelet-like morphology. *Food Struct.* **2015**, *3*, 30–34. [CrossRef]
31. Doan, C.D.; To, C.M.; De Vrieze, M.; Lynen, F.; Danthine, S.; Brown, A.; Dewettinck, K.; Patel, A.R. Chemical profiling of the major components in natural waxes to elucidate their role in liquid oil structuring. *Food Chem.* **2017**, *214*, 717–725. [CrossRef] [PubMed]
32. Abdallah, D.J.; Lu, L.; Weiss, R.G. Thermoreversible Organogels from Alkane Gelators with One Heteroatom. *Chem. Mater.* **1999**, *11*, 2907–2911. [CrossRef]
33. Clarkson, C.E.; Malkin, T. 139. Alternation in long-chain compounds. Part II. An X-ray and thermal investigation of the triglycerides. *J. Chem. Soc.* **1934**, 666–671. [CrossRef]
34. Yılmaz, E.; Öğütcü, M.; Arifoglu, N. Assessment of Thermal and Textural Characteristics and Consumer Preferences of Lemon and Strawberry Flavored Fish Oil Organogels. *J. Oleo Sci.* **2015**, *64*, 1049–1056. [CrossRef]

35. Doan, C.D.; Tavernier, I.; Bin Sintang, M.D.; Danthine, S.; Van de Walle, D.; Rimaux, T.; Dewettinck, K. Crystallization and Gelation Behavior of Low- and High Melting Waxes in Rice Bran Oil: A Case-Study on Berry Wax and Sunflower Wax. *Food Biophys.* **2017**, *12*, 97–108. [CrossRef]
36. Chen, X.-W.; Chen, Y.-J.; Wang, J.-M.; Guo, J.; Yin, S.-W.; Yang, X.-Q. Tunable volatile release from organogel-emulsions based on the self-assembly of β-sitosterol and γ-oryzanol. *Food Chem.* **2017**, *221*, 1491–1498. [CrossRef]
37. Yin, W.; Washington, M.; Ma, X.; Yang, X.; Lu, A.; Shi, R.; Zhao, R.; Wang, X. Consumer acceptability and sensory profiling of sesame oils obtained from different processes. *Grain Oil Sci. Technol.* **2020**, *3*, 39–48. [CrossRef]
38. Chen, X.-W.; Guo, J.; Wang, J.-M.; Yin, S.-W.; Yang, X.-Q. Controlled volatile release of structured emulsions based on phytosterols crystallization. *Food Hydrocoll.* **2016**, *56*, 170–179. [CrossRef]

 *foods*

Article

Comparative Study of Chemical Compositions and Antioxidant Capacities of Oils Obtained from 15 Macadamia (*Macadamia integrifolia*) Cultivars in China

Xixiang Shuai [1], Taotao Dai [2], Mingshun Chen [1], Ruihong Liang [1], Liqing Du [3], Jun Chen [1] and Chengmei Liu [1,*]

[1] State Key Laboratory of Food Science and Technology, Nanchang University, Nanchang 330047, China; shuaixixiang1989@163.com (X.S.); chenshun1221@163.com (M.C.); liangruihong@ncu.edu.cn (R.L.); chen-jun1986@hotmail.com (J.C.)

[2] Agro-Products Processing Science and Technology Research Institute, Guangxi Academy of Agricultural Sciences, Nanning 530007, China; ncubamboo@163.com

[3] South Subtropical Crop Research Institute, China Academy of Tropical Agricultural Sciences, Zhanjiang 524091, China; duliqing927618@163.com

* Correspondence: liuchengmei@ncu.edu.cn

Abstract: The planting area of macadamia in China accounted for more than one third of the world's planted area. The lipid compositions, minor components, and antioxidant capacities of fifteen varieties of macadamia oil (MO) in China were comparatively investigated. All varieties of MO were rich in monounsaturated fatty acids, mainly including oleic acid (61.74–66.47%) and palmitoleic acid (13.22–17.63%). The main triacylglycerols of MO were first time reported, including 19.2–26.1% of triolein, 16.4–18.2% of 1-palmitoyl-2,3-dioleoyl-glycerol, and 11.9–13.7% of 1-palmitoleoyl-2-oleoyl-3-stearoyl-glycerol, etc. The polyphenol, α-tocotrienol and squalene content varied among the cultivars, while Fuji (791) contained the highest polyphenols and squalene content. Multiple linear regression analysis indicated the polyphenols and squalene content positively correlated with the antioxidant capacity. This study can provide a crucial directive for the breeding of macadamia and offer an insight into industrial application of MO in China.

Keywords: macadamia oil; cultivars; minor components; antioxidant capacity; triacylglycerols

Citation: Shuai, X.; Dai, T.; Chen, M.; Liang, R.; Du, L.; Chen, J.; Liu, C. Comparative Study of Chemical Compositions and Antioxidant Capacities of Oils Obtained from 15 Macadamia (*Macadamia integrifolia*) Cultivars in China. *Foods* **2021**, *10*, 1031. https://doi.org/10.3390/foods 10051031

Academic Editors: Qiang Wang and Aimin Shi

Received: 5 April 2021
Accepted: 5 May 2021
Published: 10 May 2021

Publisher's Note: MDPI stays neutral with regard to jurisdictional claims in published maps and institutional affiliations.

Copyright: © 2021 by the authors. Licensee MDPI, Basel, Switzerland. This article is an open access article distributed under the terms and conditions of the Creative Commons Attribution (CC BY) license (https:// creativecommons.org/licenses/by/ 4.0/).

1. Introduction

The macadamia (*Macadamia integrifolia*) is an evergreen native tree indigenous to the coastal rainforests of Australia [1]. The macadamia fruit is made up of the husk, shell, and kernel [2]. The kernel is a rich source of lipids, proteins and important micronutrients [3,4]. However, its chemical composition may vary greatly influenced by the cultivar, kernel maturity, geographical location and growth conditions. Castilho Maro, et al. [5] studied the chemical composition of 22 macadamia varieties in Brazil, and the results showed lipid contents of macadamia kernel ranging from 33.13% to 64.28%, and protein from 8.56% to 19.24%. Entelman, et al. [6] characterized eight macadamia varieties in Sao Paulo, and reported that lipid content of macadamia kernel in the range of 65.21–68.48%. Kaijser, et al. [7] investigated the chemical composition of four macadamia cultivars in New Zealand, and found that lipid content of macadamia kernel was 69.1–78.4%.

The minor components of macadamia oil (MO) may also influenced by cultivar and location. Mereles, et al. [8] found the difference in the contents of polyphenols (77.9–96.3 mg gallic acid equivalent/kg) and α-tocopherol (0.2–18.4 mg/100 g) of three macadamia cultivars grown in Paraguay. Gong, et al. [9] reported that the composition and content of tocopherol (22.3–49.7 mg/kg oil) and sterol (1952–2571 mg/kg oil) of 3 MO from the U.S. market were different. Wall [10] indicated tocotrienols and squalene was largely effect by varieties, their content in seven macadamia cultivars produced in Hawaii were 31–92 and 72–171 µg/g oil, respectively.

The introduction of macadamia from Australia into China began in the 1970s. By the end of 2018, the cultivated area of macadamia in China has exceeded 3012.06 km^2, which accounted for more than one third of the world's planted area [11]. Up to now, China has become the largest and fastest growing macadamia plantation country in the world. The geographical environment of China is quite different from that of the above mentioned regions, however, the chemical compositions and antioxidant capacities of MOs cultivated in China were never reported.

In this study, the compositions of macadamia kernel, and the fatty acids, triacylglycerols, minor components, as well as antioxidant activity of MO of 15 different varieties of macadamia in China were comparatively investigated. In addition, the correlation between the main minor components and the antioxidant capacity were analyzed by multiple linear regression (MLR) analysis. This study are expected to provide a crucial directive for the industrial application of MO in China.

2. Materials and Methods

2.1. Material

Fifteen macadamia cultivars (1 = Hinde (H2), 2 = Fuji (791), 3 = Purvis (294), 4 = HAES 816, 5 = Keauhou (246), 6 = Beaumont (695), 7 = Pahala (788), 8 = HAES 863, 9 = A4, 10 = A16, 11 = HAES 344, 12 = Keaau (660), 13 = Makai (800), 14 = GUANG 11, 15 = own choice (OC)) were collected from a commercial plantation and harvested during the 2019 in September. The selected plantation is located in Lincang City (longitude: 99°26′39.2″, latitude: 14°11′43.6″), Yunnan Province, which belongs to the mountainous area of low hot valley and is the largest macadamia-cultivated regions in China. Macadamias were collected using the diagonal method. A total of 10 plants with same growth potential and without disease and pest were randomly selected in ripening period, on the canopy of which 500 g of macadamias were collected, respectively, at lower, middle, higher, inner and outer parts. The samples were fully mixed, packaged with valve bags and labeled, then taken back to laboratory. The fresh and mature macadamias were selected and its peel were removed to obtain the kernel with shell, then it was dried using hot air drying. The shell of dried fruits were removed, then the kernels were oxygen insulation packed and stored at −20 °C for further analyses.

Standards of 37 fatty acid methyl esters, eight tocopherols, phytosterols, squalane, and 5α-cholestane were purchased from Sigma-Aldrich Co., Ltd. (Shanghai, China). Further, 2,2-Diphenyl-1-picrylhydrazyl (DPPH), 2,2′-Azino-bis (3-ethylbenzothiazoline-6-sulfonic acid) diammonium salt (ABTS), chromatographic grade methanol, ethanol, *n*-hexane, and isopropanol were purchased from the National Institute for the Control of Pharmaceutical and Biological Products (Beijing, China). All other reagents were of analytical reagent grade.

2.2. Proximate Analysis of Dried Kernel

The ash, moisture, crude protein and lipid content of dried kernel were measured by the hot air drying at 550 °C (AOAC Method 923.03), hot air drying at 120 °C (AOAC Method 990.19), Kjeldahl (N × 6.25) (AOAC Method 979.09) and Soxhlet extraction (AOAC Method 963.15), respectively [12].

2.3. Extraction of MO

Macadamias were collected and pretreated according to the method reported by Li, et al. [13]. The dried macadamia kernels were crushed into slurry, and sifted through a 40-mesh sieve. Then, the MOs were extracted from slurry. In brief, the slurry (100 g) and *n*-hexane (600 mL) were mixed at 50 °C for 4.0 h under 500 r/min. Then, the mixture was filtered with a Brinell funnel. Finally, the *n*-hexane was removed by rotary evaporator (Hei-vap Precision, Heidolph Co., Schwabach, Germany) at 50 °C to obtain the MO. The composition and content of MO were further analyzed.

2.4. Fatty Acid and Triacylglycerol Composition Analysis

Briefly, 250 mg of MO was mixed with 10.0 mL of *n*-hexane. Then, the 0.3 mL of KOH-CH$_3$OH solution (2.0 mol/L) was added for the methylation reaction. This reaction was performed at room temperature for 30.0 min. Then, 8.0 mL of NaHSO4 solution (0.1 mol/L) was added to neutralize the excessive KOH. Na$_2$SO$_4$ was used to absorb any trace water remaining in the *n*-hexane part. The obtained fatty acid methyl esters were analyzed by a gas chromatograph (GC) system (Agilent 7890A, Agilent Technologies, Santa Clara, CA, USA) equipped with a BPX capillary column (0.25 μm, 60.0 m × 0.22 mm) and a flame ionization detector (FID) (Agilent Technologies, USA). The following parameters were used: flow rate of nitrogen = 1.0 mL/min; the initial temperature of the column = 60 °C, temperature programming to 170 °C = 10 °C/min, continued temperature programming to 230 °C = 3 °C/min, then held for 15.0 min, injection volume = 1.0 μL, injector temperature = 225 °C, and detector temperature = 250 °C. Fatty acid composition of oils were identified by comparing with the relative retention time of the fatty acid methyl ester peaks with standards of fatty acid methyl esters, and the results were expressed as a percentages of each fatty acid in the total.

Triacylglycerol composition of MO was analyzed using the method of American Oil Chemists' Society [14]. Triacylglycerol was identified by comparing the retention time, carbon numbers of triacylglycerol standards, and the results were expressed as relative proportion.

2.5. Tocopherols Analysis

Tocopherols were analyzed using a high-performance liquid chromatographic system (Agilent 1260, Agilent Technologies, USA) equipped with a C$_{18}$ column (4.6 mm × 250 mm, 5 μm, Agilent Technologies, USA) and ultraviolet detector. Initially, 0.1 g of MO was mixed with 10.0 mL of *n*-hexane, then filtered by a 0.45 μm organic filter. The following parameters were used: mobile phase = *n*-hexane/methanol/isopropanol (92.5/7.4/0.1, *v/v/v*), flow rate = 1.0 mL/min, column temperature = 30 °C, injection volume = 20.0 μL, and determining wavelength = 294 nm. Tocopherol and tocotrienol of MOs were identified and quantified by the retention time and standard curve.

2.6. Phytosterols Analysis

Phytosterols analysis were analyzed using a GC system equipped with a DB-5MS capillary column (0.25 μm, 30.0 m × 0.25 mm) and a FID (Agilent 7890A, Agilent Technologies, USA). First of all, 250 mg of MO was mixed with 1.0 mL of 5α-cholestane (0.5 mg/mL) and 3.0 mL of KOH-methanol (2.0 mol/L). The mixture was saponified at 85 °C for 1 h, then cooled to room temperature. 5.0 mL of *n*-hexane and 2.0 mL of distilled water were added to extract the phytosterols and repeated three times. The extract was dried by nitrogen and then silylated using 200 μL silylation reagents (pyridine/hexamethyldisilane/trimethylchlorosilane, 9/3/1, *v/v/v*) at 75 °C for 30 min. Then the silylated samples was cooled to room temperature and filtered by 0.22 μm organic membrane. After that, the filtered sample was determined using the following parameters: flow rate of helium = 3.0 mL/min; the initial temperature of the column = 230 °C, temperature programming to 280 °C = 10 °C/min, continued temperature programming to 290 °C = 5 °C/min, then held for 35 min, injection volume = 1.0 μL, injector temperature = 290 °C, and detector temperature = 300 °C. Phytosterols were identified and quantified by the retention time and internal standard substance.

2.7. Squalene Analysis

Squalene contents were determined using a GC system equipped with a HP-5 capillary column (0.25 μm, 30 m × 0.32 mm) and a FID (Agilent 7890A, Agilent Technologies, USA). A total 1.0 g of MO was mixed with 0.3 mL of squalane standard (1.0 mg/mL) and 50.0 mL of KOH-ethanol (1.0 mol/L). The mixture was saponified at 85 °C for 1.0 h. Then, distilled water (50.0 mL) was added to remove some water-soluble component, and cooled to room temperature. Then, 50.0 mL of *n*-hexane was added to extract the squalene and repeated three times. The combined *n*-hexane extracts were washed to pH 7.0 using 10% ethanol solution. The *n*-hexane was removed by rotary evaporator (Hei-vap Precision, Heidolph Co., Germany) at 30 °C to obtain the squalene samples. Finally, the squalene samples were dissolved in *n*-hexane again and reached to 10.0 mL. Then the sample filtered by 0.22 μm was determined. The following parameters were then used for measurement: flow rate of nitrogen = 1.0 mL/min; the initial temperature of the column = 160 °C, temperature programming to 220 °C = 15 °C/min, continued temperature programming to 280 °C = 5 °C/min (held for 20 min), continued temperature programming to 300 °C = 5 °C/min (held for 2 min), injection volume = 1.0 μL, injector temperature = 250 °C, and detector temperature = 300 °C. Squalene were identified and quantified by comparing the standards and internal standard of squalane.

2.8. The Total Polyphenol Content Analysis

The total polyphenol content was measured using the Folin–Ciocalteu reagent method based on the procedure of Gao, et al. [15] with some modifications. In brief, 4.0 g of MO was mixed with 3.0 mL of methanol using a vortex mixer (VX200-T, MET, USA) and placed in the dark to extract the total polyphenol. The extract processing was repeated three times, and the supernatants were combined. The supernatant (0.2 mL) was mixed with 0.8 mL of distilled water and 1.0 mL of Folin-Ciocalteu reagent and incubated for 5 min at room temperature. Then, 1.0 mL of sodium carbonate (7.5%) were added and incubated in the dark for 1.5 h. Afterwards, the absorbance was determined by UV-vis spectrophotometer (UV1700; Rangqi, Shanghai, China) at 760 nm. The results were expressed as mg of gallic acid equivalents (GAE) per kg of MO (mg/kg).

2.9. Mineral Composition Analysis

The mineral composition of the MO was detected according to the Inductively Coupled Plasma Mass Spectrometry (ICP-MS) method of Policarpi, et al. [16] with some modifications. 1.0 mL of hydrogen peroxide and 3.0 mL of nitric acid were added to 0.5 g of MO in closed microwave digestion tank. Then the digestion was according to the following procedure: microwave power was 1500 W, temperature programming from 0 to 120 °C within 5 min (held for 5 min), continued temperature programming to 200 °C within 5 min (held for 30 min). A total of 0.1 mg/L of arsenic was added as an internal standard substance. A multi-element stock standard solution containing all the analytical mineral elements was used to prepare a standard curve. The digested MO were dissolved in deionized water and analyzed using ICP-MS (Agilent 7900, Agilent Technologies, USA).

2.10. Determination of the Antioxidant Activity

Three different antioxidant models (DPPH, ABTS, and Ferric reducing ability of plasma (FRAP)) were employed to evaluate the antioxidant activity of MO (methanol extract mentioned above in Section 2.8) according to previous reports [15,17,18]. The antioxidant capacity of MO was expressed as Vitamin E equivalent (V_E, μmol/kg).

2.11. Statistical Analysis

All the experiments were repeated in triplicate, and the date were expressed as means ± standard deviations. Statistical analysis was performed on SPSS 25.0 (SPSS Inc., Chicago, IL, USA). A comparison of the means was performed by Tukey's test using one-way analysis of variance. MLR analysis using a stepwise method was performed to understand the correlations between antioxidant capacity assays and bioactive components. The use probability of F values *p*-to-enter and *p*-to-remove new variables into statistical model were $p < 0.05$ and $p > 0.10$, respectively.

3. Results and Discussion

3.1. Compositions of Macadamia Kernels

The compositions (lipid, crude protein, ash, and moisture content) of fifteen different cultivars of macadamia kernels were shown in Table 1. The protein content of macadamia kernels was averagely 8.07% with the highest value of 9.04% (Fuji (791)). The lipid contents of macadamia kernel ranged from 73.55% to 78.56% and the average was 75.98%. The GUANG 11 (14) contained the highest amount of oil compared to other varieties. The oil content in Chinese macadamia kernels was similar to that in New Zealand (69.1–78.4%) [7], but higher than that in Brazil (33.13% to 68.48%) [5,6]. Compared to some common edible nut seeds such as almond (~43.36%), cashew nut (~43.71%), walnut (~64.50%) [19], the macadamia kernels have a higher oil content, which can be a great potential oil sources in food industry. However, to be a good vegetable oil, it also needs to have the appropriate fatty acid composition, minor components, etc. Therefore, these components were analyzed in the next sections.

Table 1. Chemical composition of the macadamia kernels [1].

No.	Moisture Content	Ash Content	Protein Content	Oil Content
1	1.62 ± 0.03 [a]	1.25 ± 0.01 [b]	8.55 ± 0.08 [c]	75.02 ± 0.11 [b]
2	1.65 ± 0.06 [a,b]	1.13 ± 0.07 [a]	9.04 ± 0.13 [d,e]	76.11 ± 0.22 [e]
3	1.55 ± 0.11 [a]	1.21 ± 0.04 [a,b]	7.88 ± 0.16 [b]	75.45 ± 0.19 [b,c]
4	1.63 ± 0.00 [a]	1.16 ± 0.05 [a]	8.21 ± 0.16 [c]	77.19 ± 0.04 [f]
5	1.74 ± 0.01 [c]	1.33 ± 0.01 [c]	8.64 ± 0.13 [c,d]	73.55 ± 0.06 [a]
6	1.66 ± 0.10 [a]	1.43 ± 0.09 [c,d]	7.23 ± 0.23 [a,b]	77.32 ± 0.05 [c]
7	1.70 ± 0.05 [a,b,c]	1.36 ± 0.03 [c]	7.83 ± 0.14 [b]	75.71 ± 0.22 [b,d]
8	1.86 ± 0.16 [b,c]	1.14 ± 0.06 [a]	8.83 ± 0.07 [d]	75.37 ± 0.06 [c]
9	1.70 ± 0.03 [b]	1.48 ± 0.03 [d]	7.74 ± 0.10 [b]	76.33 ± 0.21 [e]
10	1.68 ± 0.06 [a,b,c]	1.39 ± 0.02 [c]	7.66 ± 0.08 [a,b]	76.16 ± 0.25 [e]
11	1.65 ± 0.02 [a]	1.47 ± 0.03 [d]	7.55 ± 0.10 [b]	75.94 ± 0.15 [e]
12	1.66 ± 0.10 [a]	1.24 ± 0.10 [b]	7.17 ± 0.25 [a]	75.32 ± 0.23 [b,c]
13	1.73 ± 0.11 [a,b,c]	1.28 ± 0.05 [b,c]	7.94 ± 0.16 [b,c]	75.77 ± 0.01 [d,e]
14	1.71 ± 0.03 [b,c]	1.17 ± 0.01 [a]	8.94 ± 0.26 [c,d,e]	78.36 ± 0.36 [g]
15	1.63 ± 0.01 [a]	1.26 ± 0.02 [b]	7.78 ± 0.07 [b]	76.04 ± 0.33 [e]
Max	1.86 ± 0.16 [b,c]	1.48 ± 0.03 [d]	9.04 ± 0.13 [d,e]	78.36 ± 0.36 [g]
Min	1.55 ± 0.11 [a]	1.13 ± 0.07 [a]	7.17 ± 0.25 [a]	73.55 ± 0.06 [a]
Average	1.68 ± 0.05 [a,b,c]	1.29 ± 0.04 [b,c]	8.07 ± 0.14 [b,c]	75.98 ± 0.13 [e]

[1] Values are means ± SD. Different letters in a column indicate significant differences ($p < 0.05$).

3.2. Fatty Acid and Triacylglycerol

3.2.1. Fatty Acid Composition and Content

The fatty acid composition is an important parameter of the quality and authenticity of oil, which can be used to detect the frauds of oil. Fatty acid composition and content of 15 different MO were shown in Table 2. The oil samples contained twelve fatty acids, and the major fatty acids were unsaturated fatty acids (UFAs). The sum of UFAs (including palmitoleic acid (C16:1), oleic acid (C18:1), linoleic acid (C18:2), linolenic acid (C18:3), eicosenoic acid (C20:1) and docosenoic acid (C22:1)) accounted for more than 83% of the total fatty acid content, where Fuji (791) (2) had the highest UFAs value (85.9%). In addition, monounsaturated fatty acids (MUFAs) accounted for 96.67–98.26% of UFAs, mainly consist of C18:1 (61.74–66.47% of fatty acid) and C16:1 (13.22–17.63% of fatty acid). The fatty acid composition was similar to previous reports [1,7,19–21]. However, the contents were different, which might relate to the cultivar, kernel maturity, geographical location and growth conditions. It was reported that diets containing high oleic acid and MUFA contents reduce low-density lipoprotein levels and decrease the risk of cardiovascular diseases. Moreover, vegetable oils with high oleic acid content are highly desirable in terms of thermal stability and longer shelf life [22,23]. Therefore, MO has potential as a healthy vegetable oil.

3.2.2. Triacylglycerol Composition

The triacylglycerol composition in the oil is also an important index of quality which is widely used in the industry to control the purity of oil and identify authenticity [24]. However, the composition and content of triacylglycerol of MO has not been yet reported elsewhere, which were analyzed in this study and presented in Table 3. Twenty kinds of triacylglycerols in MO were identified and quantified, namely 1,3-dimyristoyl-2-oleoyl-glycerol (MOM), 1,2-dipalmitoyl-3-palmitoleyl-glycerol (PPP_O), 1-myristoyl-2-oleoyl-3-palmitoyl-glycerol (MOP), 1-myristoyl-2-linoleoyl-3-palmitoyl-glycerol (MLP), 1,3-dipalmitoyl-2-oleyl-glycerol (POP), 1-myristoyl-2,3-dioleoyl-glycerol (MOO), 1-palmitoyl-2-linoleoyl-3-palmitoleyl-glycerol (PLP_O), 1-palmitoyl-2,3-distearoyl-glycerol (PSS), 1-palmitoleyl-2-oleyl-3-stearoyl-glycerol (P_OOS), 1-palmitoyl-2,3-dioleoyl-glycerol (POO), 1-palmitoleyl-2-linoleoyl-3-stearoyl-glycerol (P_OLS), 1-palmitoyl-2-linoleoyl-3-oleyl-glycerol (PLO), 1-palmitoyl-2,3-dilinoleoyl-glycerol (PLL), 1-stearoyl-2-linoleoyl-3-oleyl-glycerol (SLO), 1,3-dioleoyl-2-linoleoyl-glycerol (OLO), 1-stearoyl-2-oleyl-3-arachidyl-glycerol (SOA), 1-arachidyl-2,3-dioleyl-glycerol (AOO), 1,3-distearoyl-2-oleyl-glycerol (SOS), 1-stearoyl-2,3-dioleyl-glycerol (SOO), and triolein (OOO). The main triacylglycerol included OOO (19.18–26.14%), POO (16.36–18.19%), P_OOS (11.87–13.65%), POP (6.89–8.96%), MOO (6.08–8.46%) and SOO (4.81–6.93%), which accounted for more than 70% of total triacylglycerol. In addition, as shown in Table 3, the triacylglycerol contents also had a certain difference among cultivars. The OOO of the Fuji (791) (2) had the highest value (26.14%).

Both fatty acid and triacylglycerol compositions indicated that MO is a good source of UFAs. According to the Standards issued by the Food and Agriculture Organization of the United Nations, the MUFAs content of healthy edible oils should be higher than 75% [25]. Thus, MO has potential as a dietary resource of plant oil.

3.3. Minor Components

The minor components including polyphenols, tocopherols, squalene, minerals and phytosterols of MO from different cultivars were determined, which were important quality and nutritional characteristics of vegetable oil [26,27].

3.3.1. Polyphenols Content

Polyphenols are important minor components in vegetable oils because they confer the sensory and nutritional characteristics [28]. The content of polyphenol in MOs was shown in Table 4. In general, the content of polyphenols from different cultivars showed large differences. Among the 15 cultivars, the polyphenols contents ranged from 19.74 to 123.40 GAE mg/kg. The polyphenols content of Fuji (791) (2) was about 6 times higher than that of Hinde (H2) (1). Quinn and Tang [29] reported the polyphenols content of MO from Hawaii was 48.7 GAE mg/kg, which is within the scope of our results. However, the polyphenol contents of most cultivars in our study were higher than the report of Cicero, et al. [30], who indicated the polyphenols content of MO from Brazilian was only 2.36 GAE mg/kg. These differences may be related to the macadamia growing conditions, variety and location.

3.3.2. Tocopherol Content

The profile of tocopherols in 15 Chinese macadamia cultivars was shown in Table 4. It was worth noting that only α-tocotrienol was identified in the MO and the content ranged from 27.9 to 53.1 mg/kg. The highest content of α-tocotrienols was observed in the A4 (9), while the lowest content in HAES 863 (8). The profile of tocopherols may be related to the planting area and cultivars. For example, in Hawaii, the total tocopherol content of MO was 9–25 mg/g [28], while no tocopherol was detected in New Zealand [7]. In addition, seven different macadamia cultivars from the island of Hawaii were determined, and α-tocotrienol was 15.91–46.83 mg/g, ε-tocotrienol ranged from 3.00–17.66 mg/g, and γ-tocotrienol was 8.75–34.28 mg/g of oil, whereas the α- and γ-tocopherol compounds were only detected in two cultivars [10]. Compared to the tocotrienols contents of 21 species of plant oils reported by Gruszka and Kruk [31], the MOs in our study had higher content of α-tocotrienol. As we all know, tocotrienols are more powerful antioxidants than tocopherols, and showed better cholesterol-lowering and anti-cancer properties [10,32]. In addition, it can quickly penetrate the skin and reduce oxidative stress caused by ultraviolet light [10]. Therefore, MO may be used in functional foods and cosmetics.

3.3.3. Squalene Content

Squalene is a triterpene precursor to the biosynthesis of vitamin D, steroid hormones, and cholesterol in the human body, which is beneficial to health [33]. As shown in Table 4, the squalene contents of MO ranged from 91.15 to 268.08 mg/kg. The highest content of squalene was observed in the Fuji (791) (2), while the lowest content in the Hinde (H2) (1). The results were in accordance with [10,19], who reported the contents of MO ranged from 72–185 mg/kg, but significantly higher than cold-pressed MO (22.9 mg/kg), which was sold in the market of Brazilian [29]. In addition, the squalene content of MO was relatively higher than nut oils such as hazelnuts (186.4 mg/kg), peanuts (98.3 mg/kg), almonds (95.0 mg/kg), and walnuts (9.4 mg/kg) [20]. Squalene, like tocotrienol, is a potent and stable antioxidant [10]. In addition, research indicated that long-term consumption of squalene-rich vegetable oil can inhibit forming tumors and less often develop various cancers [34]. Therefore, MO was a potential source of dietary squalene.

3.3.4. Phytosterols Content

Phytosterols, which was usually considered as a nutritional supplement for fats and oils, and was one of the important bioactive classes of constitutes in nut oils [35]. The profile of phytosterols in 15 Chinese macadamia varieties was shown in Table 5. Eight kinds of phytosterols were identified and quantified, namely β-sitosterol (1248.8 to 1613.7 mg/kg), Δ^5-avenasterol (170.0 to 362.1 mg/kg), campesterol (82.4 to 114.1 mg/kg), Δ^7-stigmastenol (7.8 to 48.6 mg/kg), Δ^7-avenasterol (7.6 to 41.3 mg/kg), clerosterol (11.4 to 16.3 mg/kg), $\Delta^{5,24}$-stigmastadienol (7.8 to 19.8 mg/kg), β-stiostanol (5.4 to 9.6 mg/kg). The total content of phytosterols ranged from 1569.50 to 2064.54 mg/kg. Among the identified phytosterols, β-sitosterol, Δ^5-avenasterol and campesterol were the major, which accounted for more than 92% of the total phytosterols. Furthermore, β-sitosterol content was the highest, which was similar to previous reports [9,36]. It was worth noting that phytosterol content in MO from different cultivars had certain differences but not significant. Thus, the detailed phytosterol profile can be used as a fingerprint to identify the vegetable oils.

3.3.5. Mineral Content

The composition and content of minerals also affects the quality of some oils. Although minerals in macadamia nuts/kernel have been widely reported, those in MO is very rare, especially among different varieties. Seven kinds of minerals (Mg, Ca, Zn, As, Pb, K and Na) were determined, but only four kinds of minerals (Mg, Ca, K and Na) were identified and quantified, ranged from 8.13 to 49.90, 0 to 28.50, 13.75 to 51.80 and 0.41 to 2.51 mg/kg, respectively (Table 4). These minerals in oil were beneficial to the health of adults as suggested by dietary recommended intake guidelines. For example, Ca is associated with bone health and prevention of osteoporosis, and Mg is associated with activation of enzymatic systems [37]. Moreover, Aslanabadi, et al. [38] reported that long-term consumption of natural mineral foods rich in Ca, Mg, and bicarbonate can reduce cholesterol and low density lipoprotein. Hence, the relatively high minerals content can promote the quality and nutrition of MO.

3.4. In Vitro Antioxidant Capacity

The antioxidant capacity of MOs determined using different models was shown in Table 4. The antioxidant capacity analyzed by DPPH, ABTS and FRAP model was 126–359, 136–324, and 255–1086 μmol V_E/kg, respectively. Among the three models, the antioxidant activity had the same tendency, but presented different absolute values. The value difference may be due to the different chemical mechanisms of the antioxidant models [39,40]. Fuji (791) (2) showed the highest antioxidant capacity, while the Hinde (H2) (1) exhibited the lowest. It was noted that the study only analyzed the characteristics of different varieties of MO in one region for one year, but the growth conditions had a great influence on the composition and content of minor components. In order to obtain more detailed information, the team will continue to conduct long-term monitoring of MO from different regions, years and varieties in China. According to the references, the differences in antioxidant capacity of materias may be contributed to polyphenols, squalene, phytosterols, etc. [15,17,18,39]. Therefore, it is necessary to clarify the relationship between antioxidant capacities and minor compounds.

3.5. Correlations between Antioxidant Capacity and Minor Components

Bivariate correlations analysis as a common method was utilized to investigate the correlation between two continuous variables and calculate the correlation coefficient (r) [15,17]. The correlation between antioxidant capacities and minor compounds was shown in Table 6. The antioxidant capacity showed a significant positive correlation with the polyphenols (r was 0.938–0.965, $p < 0.01$) and squalene content (r was 0.870–0.927, $p < 0.01$). This result was similar with previous studies. Gao, et al. [17] reported that the oxidative stability index of walnut oil was significantly correlated with polyphenols (r = 0.873, $p < 0.01$), squalene (r = 0.701, $p < 0.01$), and stigmasterol (r = 0.770, $p < 0.01$). Shi, et al. [41] also reported that remarkable positive correlations were observed between squalene and measured by DPPH, ABTS and FRAP (r was 0.82–0.91), and the correlations between polyphenol and antioxidant capacity were also positive (r was 0.65–0.67). The analysis strongly supported the positive contribution of polyphenols and squalene to the antioxidant capacity of MOs.

3.6. MLR Analysis

MLR analysis can be considered as a useful tool to analyze a phenomenon (antioxidant capacity) associated with multiple factors (minor components) [15,17]. In this study, a regression model was defined: Y as the dependent variable and X as independent variables (minor components). Linear models were constructed in the form as: $Y = M_0 + M_1X_1 + M_2X_2 + M_3X_3 + \ldots + M_nX_n$.

Where, Y represents the predicted response, M_0 denotes the unstandardized constant coefficient, M_1, M_2, M_3 and M_n refer to partial correlation coefficients, and X_1, X_2, X_3 and X_n represent the independent variables [15,17].

Based on the analysis of a stepwise method of MLR, polyphenols and squalene exhibited a good correlation. As shown in Table 7, the DPPH-based model showed the highest value of adjusted R^2 (0.937), with the constant of the unstandardized coefficients was 35.671, and the partial correlation coefficients (R) of the predicted equation was 0.546 and 0.467 for polyphenols and squalene, respectively (Y = 35.671 + 0.546 polyphenols + 0.467 squalene). Moreover, polyphenols (R = 0.621) and squalene (R = 0.376) also affected ABTS radical scavenging activity and polyphenols (R = 0.965) influenced FRAP reducing capacity. In accordance with the correlation results mentioned in Section 3.5, polyphenols (r was 0.938–0.965) and squalene (r was 0.870–0.927) showed high regression coefficients and significant correlations ($p < 0.01$) in all antioxidant model. Taking together, the polyphenols and squalene were considered as an indicator of the antioxidant capacity of the MO. This result can provide a valuable judge for botanists and consumers.

Table 2. Fatty acid compositions (%) of oils extracted from 15 macadamia cultivars [1].

No.	C14:0 *	C16:0	C16:1	C18:0	C18:1	C18:2	C18:3	C20:0	C20:1	C22:0	C22:1	C24:0	SFA	MUFA	PUFA	UFA
1	0.72 ± 0.02 [j]	8.22 ± 0.00 [k]	16.44 ± 0.01 [j]	3.24 ± 0.00 [e,f]	63.22 ± 0.04 [c]	1.84 ± 0.01 [e]	0.21 ± 0.00 [b]	2.54 ± 0.00 [a]	2.25 ± 0.00 [b]	0.76 ± 0.02 [a]	0.19 ± 0.00 [a]	0.36 ± 0.03 [a,b]	15.84 ± 0.02 [g]	82.10 ± 0.01 [e]	2.05 ± 0.00 [c]	84.15 ± 0.00 [e]
2	0.52 ± 0.01 [g]	6.81 ± 0.01 [a]	13.73 ± 0.04 [c]	2.61 ± 0.01 [b]	66.47 ± 0.01 [l]	2.07 ± 0.00 [h]	0.26 ± 0.00 [e]	2.75 ± 0.00 [e]	3.04 ± 0.02 [m]	0.99 ± 0.00 [f]	0.33 ± 0.02 [g]	0.42 ± 0.00 [d]	14.10 ± 0.01 [a]	83.57 ± 0.02 [l]	2.33 ± 0.00 [d]	85.90 ± 0.01 [l]
3	0.50 ± 0.00 [f]	7.45 ± 0.01 [f]	15.64 ± 0.01 [k]	3.22 ± 0.00 [e]	64.79 ± 0.02 [g]	1.16 ± 0.00 [a]	0.31 ± 0.00 [h]	2.79 ± 0.02 [f]	2.55 ± 0.00 [h]	0.92 ± 0.00 [d]	0.27 ± 0.00 [e]	0.39 ± 0.00 [b]	15.27 ± 0.00 [d]	83.25 ± 0.01 [k]	1.47 ± 0.00 [a]	84.72 ± 0.00 [i]
4	0.51 ± 0.03 [f,g]	7.81 ± 0.00 [h]	14.62 ± 0.00 [g]	3.71 ± 0.03 [l]	65.17 ± 0.02 [i]	1.47 ± 0.00 [c]	0.20 ± 0.00 [a]	2.94 ± 0.00 [h]	2.22 ± 0.00 [a]	0.82 ± 0.03 [b]	0.18 ± 0.03 [a]	0.36 ± 0.00 [a]	16.15 ± 0.03 [j]	82.19 ± 0.01 [e]	1.67 ± 0.00 [b]	83.86 ± 0.00 [c]
5	0.50 ± 0.00 [f]	8.24 ± 0.02 [k]	17.11 ± 0.01 [l]	3.99 ± 0.00 [n]	61.74 ± 0.04 [a]	1.17 ± 0.00 [a]	0.28 ± 0.00 [f]	3.14 ± 0.01 [k]	2.29 ± 0.00 [c]	0.92 ± 0.01 [d]	0.21 ± 0.00 [b]	0.41 ± 0.01 [c]	17.20 ± 0.01 [l]	81.35 ± 0.01 [c]	1.45 ± 0.00 [a]	82.80 ± 0.00 [a]
6	0.40 ± 0.01 [c]	7.62 ± 0.00 [g]	13.22 ± 0.02 [a]	3.85 ± 0.00 [m]	64.71 ± 0.10 [f]	2.53 ± 0.01 [j]	0.26 ± 0.01 [e]	3.01 ± 0.00 [j]	2.78 ± 0.01 [k]	0.92 ± 0.00 [d]	0.29 ± 0.00 [f]	0.41 ± 0.00 [c]	16.21 ± 0.00 [j]	81.00 ± 0.00 [a]	2.79 ± 0.00 [f]	83.79 ± 0.00 [c]
7	0.43 ± 0.00 [d]	7.80 ± 0.01 [h]	14.64 ± 0.00 [g]	3.58 ± 0.00 [k]	64.94 ± 0.03 [h]	1.44 ± 0.01 [c]	0.25 ± 0.01 [e]	2.97 ± 0.01 [i]	2.42 ± 0.01 [f]	0.82 ± 0.00 [b]	0.22 ± 0.00 [c]	0.42 ± 0.01 [c,d]	16.02 ± 0.00 [i]	82.22 ± 0.02 [f]	1.69 ± 0.00 [b]	83.91 ± 0.01 [d]
8	0.46 ± 0.01 [e]	7.94 ± 0.03 [i]	14.46 ± 0.00 [f]	3.18 ± 0.01 [d]	64.63 ± 0.00 [f]	2.02 ± 0.00 [g]	0.23 ± 0.00 [d]	2.73 ± 0.01 [d]	2.76 ± 0.00 [j]	0.89 ± 0.02 [c]	0.29 ± 0.01 [f]	0.42 ± 0.00 [c,d]	15.62 ± 0.02 [f]	82.14 ± 0.00 [e]	2.25 ± 0.00 [d]	84.39 ± 0.00 [g]
9	0.30 ± 0.01 [a]	8.29 ± 0.02 [l]	15.16 ± 0.03 [i]	3.41 ± 0.01 [i]	63.98 ± 0.00 [d]	1.86 ± 0.00 [e]	0.22 ± 0.00 [c]	2.91 ± 0.01 [f]	2.34 ± 0.00 [d]	0.91 ± 0.00 [c,d]	0.21 ± 0.00 [b]	0.41 ± 0.00 [c]	15.96 ± 0.01 [h]	81.69 ± 0.01 [d]	2.08 ± 0.00 [c]	83.77 ± 0.00 [c]
10	0.34 ± 0.00 [b]	7.37 ± 0.00 [d]	13.57 ± 0.01 [b]	4.08 ± 0.00 [o]	64.93 ± 0.01 [h]	2.17 ± 0.00 [i]	0.22 ± 0.01 [b,c]	3.17 ± 0.00 [l]	2.49 ± 0.00 [g]	0.97 ± 0.02 [e,f]	0.25 ± 0.00 [d]	0.44 ± 0.03 [c,d,e]	16.37 ± 0.01 [k]	81.24 ± 0.00 [b]	2.39 ± 0.00 [e]	83.63 ± 0.00 [b]
11	0.63 ± 0.00 [i]	7.25 ± 0.00 [c]	14.06 ± 0.00 [d]	2.69 ± 0.03 [c]	65.55 ± 0.01 [k]	1.98 ± 0.00 [f]	0.29 ± 0.00 [g]	2.63 ± 0.00 [c]	3.13 ± 0.02 [n]	0.94 ± 0.01 [e]	0.40 ± 0.03 [h]	0.44 ± 0.00 [e]	14.58 ± 0.01 [c]	83.14 ± 0.02 [j]	2.27 ± 0.00 [d]	85.41 ± 0.01 [j]
12	0.81 ± 0.02 [k]	6.98 ± 0.01 [b]	14.89 ± 0.01 [h]	3.55 ± 0.01 [j]	65.47 ± 0.02 [k]	1.37 ± 0.01 [b]	0.27 ± 0.01 [e,f]	2.91 ± 0.00 [f]	2.36 ± 0.00 [e]	0.83 ± 0.01 [b]	0.21 ± 0.01 [b]	0.34 ± 0.03 [a]	15.42 ± 0.02 [e]	82.93 ± 0.02 [i]	1.64 ± 0.01 [b]	84.57 ± 0.02 [h]
13	0.55 ± 0.01 [h]	7.40 ± 0.00 [e]	14.70 ± 0.02 [g]	2.54 ± 0.00 [a]	65.26 ± 0.00 [j]	1.75 ± 0.01 [d]	0.31 ± 0.01 [h]	2.56 ± 0.00 [b]	3.21 ± 0.01 [o]	0.92 ± 0.00 [d]	0.38 ± 0.00 [h]	0.44 ± 0.02 [d,e]	14.41 ± 0.01 [b]	83.55 ± 0.01 [l]	2.06 ± 0.01 [c]	85.61 ± 0.01 [k]
14	0.82 ± 0.00 [k]	7.47 ± 0.00 [f]	14.27 ± 0.00 [e]	3.31 ± 0.00 [g]	64.23 ± 0.01 [e]	1.99 ± 0.00 [f]	0.29 ± 0.00 [g]	3.02 ± 0.01 [j]	2.84 ± 0.01 [l]	1.05 ± 0.00 [g]	0.31 ± 0.00 [g]	0.39 ± 0.00 [b]	16.06 ± 0.00 [i]	81.65 ± 0.01 [d]	2.28 ± 0.00 [d]	83.93 ± 0.00 [d]
15	0.42 ± 0.01 [c,d]	8.00 ± 0.02 [j]	17.63 ± 0.01 [m]	3.25 ± 0.00 [f]	62.51 ± 0.00 [b]	1.37 ± 0.00 [b]	0.36 ± 0.01 [i]	2.76 ± 0.01 [e,f]	2.31 ± 0.00 [d]	0.82 ± 0.00 [b]	0.22 ± 0.01 [b,c]	0.36 ± 0.01 [a]	15.61 ± 0.01 [f]	82.67 ± 0.01 [h]	1.73 ± 0.00 [b]	84.40 ± 0.00 [g]
Max	0.82 ± 0.00 [k]	8.29 ± 0.02 [l]	17.63 ± 0.01 [m]	4.08 ± 0.00 [o]	66.47 ± 0.01 [l]	2.53 ± 0.01 [j]	0.36 ± 0.01 [i]	3.17 ± 0.00 [k]	3.21 ± 0.01 [n]	1.05 ± 0.00 [g]	0.40 ± 0.03 [h]	0.44 ± 0.00 [e]	17.20 ± 0.01 [l]	83.57 ± 0.02 [l]	2.79 ± 0.00 [f]	85.90 ± 0.01 [l]
Min	0.30 ± 0.01 [a]	6.81 ± 0.01 [a]	13.22 ± 0.02 [a]	2.54 ± 0.00 [a]	61.74 ± 0.04 [a]	1.16 ± 0.00 [a]	0.20 ± 0.00 [a]	2.54 ± 0.00 [a]	2.22 ± 0.00 [a]	0.76 ± 0.02 [a]	0.18 ± 0.03 [a]	0.36 ± 0.00 [a]	14.10 ± 0.01 [a]	81.00 ± 0.00 [a]	1.45 ± 0.00 [a]	82.80 ± 0.00 [a]

[1] Values are means ± SD. Different letters in a column indicate significant differences ($p < 0.05$). * C14:0, myristic acid; C16:0, palmitic acid; C16:1, palmitoleic acid; C18:0, stearic acid; C18:1, oleic acid; C18:2, linoleic acid; C18:3, linolenic acid; C20:0, eicosanoic acid; C20:1, eicosenoic acid; C22:0, docosanoic acid; C22:1, docosenoic acid; C24:0, lignoceric acid. SFA, saturated fatty acids (C14:0 + C16:0 + C18:0 + C20:0 + C22:0 + C24:0); MUFA, mono-unsaturated fatty acids (C16:1 + C18:1 + C20:1 + C22:1); PUFA, poly-unsaturated fatty acids (C18:2+ C18:3); UFA, unsaturated fatty acids (MUFA + PUFA).

Table 3. Triacylglycerols [%(w/w)] of oils extracted from 15 macadamia cultivars [1].

No.	1	2	3	4	5	6	7	8	9	10	11	12	13	14	15
MOM *	0.25 ± 0.00 [f]	0.19 ± 0.00 [d,e]	0.19 ± 0.00 [d,e]	0.17 ± 0.00 [d]	0.20 ± 0.00 [e]	0.11 ± 0.00 [a]	0.13 ± 0.01 [b]	0.15 ± 0.00 [c]	0.10 ± 0.00 [a]	0.11 ± 0.00 [a]	0.20 ± 0.00 [e]	0.28 ± 0.01 [g]	0.19 ± 0.00 [d,e]	0.26 ± 0.00 [f]	0.17 ± 0.01 [c,d]
PPP$_O$	0.81 ± 0.02 [h]	0.51 ± 0.00 [b]	0.60 ± 0.00 [e]	0.61 ± 0.00 [e]	0.72 ± 0.01 [g]	0.52 ± 0.00 [b]	0.57 ± 0.00 [d]	0.60 ± 0.00 [e]	0.56 ± 0.01 [c,d]	0.47 ± 0.00 [a]	0.56 ± 0.00 [c]	0.63 ± 0.00 [f]	0.55 ± 0.00 [c]	0.70 ± 0.00 [g]	0.60 ± 0.00 [e]
MOP	2.26 ± 0.01 [k]	1.63 ± 0.00 [d]	1.87 ± 0.00 [g]	1.75 ± 0.00 [f]	2.16 ± 0.01 [j]	1.44 ± 0.00 [b]	1.59 ± 0.01 [c,d]	1.61 ± 0.01 [d]	1.59 ± 0.00 [c]	1.38 ± 0.01 [a]	1.65 ± 0.00 [d]	1.94 ± 0.01 [h]	1.70 ± 0.00 [e]	1.94 ± 0.00 [h]	1.98 ± 0.00 [i]
MLP	1.28 ± 0.01 [h]	0.96 ± 0.00 [c]	1.15 ± 0.00 [g]	1.07 ± 0.01 [e]	1.41 ± 0.00 [i]	0.88 ± 0.00 [a]	0.95 ± 0.00 [c]	0.88 ± 0.01 [a]	1.12 ± 0.00 [f]	0.92 ± 0.00 [b]	0.89 ± 0.00 [a]	0.99 ± 0.01 [d]	1.09 ± 0.00 [e]	0.95 ± 0.00 [c]	1.64 ± 0.00 [j]
PLP$_O$	2.18 ± 0.00 [h]	1.54 ± 0.02 [a]	1.84 ± 0.01 [d]	2.06 ± 0.00 [g]	2.28 ± 0.02 [i]	1.94 ± 0.02 [e]	2.03 ± 0.00 [f]	2.07 ± 0.02 [g]	2.23 ± 0.02 [i]	1.88 ± 0.02 [d]	1.61 ± 0.01 [b]	1.63 ± 0.01 [b]	1.69 ± 0.01 [c]	1.84 ± 0.01 [d]	1.95 ± 0.00 [e]
POP	8.96 ± 0.01 [l]	6.94 ± 0.02 [a]	8.00 ± 0.01 [i]	7.87 ± 0.00 [g]	8.89 ± 0.02 [k]	7.00 ± 0.01 [b]	7.77 ± 0.00 [f]	7.88 ± 0.02 [g]	7.91 ± 0.01 [g,h]	6.89 ± 0.01 [a]	7.01 ± 0.01 [b]	7.46 ± 0.00 [d]	7.15 ± 0.01 [c]	7.62 ± 0.01 [e]	8.07 ± 0.00 [j]
MOO	7.98 ± 0.01 [l]	6.35 ± 0.01 [d]	7.49 ± 0.00 [k]	6.96 ± 0.00 [i]	8.24 ± 0.01 [m]	6.08 ± 0.01 [a]	6.81 ± 0.00 [h]	6.68 ± 0.01 [g]	7.23 ± 0.01 [j]	6.23 ± 0.02 [c]	6.13 ± 0.00 [b]	6.61 ± 0.01 [f]	6.70 ± 0.01 [g]	6.24 ± 0.01 [c]	8.46 ± 0.00 [e]
PSS	1.50 ± 0.00 [f]	1.10 ± 0.00 [a,b]	1.33 ± 0.00 [c]	1.60 ± 0.00 [h]	1.85 ± 0.00 [j]	1.61 ± 0.00 [h]	1.55 ± 0.00 [g]	1.41 ± 0.00 [e]	1.59 ± 0.00 [h]	1.67 ± 0.00 [i]	1.12 ± 0.00 [b]	1.41 ± 0.00 [e]	1.08 ± 0.00 [a]	1.40 ± 0.00 [e]	1.36 ± 0.00 [d]
P$_O$OS	13.00 ± 0.01 [g]	12.12 ± 0.00 [b]	12.51 ± 0.01 [c]	13.11 ± 0.00 [h]	12.72 ± 0.01 [f]	13.22 ± 0.01 [j]	13.26 ± 0.00 [k]	13.19 ± 0.01 [i]	13.65 ± 0.01 [l]	12.71 ± 0.01 [f]	12.67 ± 0.00 [e]	11.87 ± 0.00 [a]	12.61 ± 0.00 [d]	12.54 ± 0.01 [c]	12.47 ± 0.00 [c]
P$_O$LS	2.38 ± 0.00 [h]	2.01 ± 0.00 [d]	2.34 ± 0.00 [g]	2.61 ± 0.00 [k]	2.87 ± 0.00 [l]	2.00 ± 0.00 [d]	2.45 ± 0.00 [i]	2.31 ± 0.00 [f]	2.32 ± 0.00 [f]	2.54 ± 0.00 [j]	1.54 ± 0.00 [a]	2.21 ± 0.00 [e]	1.60 ± 0.00 [b]	1.95 ± 0.00 [c]	2.21 ± 0.00 [e]
POO	18.00 ± 0.02 [i]	17.26 ± 0.00 [e]	18.19 ± 0.01 [j]	17.00 ± 0.00 [d]	17.55 ± 0.01 [g]	16.43 ± 0.01 [b]	17.37 ± 0.00 [f]	17.81 ± 0.01 [h]	16.91 ± 0.01 [c]	16.36 ± 0.01 [a]	17.36 ± 0.00 [f]	17.54 ± 0.01 [g]	17.27 ± 0.00 [e]	16.98 ± 0.01 [d]	18.01 ± 0.00 [i]
PLO	2.60 ± 0.00 [i]	2.29 ± 0.00 [d]	2.60 ± 0.00 [i]	2.48 ± 0.00 [g]	2.79 ± 0.00 [k]	2.10 ± 0.00 [b]	2.78 ± 0.00 [k]	2.43 ± 0.00 [e]	3.02 ± 0.00 [l]	2.45 ± 0.00 [f]	2.21 ± 0.00 [c]	2.45 ± 0.00 [f]	2.59 ± 0.00 [h]	1.90 ± 0.00 [a]	2.73 ± 0.00 [j]
PLL	0.99 ± 0.00 [h]	1.00 ± 0.00 [h,i]	0.61 ± 0.00 [a]	0.72 ± 0.00 [c]	0.66 ± 0.00 [b]	1.30 ± 0.00 [k]	0.75 ± 0.00 [d]	1.05 ± 0.00 [j]	0.95 ± 0.00 [f]	0.97 ± 0.00 [g]	1.01 ± 0.00 [i]	0.73 ± 0.00 [c]	0.91 ± 0.00 [e]	1.01 ± 0.00 [i]	0.76 ± 0.00 [d]
SOS	1.13 ± 0.00 [c]	1.12 ± 0.00 [c]	1.18 ± 0.00 [d]	1.35 ± 0.00 [h]	1.52 ± 0.00 [k]	1.59 ± 0.00 [m]	1.39 ± 0.00 [j]	1.28 ± 0.00 [g]	1.37 ± 0.00 [i]	1.57 ± 0.00 [l]	1.08 ± 0.00 [b]	1.25 ± 0.00 [f]	1.03 ± 0.00 [a]	1.36 ± 0.00 [h,i]	1.20 ± 0.00 [e]
SOO	5.25 ± 0.00 [c]	5.06 ± 0.00 [b]	5.71 ± 0.01 [h]	6.35 ± 0.00 [l]	6.13 ± 0.00 [j]	6.93 ± 0.00 [m]	6.24 ± 0.00 [k]	5.68 ± 0.00 [g]	5.71 ± 0.00 [h]	6.98 ± 0.00 [n]	5.56 ± 0.00 [e]	6.04 ± 0.00 [i]	4.81 ± 0.00 [a]	5.64 ± 0.00 [f]	5.31 ± 0.00 [d]
OOO	21.30 ± 0.02 [b]	26.14 ± 0.00 [m]	22.90 ± 0.02 [e]	23.28 ± 0.00 [g]	19.18 ± 0.01 [a]	23.97 ± 0.02 [h]	23.10 ± 0.00 [f]	22.86 ± 0.02 [e]	21.62 ± 0.01 [c]	24.12 ± 0.01 [i]	25.22 ± 0.00 [k]	24.61 ± 0.01 [j]	25.75 ± 0.00 [l]	24.16 ± 0.02 [i]	21.65 ± 0.00 [d]
SLO	3.24 ± 0.00 [a]	4.27 ± 0.00 [l]	4.14 ± 0.00 [k]	3.65 ± 0.00 [e]	3.93 ± 0.00 [i]	3.40 ± 0.00 [b]	3.68 ± 0.00 [f]	3.68 ± 0.00 [f]	4.29 ± 0.00 [m]	3.55 ± 0.00 [c]	4.15 ± 0.00 [k]	4.05 ± 0.00 [j]	3.62 ± 0.00 [d]	3.75 ± 0.00 [g]	3.88 ± 0.00 [h]
OLO	1.42 ± 0.00 [g]	1.93 ± 0.00 [l]	0.89 ± 0.00 [b]	1.20 ± 0.00 [e]	0.77 ± 0.00 [a]	2.10 ± 0.00 [m]	1.09 ± 0.00 [c]	1.66 ± 0.00 [i]	1.51 ± 0.00 [h]	1.94 ± 0.00 [l]	1.87 ± 0.00 [k]	1.22 ± 0.00 [f]	1.65 ± 0.00 [i]	1.76 ± 0.00 [j]	1.13 ± 0.00 [d]
SOA	2.89 ± 0.00 [a]	3.67 ± 0.02 [g]	3.49 ± 0.02 [c]	3.56 ± 0.01 [e]	3.54 ± 0.00 [e]	3.92 ± 0.01 [j]	3.63 ± 0.00 [f]	3.44 ± 0.00 [b]	3.52 ± 0.00 [c,d]	4.04 ± 0.00 [k]	3.86 ± 0.02 [i]	3.85 ± 0.00 [i]	3.71 ± 0.01 [h]	4.12 ± 0.01 [l]	3.45 ± 0.00 [b]
AOO	2.29 ± 0.00 [a]	3.67 ± 0.03 [l]	2.78 ± 0.02 [f]	2.39 ± 0.00 [b]	2.30 ± 0.00 [a]	3.15 ± 0.01 [j]	2.66 ± 0.00 [d]	3.09 ± 0.00 [i]	2.59 ± 0.00 [c]	3.00 ± 0.00 [h]	4.05 ± 0.02 [m]	2.97 ± 0.00 [g]	4.08 ± 0.01 [m]	3.54 ± 0.01 [k]	2.75 ± 0.00 [e]

[1] Values are means ± SD. Different letters in a column indicate significant differences ($p < 0.05$). * O: Oleic acid; P: Palmitic acid; L: Linoleic acid; S: Stearic acid; P$_O$: Palmitoleic acid; A: Arachidic acid; M: Myristic acid.

Table 4. The minor components content (mg/kg of oil) and antioxidant capacity (μmol Vitamin E /kg) of oils extracted from 15 macadamia cultivars [1].

No.	Antioxidant Capacity			Polyphenols	α-Tocotrienols	Squalene	Minerals			
	DPPH *	ABTS	FRAP				Mg	Ca	K	Na
1	126.64 ± 9.76 [a]	255.17 ± 5.40 [a]	136.26 ± 0.92 [a]	19.74 ± 1.83 [a]	52.3 ± 1.11 [k]	91.15 ± 1.38 [a]	12.21 ± 0.51 [c]	N.D. [a]	22.28 ± 0.20 [d]	1.20 ± 0.01 [h]
2	359.36 ± 5.67 [g]	1086.94 ± 18.05 [i,j]	324.72 ± 3.42 [k]	123.40 ± 1.67 [k]	46.6 ± 0.26 [i]	268.08 ± 1.52 [n]	17.05 ± 0.08 [f]	N.D. [a]	19.42 ± 0.16 [d]	0.75 ± 0.02 [d]
3	303.80 ± 7.51 [f]	1036.79 ± 33.73 [i]	246.63 ± 0.69 [h]	77.17 ± 2.67 [h]	33.5 ± 0.25 [d]	252.07 ± 0.59 [l]	15.54 ± 0.65 [e]	1.35 ± 0.01 [c]	17.14 ± 0.46 [b]	0.86 ± 0.01 [e]
4	247.50 ± 9.76 [d,e]	629.39 ± 21.88 [f]	207.76 ± 2.28 [f]	52.64 ± 1.33 [e]	46.9 ± 0.26 [i]	232.13 ± 0.66 [i]	8.84 ± 0.09 [b]	N.D. [a]	18.58 ± 0.33 [c]	0.82 ± 0.02 [e]
5	307.56 ± 1.30 [f]	1083.09 ± 28.29 [i,j]	270.58 ± 1.39 [j]	96.16 ± 3.50 [j]	30.0 ± 0.72 [b]	264.07 ± 0.79 [m]	14.00 ± 0.18 [d]	1.21 ± 0.00 [b]	24.81 ± 1.46 [e]	0.61 ± 0.02 [c]
6	179.19 ± 19.42 [b,c]	438.81 ± 9.85 [d]	160.56 ± 1.59 [c]	50.17 ± 1.50 [d,e]	33.7 ± 0.17 [d]	163.89 ± 0.64 [d]	14.42 ± 0.19 [d]	1.32 ± 0.01 [c]	17.88 ± 0.45 [b]	1.05 ± 0.06 [f]
7	171.68 ± 14.02 [b]	339.28 ± 19.01 [c]	158.13 ± 3.31 [c]	40.97 ± 0.50 [c]	35.6 ± 0.00 [e]	182.79 ± 0.95 [e]	8.13 ± 0.11 [a]	N.D. [a]	13.75 ± 0.44 [a]	0.55 ± 0.02 [c]
8	155.91 ± 8.36 [b]	326.16 ± 2.31 [c]	156.74 ± 2.11 [c]	24.22 ± 1.83 [b]	27.9 ± 0.29 [a]	158.62 ± 0.64 [c]	18.94 ± 0.26 [g]	1.44 ± 0.02 [d]	23.77 ± 0.83 [e]	0.55 ± 0.01 [c]
9	191.20 ± 26.72 [b,c]	520.60 ± 19.69 [e]	177.56 ± 0.35 [d]	40.97 ± 1.83 [c]	53.1 ± 1.21 [k]	189.79 ± 0.68 [f]	22.25 ± 0.45 [h]	1.55 ± 0.01 [e]	23.41 ± 0.40 [e]	0.41 ± 0.02 [a]
10	225.73 ± 16.92 [d]	618.59 ± 13.18 [f]	183.11 ± 2.11 [e]	50.40 ± 0.17 [d]	45.1 ± 0.87 [h]	229.22 ± 0.74 [h]	15.65 ± 0.47 [e]	N.D. [a]	21.38 ± 0.88 [d]	0.51 ± 0.01 [b]
11	221.23 ± 13.28 [d]	579.24 ± 34.13 [a]	178.26 ± 1.59 [d,e]	50.28 ± 0.67 [d,e]	32.2 ± 0.52 [c]	208.77 ± 0.39 [g]	49.90 ± 0.70 [l]	28.50 ± 0.88 [i]	48.30 ± 2.33 [i]	1.01 ± 0.03 [f]
12	259.51 ± 6.00 [e]	669.52 ± 1.54 [g]	227.89 ± 4.59 [g]	55.59 ± 1.17 [f]	40.9 ± 0.21 [g]	240.16 ± 0.69 [j]	46.10 ± 0.14 [k]	21.00 ± 0.56 [h]	51.80 ± 0.90 [i]	2.51 ± 0.01 [k]
13	300.80 ± 16.90 [f]	819.20 ± 14.88 [h]	234.48 ± 6.10 [g]	64.43 ± 0.67 [g]	38.3 ± 0.51 [f]	244.71 ± 0.66 [k]	34.40 ± 0.16 [i]	17.70 ± 0.22 [f]	34.90 ± 0.17 [g]	1.85 ± 0.00 [i]
14	153.66 ± 22.91 [a,b]	280.64 ± 26.77 [a,b]	143.55 ± 3.08 [b]	19.74 ± 1.83 [a]	52.3 ± 1.11 [k]	91.15 ± 1.38 [a]	38.10 ± 0.29 [j]	18.70 ± 0.14 [g]	43.90 ± 0.17 [h]	1.17 ± 0.00 [g]
15	306.81 ± 10.90 [f]	1039.11 ± 8.13 [i]	263.64 ± 3.66 [i]	123.40 ± 1.67 [k]	46.6 ± 0.26 [i]	268.08 ± 1.52 [n]	33.40 ± 1.14 [i]	18.20 ± 0.21 [g]	32.30 ± 1.54 [f]	1.59 ± 0.02 [i]
Max	359.36 ± 5.67 [g]	1086.94 ± 18.05 [i,j]	324.72 ± 3.42 [k]	123.40 ± 1.67 [k]	53.1 ± 1.21 [k]	268.08 ± 1.52 [n]	49.90 ± 0.70 [l]	28.50 ± 0.88 [i]	51.80 ± 0.90 [i]	2.51 ± 0.01 [k]
Min	126.64 ± 9.76 [a]	255.17 ± 5.40 [a]	136.26 ± 0.92 [a]	19.74 ± 1.83 [a]	27.9 ± 0.29 [a]	91.15 ± 1.38 [a]	8.13 ± 0.11 [a]	N.D. [a]	13.75 ± 0.44 [a]	0.41 ± 0.02 [a]

[1] Values are means ± SD. Different letters in a column indicate significant differences ($p < 0.05$). N.D., not detected. * DPPH: 2,2-diphenyl-1-picrylhydrazyl radical scavenging ability. ABTS: 2,2'-azino-bis (3-ethylbenzothiazoline-6-sulfonic acid) diammonium salt radical scavenging ability. FRAP: ferric reducing ability of plasma.

Table 5. Phytosterols content (mg/kg of oil) of oils extracted from 15 macadamia cultivars [1].

No.	Campesterol	Clerosterol	β-Sitosterol	β-Stiostanol	Δ5-Avenasterol	Δ5,24-Stigmastadienol	Δ7-Stigmastenol	Δ7-Avenasterol	Total Phytosterols
1	88.36 ± 1.13 c	15.50 ± 0.77 c,d	1533.93 ± 20.45 f	8.66 ± 0.49 e	291.75 ± 2.80 j	12.27 ± 0.74 d	30.71 ± 0.25 f	25.59 ± 0.25 g	2006.77 ± 19.65 d
2	103.45 ± 1.85 g	12.88 ± 0.60 a,b	1248.78 ± 23.80 a	9.59 ± 0.83 e	170.01 ± 0.65 a	9.39 ± 0.54 b	7.82 ± 0.07 a	7.58 ± 0.10 a	1569.50 ± 25.50 a
3	88.58 ± 0.62 c	11.69 ± 0.50 a	1354.52 ± 15.65 b	7.29 ± 0.09 c,d	263.07 ± 0.84 g	7.96 ± 0.21 a	18.09 ± 0.38 c,d	16.19 ± 0.16 c	1767.39 ± 15.69 a,b
4	82.36 ± 0.84 a,b	16.26 ± 0.13 d	1613.7 ± 32.24 g	6.94 ± 0.15 c	250.74 ± 1.40 e	10.96 ± 0.38 c,d	18.60 ± 0.27 d	15.30 ± 0.27 b	2014.86 ± 34.65 d
5	81.45 ± 0.78 a	11.41 ± 0.31 a	1357.07 ± 35.69 b	6.22 ± 0.02 b	343.78 ± 2.11 l	10.17 ± 0.70 b,c	34.95 ± 0.05 g	25.64 ± 0.54 g	1870.69 ± 36.51 b,c
6	114.13 ± 1.39 i	14.39 ± 0.52 b,c	1422.23 ± 17.78 c,d	8.84 ± 0.65 e	338.72 ± 1.20 k	14.61 ± 0.21 e	35.93 ± 0.14 h	32.11 ± 0.24 h	2014.28 ± 15.60 d
7	93.57 ± 0.05 d	13.85 ± 0.23 b	1470.71 ± 17.61 d,e	7.13 ± 0.32 c	191.22 ± 0.41 c	10.63 ± 0.76 c,d	30.73 ± 0.25 f	19.54 ± 0.05 e	1837.38 ± 17.64 b
8	97.75 ± 0.78 f	14.33 ± 0.34 b,c	1441.08 ± 10.47 d	7.74 ± 0.28 c	266.99 ± 3.81 g,h	12.35 ± 0.55 d	17.96 ± 0.06 c	19.13 ± 0.09 d	1877.33 ± 14.87 b,c
9	95.86 ± 0.15 e	13.48 ± 0.66 b	1443.81 ± 21.86 d	6.97 ± 0.27 c	362.08 ± 1.79 g	19.79 ± 0.80 h	47.95 ± 0.19 j	41.34 ± 0.49 i	2031.28 ± 22.79 d
10	90.98 ± 1.42 c	13.49 ± 0.60 b	1510.38 ± 14.65 f	8.34 ± 0.11 e	333.57 ± 2.80 d	18.38 ± 0.36 g	48.57 ± 0.61 j	40.83 ± 0.08 i	2064.54 ± 15.06 d
11	107.41 ± 0.95 h	13.39 ± 0.24 b	1441.48 ± 16.65 d	7.74 ± 0.12 d	257.25 ± 2.37 f	15.92 ± 0.35 f	47.42 ± 0.29 j	31.40 ± 0.35 h	1922.02 ± 16.98 c,d
12	93.72 ± 0.21 d	11.71 ± 0.42 a	1329.66 ± 15.40 b	7.62 ± 0.20 d	178.90 ± 0.65 b	9.37 ± 0.48 b	30.49 ± 0.73 f	20.69 ± 0.87 f	1682.16 ± 15.02 a
13	91.40 ± 1.96 c,d	12.63 ± 0.53 ab	1396.40 ± 6.51 c	7.72 ± 0.05 d	280.42 ± 2.73 i	9.02 ± 0.19 b	16.55 ± 0.12 b	16.72 ± 0.43 c	1830.86 ± 6.34 b
14	92.06 ± 0.44 d	13.10 ± 0.61 b	1324.06 ± 9.15 b	7.22 ± 0.08 c	248.04 ± 1.01 e	7.84 ± 0.32 a	24.45 ± 0.40 e	21.12 ± 0.65 f	1737.89 ± 11.44 a,b
15	83.90 ± 0.64 b	12.41 ± 0.40 a	1348.70 ± 28.15 b	5.39 ± 0.27 a	272.42 ± 1.17 h	12.82 ± 1.70 d	39.17 ± 0.26 i	31.83 ± 0.11 h	1806.64 ± 32.35 b
Max	114.13 ± 1.39 i	16.26 ± 0.13 d	1613.7 ± 32.24 g	9.59 ± 0.83 e	362.08 ± 1.79 g	18.38 ± 0.36 g	48.57 ± 0.61 j	41.34 ± 0.49 i	2064.54 ± 15.06 d
Min	81.45 ± 0.78 a	11.41 ± 0.31 a	1248.78 ± 23.80 a	5.39 ± 0.27 a	170.01 ± 0.65 a	7.84 ± 0.32 a	7.82 ± 0.07 a	7.58 ± 0.10 a	1569.50 ± 25.50 a

[1] Values are means ± SD. Different letters in a column indicate significant differences ($p < 0.05$).

Table 6. Correlations between minor components and antioxidant capacity.

	Polyphenols	Tocotrienols	Squalence	Campesterol	Clerosterol	β-Sitosterol	β-Stiostanol	Δ5-Avenasterol	Δ5,24-Stigmastadienol	Δ7-Stigmastenol	Δ7-Avenasterol
DPPH	0.939 a	−0.111	0.927 a	−0.225	−0.605 b	−0.525 b	−0.144	−0.219	−0.327	−0.307	−0.348
FRAP	0.965 a	−0.060	0.870 a	−0.217	−0.584 b	−0.589 b	−0.106	−0.266	−0.354	−0.360	−0.397
ABTS	0.938 a	−0.142	0.899 a	−0.287	−0.647 a	−0.519 b	−0.242	−0.063	−0.260	−0.198	−0.226

[a] Correlation is significant at the 0.01 level (2-tailed). [b] Correlation is significant at the 0.05 level (2-tailed).

Table 7. Equation, variable, and regression coefficient in the prediction of the antioxidant capacity by MLR.

Dependent Variable	R^2	Adjusted R^2	Variable	R	Standard Error	t	Significance (Two Tails p) [a]	Equation
DPPH	0.946	0.937	(Constant)	35.671	20.695	1.724	0.110	$Y = 35.671 + 0.546$ (polyphenols) $+ 0.467$ (squalene)
			polyphenols	0.546	0.306	4.374	0.001	
			squalene	0.467	0.158	3.741	0.003	
ABTS	0.920	0.907	(Constant)	−142.676	106.934	−1.334	0.207	$Y = -142.676 + 0.621$ (polyphenols) $+ 0.376$ (squalene)
			polyphenols	0.621	1.583	4.111	0.001	
			squalene	0.376	0.816	2.491	0.028	
FRAP	0.932	0.927	(Constant)	99.412	8.769	11.337	0.000	$Y = 99.412 + 0.965$ (polyphenols)
			polyphenols	0.965	0.138	13.346	0.000	

[a] $p < 0.05$, significant regression.

4. Conclusions

The chemical compositions and antioxidant capacities of oils obtained from 15 macadamia cultivars grown in China were analyzed. Generally, the macadamia kernel is rich in oils (average of 75.98%). The major fatty acids in MOs were monounsaturated fatty acids, mainly including oleic acid and palmitoleic acid. The main triacylglycerols were OOO, POO and P_OOS, which was first time reported for MO. Furthermore, MO contained high minor components including α-tocotrienols, phytosterols, squalene, and polyphenols. Among all the cultivars, Fuji (791) (2) contained the highest polyphenols (123.4 mg/kg), squalene (268.1 mg/kg) and exhibited the highest antioxidant capacity. As correlation analysis indicated, polyphenols and squalene content positively correlated with the antioxidant capacity of MOs. In addition, the antioxidant capacity can be predicted by the model of MLR. The above results indicated that MO has potential as a healthy vegetable oil and offer a valuable directive for the breeding of macadamia in China.

Author Contributions: Conceptualization, X.S., T.D., L.D., J.C. and C.L.; methodology, X.S. and T.D.; data curation, X.S. and T.D.; writing—original draft preparation, X.S., T.D. and J.C.; writing—review and editing, X.S., T.D. and J.C.; visualization, M.C. and R.L.; supervision, X.S. and C.L. All authors have read and agreed to the published version of the manuscript.

Funding: This work was supported by China Postdoctoral Science Foundation [Grant No. 2020M6832211], Postdoctoral Foundation of Guangxi Academy of Agricultural Sciences [Grant No. 2020037] and Natural Science Foundation of Hainan province [Grant No.219QN286].

Data Availability Statement: Data is contained within the article.

Conflicts of Interest: The authors declare no conflict of interest.

References

1. Aquino-Bolanos, E.N.; Mapel-Velazco, L.; Martin-del-Campo, S.T.; Chavez-Servia, J.L.; Martinez, A.J.; Verdalet-Guzman, I. Fatty acids profile of oil from nine varieties of Macadamia nut. *Int. J. Food Prop.* **2017**, *20*, 1262–1269. [CrossRef]
2. Dos Santos Penoni, E.; Pio, R.; Rodrigues, F.A.; Castilho Maro, L.A.; Costa, F.C. Analysis of fruits and nuts of macadamia walnut cultivars. *Cienc. Rural* **2011**, *41*, 2080–2083. [CrossRef]
3. Hu, W.; Fitzgerald, M.; Topp, B.; Alam, M.; O'Hare, T.J. A review of biological functions, health benefits, and possible de novo biosynthetic pathway of palmitoleic acid in macadamia nuts. *J. Funct. Foods* **2019**, *62*, 103520. [CrossRef]
4. Navarro, S.L.B.; Rodrigues, C.E.C. Macadamia oil extraction methods and uses for the defatted meal byproduct. *Trends Food Sci. Technol.* **2016**, *54*, 148–154. [CrossRef]
5. Castilho Maro, L.A.; Pio, R.; Penoni, E.d.S.; de Oliveira, M.C.; Prates, F.C.; de Oliveira Lima, L.C.; Cardoso, M.d.G. Chemical characterization and fatty acids profile in macadamia walnut cultivars. *Cienc. Rural* **2012**, *42*, 2166–2171. [CrossRef]
6. Entelman, F.A.; Scarpare Filho, J.A.; Pio, R.; da Silva, S.R.; Machado de Souza, F.B. Production and quality attributes of macadamia tree cultivars in the southwest of Sao Paulo State. *Rev. Bras. Frutic.* **2014**, *36*, 192–198. [CrossRef]
7. Kaijser, A.; Dutta, P.; Savage, G. Oxidative stability and lipid composition of macadamia nuts grown in New Zealand. *Food Chem.* **2000**, *71*, 67–70. [CrossRef]
8. Mereles, L.G.; Ferro, E.A.; Alvarenga, N.L.; Caballero, S.B.; Wiszovaty, L.N.; Piris, P.A.; Michajluk, B.J. Chemical composition of *Macadamia integrifolia* (Maiden and Betche) nuts from Paraguay. *Int. Food Res. J.* **2017**, *24*, 2599–2608.
9. Gong, Y.; Pegg, R.B.; Carr, E.C.; Parrish, D.R.; Kellett, M.E.; Kerrihard, A.L. Chemical and nutritive characteristics of tree nut oils available in the US market. *Eur. J. Lipid Sci. Technol.* **2017**, *119*, 1600520. [CrossRef]
10. Wall, M.M. Functional lipid characteristics, oxidative stability, and antioxidant activity of macadamia nut (*Macadamia integrifolia*) cultivars. *Food Chem.* **2010**, *121*, 1103–1108. [CrossRef]
11. Ma, S.; Guo, G.; Huang, K.; Hu, X.; Fu, J.; Xu, R.; Li, Z.; Zou, J. Amino acid compositions and antibacterial activities of different molecular weight macadamia nut polypeptides. *Sci. Technol. Food Ind.* **2020**, 1–9. [CrossRef]
12. AOAC. *Official Methods of Analysis of AOAC International*; Association of Official Analytical Chemists: Gaithersburg, MA, USA, 2005.
13. Li, Q.; Yin, R.; Zhang, Q.R.; Wang, X.P.; Hu, X.J.; Gao, Z.D.; Duan, Z.M. Chemometrics analysis on the content of fatty acid compositions in different walnut (*Juglans regia* L.) varieties. *Eur. Food Res. Technol.* **2017**, *243*, 2235–2242. [CrossRef]
14. AOCS. *Official Methods and Recommended Practices of the American Oil Chemist's Society*; Ce 5-86; AOCS Press: Champaign, IL, USA, 2009.
15. Gao, P.; Liu, R.; Jin, Q.; Wang, X. Comparison of solvents for extraction of walnut oils: Lipid yield, lipid compositions, minor-component content, and antioxidant capacity. *LWT Food Sci. Technol.* **2019**, *110*, 346–352. [CrossRef]
16. Policarpi, P.D.B.; Turcatto, L.; Demoliner, F.; Ferrari, R.A.; Azzolin Frescura Bascunan, V.L.; Ramos, J.C.; Jachmanian, I.; Vitali, L.; Micke, G.A.; Block, J.M. Nutritional potential, chemical profile and antioxidant activity of Chicha (*Sterculia striata*) nuts and its by-products. *Food Res. Int.* **2018**, *106*, 736–744. [CrossRef]

17. Gao, P.; Liu, R.; Jin, Q.; Wang, X. Comparative study of chemical compositions and antioxidant capacities of oils obtained from two species of walnut: *Juglans regia* and *Juglans sigillata*. *Food Chem.* **2019**, *279*, 279–287. [CrossRef]

18. Gao, P.; Liu, R.; Jin, Q.; Wang, X. Effects of processing methods on the chemical composition and antioxidant capacity of walnut (*Juglans regia* L.) oil. *LWT Food Sci. Technol.* **2021**, *135*, 109958. [CrossRef]

19. Venkatachalam, M.; Sathe, S.K. Chemical composition of selected edible nut seeds. *J. Agric. Food Chem.* **2006**, *54*, 4705–4714. [CrossRef] [PubMed]

20. Maguire, L.S.; O'Sullivan, S.M.; Galvin, K.; O'Connor, T.P.; O'Brien, N.M. Fatty acid profile, tocopherol, squalene and phytosterol content of walnuts, almonds, peanuts, hazelnuts and the macadamia nut. *Int. J. Food Sci. Nutr.* **2004**, *55*, 171–178. [CrossRef]

21. Naveed Akhtar, M.A. Asadullah Madni, 1Malik Sattar Bakhsh, Evaution of basic properties of Macsdamia nut oil. *Gomal Univ. J. Res.* **2006**, *22*, 21–27.

22. Guan, M.; Chen, H.; Xiong, X.; Lu, X.; Li, X.; Huang, F.; Guan, C. A Study on Triacylglycerol Composition and the Structure of High-Oleic Rapeseed Oil. *Engineering* **2016**, *2*, 258–262. [CrossRef]

23. Kris-Etherton, P.M.; Yu-Poth, S.; Sabate, J.; Ratcliffe, H.E.; Zhao, G.X.; Etherton, T.D. Nuts and their bioactive constituents: Effects on serum lipids and other factors that affect disease risk. *Am. J. Clin. Nutr.* **1999**, *70*, 504S–511S. [CrossRef]

24. Christopoulou, E.; Lazaraki, M.; Komaitis, M.; Kaselimis, K. Effectiveness of determinations of fatty acids and triglycerides for the detection of adulteration of olive oils with vegetable oils. *Food Chem.* **2004**, *84*, 463–474. [CrossRef]

25. FAOSTAT. New Food Balances. 2014. Available online: http://www.fao.org/faostat/en/#data/FBS: (accessed on 30 March 2021).

26. Chew, S.C. Cold-pressed rapeseed (*Brassica napus*) oil: Chemistry and functionality. *Food Res. Int.* **2020**, *131*, 108997. [CrossRef]

27. Yang, R.; Zhang, L.; Li, P.; Yu, L.; Mao, J.; Wang, X.; Zhang, Q. A review of chemical composition and nutritional properties of minor vegetable oils in China. *Trends Food Sci. Technol.* **2018**, *74*, 26–32. [CrossRef]

28. Van Hoed, V.; Ben Ali, C.; Slah, M.; Verhe, R. Quality differences between pre-pressed and solvent extracted rapeseed oil. *Eur. J. Lipid Sci. Technol.* **2010**, *112*, 1241–1247. [CrossRef]

29. Quinn, L.A.; Tang, H.H. Antioxidant properties of phenolic compounds in macadamia nuts. *J. Am. Oil Chem. Soc.* **1996**, *73*, 1585–1588. [CrossRef]

30. Cicero, N.; Albergamo, A.; Salvo, A.; Bua, G.D.; Bartolomeo, G.; Mangano, V.; Rotondo, A.; Di Stefano, V.; Di Bella, G.; Dugo, G. Chemical characterization of a variety of cold-pressed gourmet oils available on the Brazilian market. *Food Res. Int.* **2018**, *109*, 517–525. [CrossRef]

31. Gruszka, J.; Kruk, J. RP-LC for determination of plastochromanol, tocotrienols and tocopherols in plant oils. *Chromatographia* **2007**, *66*, 909–913. [CrossRef]

32. Sen, C.K.; Khanna, S.; Roy, S. Tocotrienols in health and disease: The other half of the natural vitamin E family. *Mol. Asp. Med.* **2007**, *28*, 692–728. [CrossRef] [PubMed]

33. Shimizu, N.; Ito, J.; Kato, S.; Eitsuka, T.; Miyazawa, T.; Nakagawa, K. Significance of squalene in rice bran oil and perspectives on squalene oxidation. *J. Nutr. Sci. Vitaminol.* **2019**, *65*, S62–S66. [CrossRef]

34. Sotiroudis, T.G.; Kyrtopoulos, S.A. Anticarcinogenic compounds of olive oil and related biomarkers. *Eur. J. Nutr.* **2008**, *47*, 69–72. [CrossRef]

35. Gao, P.; Jin, J.; Liu, R.; Jin, Q.; Wang, X. Chemical Compositions of Walnut (*Juglans regia* L.) Oils from Different Cultivated Regions in China. *J. Am. Oil Chem. Soc.* **2018**, *95*, 825–834. [CrossRef]

36. Maestri, D.; Cittadini, M.C.; Bodoira, R.; Martinez, M. Tree nut oils: Chemical profiles, extraction, stability, and quality concerns. *Eur. J. Lipid Sci. Technol.* **2020**, *122*, 1900450. [CrossRef]

37. Demoliner, F.; de Britto Policarpi, P.; Ramos, J.C.; Bascunan, V.; Ferrari, R.A.; Jachmanian, I.; de Francisco de Casas, A.; Vasconcelos, L.F.L.; Block, J.M. Sapucaia nut (*Lecythis pisonis Cambess*) and its by-products: A promising and underutilized source of bioactive compounds. Part I: Nutritional composition and lipid profile. *Food Res. Int.* **2018**, *108*, 27–34. [CrossRef]

38. Aslanabadi, N.; Habibi Asl, B.; Bakhshalizadeh, B.; Ghaderi, F.; Nemati, M. Hypolipidemic activity of a natural mineral water rich in calcium, magnesium, and bicarbonate in hyperlipidemic adults. *Adv. Pharm. Bull.* **2014**, *4*, 303–307. [CrossRef]

39. Gao, P.; Liu, R.; Jin, Q.; Wang, X. Comparison of different frocessing methods of iron walnut oils (*Juglans sigillata*): Lipid yield, lipid compositions, minor components, and antioxidant capacity. *Eur. J. Lipid Sci. Technol.* **2018**, *120*, 1800151. [CrossRef]

40. Lopez-Alarcon, C.; Denicola, A. Evaluating the antioxidant capacity of natural products: A review on chemical and cellular-based assays. *Anal. Chim. Acta* **2013**, *763*, 1–10. [CrossRef]

41. Shi, L.-K.; Mao, J.-H.; Zheng, L.; Zhao, C.-W.; Jin, Q.-Z.; Wang, X.-G. Chemical characterization and free radical scavenging capacity of oils obtained from *Torreya grandis* Fort. ex. Lindl. and *Torreya grandis* Fort. var. Merrillii: A comparative study using chemometrics. *Ind. Crop. Prod.* **2018**, *115*, 250–260. [CrossRef]

Review

Recent Advances for the Developing of Instant Flavor Peanut Powder: Generation and Challenges

Yue Liu, Hui Hu, Hongzhi Liu * and Qiang Wang *

Institute of Food Science and Technology, Chinese Academy of Agricultural Sciences, Key Laboratory of Agro-Products Processing, Ministry of Agriculture, Beijing 100193, China; 15552806102@163.com (Y.L.); huhui@caas.cn (H.H.)
* Correspondence: lhz0416@126.com (H.L.); wangqiang06@caas.cn (Q.W.); Tel.: +86-(10)-62818455 (H.L.); +86-(10)-62815837 (Q.W.)

Abstract: Instant flavor peanut powder is a nutritional additive that can be added to foods to impart nutritional value and functional properties. Sensory acceptability is the premise of its development. Flavor is the most critical factor in sensory evaluation. The heat treatment involved in peanut processing is the main way to produce flavor substances and involves chemical reactions: Maillard reaction, caramelization reaction, and lipid oxidation reaction. Peanut is rich in protein, fat, amino acids, fatty acids, and unsaturated fatty acids, which participate in these reactions as volatile precursors. N-heterocyclic compounds, such as the pyrazine, are considered to be the key odorants of the "baking aroma". However, heat treatment also affects the functional properties of peanut protein (especially solubility) and changes the nutritional value of the final product. In contrast, functional properties affect the behavior of proteins during processing and storage. Peanut protein modification is the current research hotspot in the field of deep processing of plant protein, which is an effective method to solve the protein denaturation caused by heat treatment. The review briefly describes the characterization and mechanism of peanut flavor during heat treatment combined with solubilization modification technology, proposing the possibility of using peanut meal as material to produce IFPP.

Keywords: instant flavor peanut powder; heat treatment; flavor; MR; functional properties; peanut meal

Citation: Liu, Y.; Hu, H.; Liu, H.; Wang, Q. Recent Advances for the Developing of Instant Flavor Peanut Powder: Generation and Challenges. *Foods* **2022**, *11*, 1544. https://doi.org/10.3390/foods11111544

Received: 31 March 2022
Accepted: 11 May 2022
Published: 24 May 2022

Publisher's Note: MDPI stays neutral with regard to jurisdictional claims in published maps and institutional affiliations.

Copyright: © 2022 by the authors. Licensee MDPI, Basel, Switzerland. This article is an open access article distributed under the terms and conditions of the Creative Commons Attribution (CC BY) license (https://creativecommons.org/licenses/by/4.0/).

1. Introduction

Oil extraction is one of the main uses of peanuts in China [1]. The peanut residue produced after pressing through an oil press is also known as peanut meal [2]. Peanut meal is the main by-product of oil extraction, and about 125,000 tons of peanut meal can be obtained each year [1,3].

The protein content of peanut meal is about 55% along with some lipids, carbohydrates, and other nutrient content, such as vitamins A, B, E, K, triterpenoids, sterols, thiamin, riboflavin, and niacin, making it an ideal source of high-quality protein [4]. Fatty acids, triterpenoids, and sterols have been shown to have potential in the fight against diabetes, Alzheimer's disease, and cancer [5]. Compared with the common protein powder raw material of soy protein, peanut protein is easier for human digestion and absorption (more than 90%), has higher nutritional potency (about equal to 59%), and has no unpleasant beany taste [6]. Additionally, peanut protein has desirable functional properties, such as mixing ability, viscosity, emulsifying ability, and water- and oil-holding capacity [7–13]. When added to food, peanut protein plays a decisive role in nutritional, organoleptic, physicochemical, and sensory properties (color, texture, flavor) [14–18]. Instant flavor peanut powder (IFPP) has low processing costs. While extracting peanut oil, we can obtain high-quality peanut protein, which effectively increases the added value of the peanut product [19,20].

The heat-treatment technology is an essential part of the peanut industry for the production and manufacture of peanut oil [6], which imparts peanut oil and meal's distinctive "baking aroma" [21]. The heat treatment of tempering, pressing, crushing, and sterilization in the traditional oil-extraction process can cause a series of complex thermochemical reactions [22], which would cause certain effects on the sensory quality of IFPP. Tempering is an effective means of imparting the typical flavor of peanut oil and IFPP. Roasting, frying, and steaming are effective methods of peanut tempering processes. During the heating process, a series of complex chemical reactions take place, including Maillard reaction (MR), caramelization reaction (CR), lipid oxidation reaction (LOR), and protein thermal degradation [23]. The abundant protein, free amino acids, and unsaturated fatty acids in peanuts become precursors for these reaction. A large number of N, O, and S heterocyclic compounds, such as pyrazine, pyrrole, pyridine, and furan derivatives, and non-heterocyclic substances, such as aldehydes and ketones, are produced during the reaction [24,25], and the interaction between these flavor substances directly affects the sensory evaluation index of the products [20], which is the root cause of the unique "baking aroma" of roasted peanut, peanut oil, peanut meal, and other peanut products. Furthermore, the heat treatments involved in the processing affect the nutritional value and functional properties of the final product [26]. At present, the traditional peanut oil extraction mostly uses a high-temperature pressing process (110–160 °C), which is shown in Figure 1 [27]. While imparting aroma to peanut oil and peanut meal, excessive temperature also affects the spatial structure and degree of polymerization of peanut proteins, resulting in denaturation of peanut proteins, specifically in terms of reduced nutritional value and functional properties (especially solubility) [18], which greatly limits the wide application of peanut proteins in food processing [8,28].

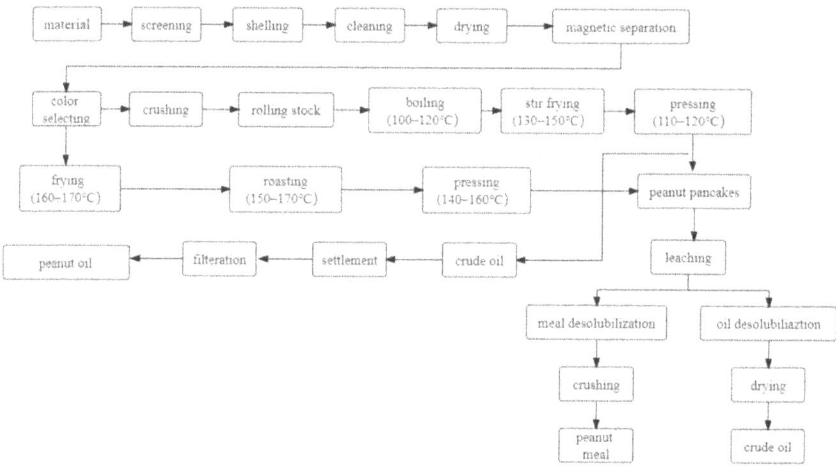

Figure 1. The traditional peanut oil production process.

Solubility is often the first consideration when developing new products and determining the functional properties of a protein component [29]. To solve the problem of severe protein denaturation during the high-temperature pressing of peanut oil, different researchers have invented new pressing and oil-extraction techniques. The focus of their work is concentrated on controlling the heat-treatment temperature during the process. The commonly used preparation methods include cold pressing, leaching, aqueous agent, membrane separation technology, and aqueous enzyme method. These technologies, with the exception of membrane separation, are well-established in the field of oil production. Zhang, Qin, and Sun et al. reviewed the process of making low-denaturing peanut protein powder under different low-temperature conditions, which was effective in reducing the degree of peanut protein denaturation [30–32]. In 2004, based on the summary of previous research results, Wang adopted the method of low-temperature cold pressing combined with No. 4 solvent leaching, and through the study of moderate adjustment of peanut oil yield gap and meal thickness [33], established the largest peanut low-temperature pressing oil–protein co-production line in China, and it has reached the international advanced level. The specific process flow is shown in Figure 2.

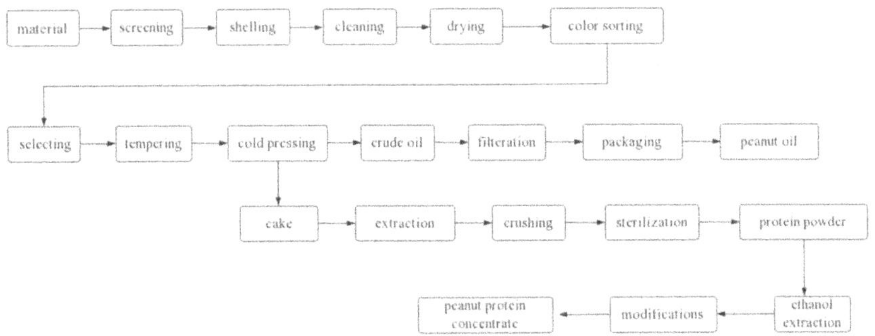

Figure 2. The route of low-temperature peanut oil and protein production.

Although these methods could effectively prevent peanut protein denaturation, basically keeping the protein content and residual oil rate in a suitable range (residual oil $\leq 0.8\%$, protein $\geq 58\%$, NSI $\geq 70.6\%$), the obtained peanut oil and peanut protein powder were significantly worse in aroma than the traditional high-temperature pressing process. This is because the flavor substances produced at low temperatures are limited. With the continuous maturation of protein technology, the modification of denatured peanut protein after heating at the protein molecular level has become possible. Modification is the artificial modification of the structure of a protein, and these modifications are related with delayed deterioration reactions, removal of toxic or inhibitory compounds, as well as incorporation of nutrients and additives through the formation of covalent bonding [34]. At the molecular level, these modifications are made by cutting the main chain of the protein molecule or by modifying the side chain groups of the protein molecule to change the protein structure and conformation at different levels to achieve optimal properties of the protein in terms of size, surface, charge, hydrophobicity/hydrophilicity ratio, and molecular flexibility [15,30]. It is feasible to achieve increased solubility by changing the composition and structure of peanut proteins. According to the different mechanisms of structural modification of proteins, the common means of solubility modification mainly include physical modification, chemical modification, and biological modification [34].

Flavor is an important quality criterion for food [35]. On the basis of ensuring the functional properties of peanut protein, it is imperative to conduct a summary work on peanut flavor and protein modification to reduce the production and economic loss of IFPP caused by flavor deficiency. In order to improve the grim situation of low utilization of peanut meal yield, through the in-depth study of the mechanism of flavor production and microscopic protein structure and function under heating conditions, the use of moderate temperature preheating, medium temperature baking, protein pretreatment to control the degree of peanut protein denaturation, and MR, CR, LOD, and other chemical reactions during the heating process is an ideal production method of IFPP. This paper reviews the research progress on the characterization and generation mechanism of peanut flavor substances during different heat-treatment processes. In addition, it also summarizes the world's applications related to the solubilization modification technology of peanut protein. This review proposes the feasibility of IFPP product development based on the mechanism of flavor substance generation and solubilization modification methods that combine the sensory and functional properties of peanut protein. Overall, this review can provide more ideas and possibilities for future research on exploring and establishing higher-quality peanut pressed oil production and flavor protein powder co-production lines.

2. Characterization of Characteristic Flavor Substances in Raw Peanuts and Heat-Treatment Peanut Products

Heat treatment is an important step in the production of IFPP, and the various aromatic substances produced during the heating process, such as pyrazine, give it a unique "baking aroma". IFPP is a new product, and there are no studies in the world to identify and analyze its flavor components. However, flavor characterization of studies for raw peanuts, roasted peanuts, peanut oil, and peanut butter is comprehensive, and IFPP has a similar processing as a by-product of peanut oil production. This paragraph summarizes a review of the research on the flavor characterization content of these four kinds of common peanut products on the market. Although a large number of volatile components have been identified from different processed peanut products, not all of them contribute to the finishing aroma of peanut products. A variety of important aroma substances as well as aroma characteristic components currently analyzed and identified from raw peanuts, roasted peanuts, peanut oil, and peanut butter are summarized in Table 1 and Figure 3. Table 1 summarizes the key flavor substances that have been identified in four different processed peanut products and their aroma descriptions. These substances have a large contribution to the overall aroma attributes of the products even if they present different aroma characteristics, and furthermore, they generally have odor activity values (OAV) ≥ 1 or have a high flavor dilution (FD). On the basis of Table 1, Figure 3 summarizes the common aroma property expressions in the four peanut products and lists the aroma active substances that contribute significantly to them, respectively, which are mostly present in these four peanut products and belong to the characteristic aroma substances of peanut products.

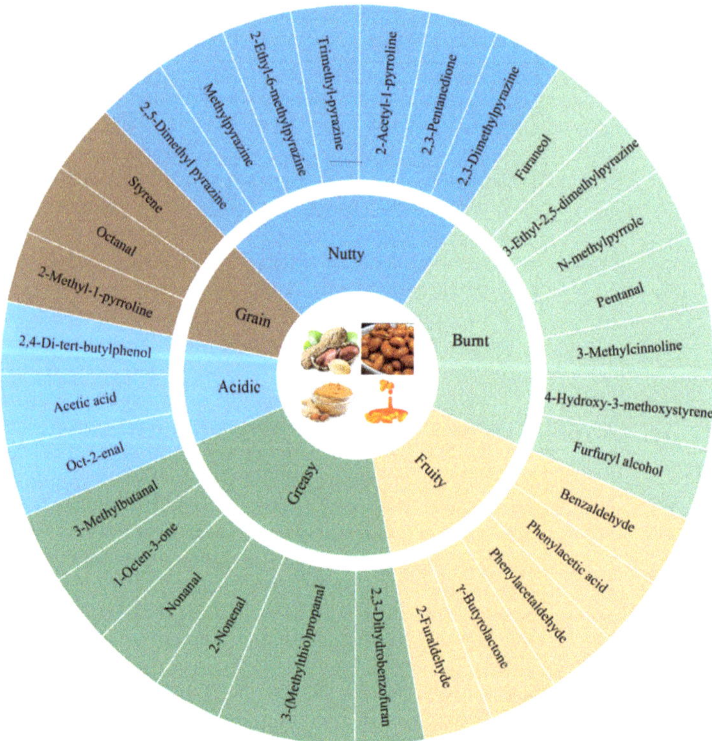

Figure 3. Raw peanuts, roasted peanuts, peanut oil, and peanut butter characteristic aroma components.

Table 1. Key flavor components in peanut heat processed products.

No.	Compounds	CAS No.	Flavor Description	Reported Flavor of Peanut Product	Reference
1	2-isobutyl-3-methoxypyrazine	24683-00-9	Bell pepper-like, earthy	Raw peanuts	[28]
2	Trans-4,5-epoxy-(E)-2-decenal		metallic	Raw peanuts	[28]
3	3-Isopropyl-2-methoxy-pyrazine	25773-40-4	Chocolate, nutty	Raw peanuts	[36]
4	Acetic acid	64-19-7	Sharp pungent	Raw peanuts	[36]
5	2-isopropyl-3-methoxypyrazine	93905-03-4	Earthy, pea-like	Raw peanuts, roasted peanut	[28]
6	Nonanal	124-19-6	Beany aroma	Raw peanuts, roasted peanuts	[36]
7	2-Acetylpyrroline	85213-22-5	Popcorn scent	Roasted peanuts, peanut oil	[36]
8	1-Octen-3-one	4312-99-6	Mushroom aroma	Roasted peanuts, raw peanut	[36,37]
9	Trans-4,5-Epoxy-(E)-2-decanal		Metallic aroma	Roasted peanuts	[23,36]
10	2-Methyl-1-pyrroline	872-32-2	Rice aroma	Roasted peanuts	[37]
11	2-Nonenal	2463-53-8	Greasy	Roasted peanuts	[17]
12	Phenylacetaldehyde	122-78-1	Fruity	Roasted peanuts	[38]
13	Phenylacetic acid	103-82-2	Honey, floral scent	Roasted peanuts	[36,37]
14	Methanethiol	74-93-1	Decomposed aroma	Roasted peanuts	[22]
15	2,3-pentanedione	600-14-6	Nutty	Roasted peanuts	[22,36]
16	3-(methylthio) propanal	3268-49-3	Musty potato, tomato	Roasted peanuts, peanut oil, raw peanut	[23,36,38]
17	3-methylbutanal	590-86-3	Fatty	Roasted peanuts	[37]
18	2-methylbutanal	96-17-3	Musty, cocoa, nutty	Roasted Peanuts, raw peanut	[36]
19	Trans-2,4-nonadienal-trans	6750-03-4	Greasy, fatty	Roasted peanuts	[38]
20	Octanal	124-13-0	Beany	Roasted peanuts, raw peanut	[16,17]
21	Pentanal	110-62-3	Fruity, nutty, berry	Roasted peanuts, raw peanut	[39]
22	N-methylpyrrole	96-54-8	Sweet, woody odor	Roasted peanuts	[38]
23	4,5-dimethyloxazole	20662-83-3	Green, sweet, vegetable	Roasted peanuts	[38]
24	Hexanal	66-25-1	Grassy, refreshing, beany	Roasted peanuts, peanut oil, raw peanut	[23,37]
25	2,3-dimethylpyrazine	5910-89-4	Nutty	Roasted peanuts, peanut oil	[23,38]
26	Oct-2-enal	2363-89-5	Fatty	Peanut oil	[38]
27	Benzaldehyde	100-52-7	Fruity	Roasted peanuts, peanut oil	[23,38]
28	2-Acetyl-3-methylpyrazine	23787-80-6	Rose aroma	Peanut oil	[38]
29	Phenethyl alcohol	60-12-8	Bitter medicinal	Peanut oil	[38]
30	Pyrrole-2-carboxaldehyde	1003-29-8	Sweet fragrance	Peanut oil	[38]
31	Gamma butyrolactone	96-48-0	Alcohol odor	Peanut oil	[38]
32	2-furaldehyde	98-01-1	Green, sweet, and vegetable	Roasted peanuts, peanut oil, peanut butter	[23,39,40]
33	Methylpyrazine	109-08-0	Nutty	Roasted peanuts, peanut oil, Peanut butter	[23,39,40]
34	2,5-dimethyl pyrazine	123-32-0	Nutty	Roasted peanuts, peanut oil, peanut butter	[40]
35	Trimethyl-pyrazine	14667-55-1	Nutty	Roasted peanuts, peanut oil, peanut butter	[40]
36	Furaneol	3658-77-3	Caramel aroma	Roasted peanuts, peanut oil, peanut butter	[36,37]
37	3-Ethyl-2,5-diMethylpyrazine	13360-65-1	Burnt aroma	Peanut oil, peanut butter	[23]
38	4-Hydroxy-3-methoxystyrene	7786-61-0	Burnt aroma	Peanut butter	[40]

In the fact that the aroma attributes of peanut products are the result of many different kinds of flavor substances, the interactions between aroma active substances such as pyrazines, furans, pyridines, aldehydes, ketones, and alcohols are key. Regarding both raw or heat-treatment peanut products, the aroma active substances are the root cause of their different aromas.

2.1. Characterization of Raw Peanut Flavor Components

Up to now, about 70 volatile compounds have been found in fresh peanuts, mainly consisting of heterocyclic and non-heterocyclic compounds, of which alkyl pyrazines and furans are the main ones: ethyl 2-ethyl-3,6-dihydroxypropionate (fruity aroma), ethyl 2-methylbutyrate (green apple aroma), ethyl 3-methylbutyrate (fruity aroma), octanal (grassy aroma), hexanal (oily aroma), butyric acid (cheese aroma), isovaleric acid (rotten aroma), pentanoic acid (rotten aroma), isovaleric acid (flower aroma), hexanoic acid (coconut oil aroma), and furfuryl alcohol (burnt aroma), with hexanal and octanal being the main sources of raw peanut bean aroma [22,37,41]. The non-heterocyclic compounds are mainly formed by lipid oxidation and degradation of hydroperoxides [41]. They have a very low aroma threshold, which can only be detected by GC-O [37]. In addition, slight differences in compounds in peanuts are caused by varietal differences, resulting in differences in flavor between peanut species. For example, Argentine peanuts have been described as musty, but Chinese peanuts exhibit a woody/earthy shell/skin flavor as well as bitter and sour flavor, while American peanuts have a sweet, aromatic intensity and are often used in the production of roasted peanuts and peanut butter [42]; Nepote found that high-oleic peanuts contained significantly higher levels of pyrazine than ordinary peanuts and contained lower levels of aldehydes. The flavor of high-oleic peanuts was more stable after prolonged storage [43]. Interestingly, differences in the harvest season of the same peanut also had an effect on peanut flavor. Autumn-ripened roasted peanuts in Taiwan found to contain higher levels of total pyrazine than spring-ripened peanuts, consistent with the general consumer perception that autumn-ripened peanuts have a richer peanut flavor [44]. The overall flavor of raw peanuts is relatively bland, with a low content of aroma active substances but rich in amino acids, sugars, fatty acids, and other flavor precursors. The differences in flavor precursors in raw peanuts directly affect the production of flavor substances during heat treatment. Settaluri found that the carbon skeleton in pyrazine was mainly derived from the degradation of sugar compounds in peanut seeds. Meanwhile, the nitrogen elements in pyrazine, pyrrole, and pyridine are mainly derived from amino acids in peanut seeds [45]; Maga found that some of the pyridines in roasted peanuts were formed from alkanals amino acids. Amino acids are considered to be the most important flavor precursors in peanut seeds [46].

2.2. Characterization of Heat-Treatment Peanuts Products

2.2.1. Roasted Peanut Flavor Components

The distinctive flavor of peanuts during roasting is one of the main factors influencing consumer choice and acceptance [47]. The flavor of roasted peanuts has been a hot topic for 50 years, with more than 300 volatile compounds identified. In 1966, Mason and Johnson first proposed that heterocyclic compounds, namely pyrazine, pyrrole, and furan, with low molecular weight produce the typical roasted peanut flavor [48]. Baker also suggested that pyrazines are the main cause of roasted flavor during peanut roasting, with 2,5-dimethylpyrazine being the most relevant to roasted peanut aroma [49]. Pyrazine is the heterocyclic compound containing two nitrogen atoms and has traditionally been considered a key element in the formation of the typical aroma of roasted peanuts. Johnson identified 46 peanut volatile components in roasted peanuts, and most of them were pyrazines [50]. Buckholz and Ho quantified roasted peanut volatiles by a variety of different extraction and isolation methods, and the results of their study indicated that several pyrazine compounds were major contributors to flavor [39]. Methylpyrazine; 2-ethyl-3-methylpyrazine; 2,4,5-trimethylpyrazine; 2,5-ethyl-3-dimethylpyrazine; 2,3-diethyl-5-methylpyrazine; and 3-ethyl-2,5-dimethylpyrazine have been identified as contributing to the roasting odor in roasted peanuts [51]. In addition, 2-acetyl-1-pyrroline, 2-propionyl-1-pyrroline, and 4-hydroxy-2,5-dimethyl-3(2h)-furanone were likewise identified as the main sources of roasting aroma, with 2-acetyl-1-pyrroline, which has a strong popcorn flavor, showing the highest OAV in roasted peanuts [36].

The formation of roasted peanut flavor substances is directly linked to peanut varieties. The importance of variety selection for roasted peanut flavor was discovered long ago. The American Runner-type peanut and the Chinese Four-grain Red were considered ideal for roasted peanut production, with a stronger roasted flavor under the same roasting process. Baker compared four different varieties of peanut by establishing a correlation between pyrazine compound content and roasted peanut flavor [49]. By combining with sensory evaluation scores, he found that peanut genotypes differed in roasted flavor and aroma regardless of roast color. Florida MDR 98 formed the highest levels of pyrazines under the same roasting conditions, followed by Florunner, Georgia Greene, and Sun Oleic, respectively [49]. Nepote roasted 16 different genotypes of normal and high-oleic peanuts at 170 °C for 30 min before subjecting the peanut genotypes to chemical and sensory data by principal component analysis and cluster analysis [43]. It was found that the high-oleic geno-types, 4896-11-C and 9399-10, showed high consumer acceptance in a questionnaire, suggesting that some of the high-oleic peanuts could be substituted for regular peanuts. And as mentioned in 2.1, high oleic peanuts have a higher stability. However, Hu found that the higher initial concentration of characteristic precursors (arginine, tyrosine, lysine, and glucose) in normal-oleic peanuts than in high-oleic peanuts was the underlying cause of the inferior aroma of high-oleic peanut oil compared to normal peanut oil [47]. In addition, the roasting temperature and time parameters had a decisive influence on the formation of flavor substances in peanuts [37].

With advances in flavor omics analysis, it has recently become apparent that pyrazines do not necessarily mean that they play an important role in the flavor of peanuts. If the concentration of the compound is below the sensory threshold, then it makes little contri-bution to the overall aroma. Surprisingly, Chetschik found that most pyrazines did not have as much impact on the overall aroma of roasted peanuts as expected [22]. Schirack found that thirty-eight compounds were the main contributors to the aroma of roasted peanuts, of which only seven were pyrazines [52]. With the continuous exploration of roasted peanut flavor, it was found that short-chain fatty aldehydes, such as pentanal, hexanal, 2-heptanal, 2-octenal, and phenylethylaldehyde, and the sulfides, such as methyl mercaptan, 2-furan mercaptan, and 2-thiophene mercaptan, were also major contributors to the flavor of roasted peanut [22,36].

In addition, some progress has been made in the study of the off-flavor of roasted peanuts. Basha and Young discovered that the protein fraction capable of producing odoriferous volatiles is a naturally occurring oleic-acid-rich lipoprotein by gel filtration of peanut proteins [53]. The production of N-methylpyrolopentane, acetone, dimethyl sulfide, 2-methyl propanol, pentanal, and hexanal was directly related to the musty, fruity, tongue-burning, and soy flavors of roasted peanuts, suggesting that in addition to lipid oxidation, thermal degradation of proteins may be a source of undesirable flavor com-pounds in roasted peanuts. In addition, fruity esters, such as 2-methylpropionate ethyl, 2-methylbutyrate ethyl, and 3-methylbutyrate ethyl, as well as high levels of short-chain organic acids, such as butyric, 3-methylbutyric, and hexanoic acids, are associated with the fruity taste of off-flavored fermentation [54]. Ethyl acetate, on the other hand, is considered to be an indicator aroma substance for roasted peanut spoilage. The interaction between different compounds resulted in differences in the aroma of roasted peanuts [55].

2.2.2. Characterization of Roasted Peanut Oil Flavor Components

Peanut oil has a strong nutty and roasted flavor that distinguishes it from other edible vegetable oils [23]. For the production of traditional concentrated peanut oil, crushed peanut seeds are roasted and pressed at around 120 °C, which develops the typical flavor of nuts and roasting. The formation pathways of flavor substances in richly flavored peanut oil are similar to those of roasted peanuts in that all amino acids, lipids, carbohydrates, and other substances in peanuts are subjected to high temperatures, resulting in a series of chemical changes.

Liu and Hu accurately identified flavor substances in peanut oil. 2,3-dimethylpyrazine (nutty); 2, 3,5-trimethylpyrazin (nutty); 3-ethyl-2,5-dimethylpyrazine (coffee); 2-acetyl-3-methylpyrazine(roasted); 2-acetylpyrrole (floral aroma); hexanal (grassy); benzaldehyde (fruit aroma); and 3-hydroxy-4,4-dimethyl-γ-butyrolactone (wine aroma) were the main aroma substances in traditional peanut oil, while methylpyrazine; 2,5-dimethylpyrazine; and 2-ethyl-5-methylpyrazine are considered to be the key volatiles contributing to the nutty and roasted flavor of peanut oil [38,47]. Liu suggested that the aldehydes mainly produced in the first and middle stages of seed frying and that the nitrogen heterocyclic compounds produced in the later stages contributed to the flavor of peanut oil. More importantly, 30 min of heating at high temperature (200 °C) is the most important stage in the formation of typical flavor of peanut oil [38].

Compared with traditional high-temperature pressing, cold-pressed peanut oil made from low-temperature conditioning (50–60 °C) and low-temperature pressing (60–70 °C) generally has a poorer flavor [23]. As early as 1972, Brown suggested that 2-heptanal; 2-octenal; 2-nonenal' and 2,4-dodecenal in low-temperature cold-pressed oil were associated with the greasy and fried flavor of freshly cold-pressed roasted peanut oil [25]. In recent years, Dun identified a total of 101 compounds in hot-pressed peanut oil and only 64 compounds in cold-pressed peanut oil [23]. Aldehydes, ketones, hydrocarbons, alcohols, acids, esters, furans, and pyrroles were detected in both hot- and cold-pressed peanut oil samples, while pyrazines and pyridines were only detected in hot-pressed peanut oil samples. The result confirmed previous experimental results that pyrazines are responsible for the roasted flavor and aroma of peanut oil, with 2,5-dimethylpyrazines showing the highest correlation with roasted peanut flavor and aroma. The volatiles of cold-pressed peanut oil contained high levels of aldehydes, mainly hexanal, nonanal, and (Z)-2-heptenal, accounting for 50.88% to 70.12% of the total volatiles, which was about three times higher than that of hot-pressed peanut oil [23]. This finding is consistent with Hu et al., who suggested that the flavor substances in low-temperature cold-pressed oil are mainly aldehydes [47]. Fatty aldehydes, in particular, produce fatty and grassy odors that bring a significant impact on the overall aroma of peanut oil. Because of the lack of high-temperature roasting, the content of heterocyclic compounds, such as pyrazine and pyrrole, were significantly low, and the peanut oil was low in compounds with roasted and toasted aromas. These results clearly indicate that the difference in volatile flavor substances in peanut oil due to the temperature and duration of high temperature between tumbling and pressing is the underlying cause of the difference in flavor between cold-pressed and hot-pressed peanut oil.

2.2.3. Characterization of Peanut Butter Flavor Components

Peanut butter is a casual food product made from fresh peanut seeds roasted and stripped of their red coating and ground. The flavor profile of peanut butter is similar to that of roasted peanuts, with a strong roasted and nutty aroma, particularly typical of roasted peanuts [40]. Peanut butter flavor formation varies by the peanut plant varieties, picking treatment, roasting processing, and storage conditions factors [56]. The roasting process is the most important measure in the formation of the characteristic aroma of roasted peanuts and peanut butter. The peanut seeds are roasted at high temperatures and undergo a series of chemical reactions, including the MR, CLO, and protein thermal degradation, similar to the aroma presentation mechanism of roasted peanuts. After roasting, the peanuts are ground, and the volatile aroma active ingredients are further emitted, endowing peanut butter with a more intense flavor than roasted peanuts. The MR is generally considered to be the main source of flavoring substances in peanut butter. Heterocyclic compounds, such as pyrazines, furans, and pyrroles, produced by the MR, contribute to the roasted, nutty flavor of peanut butter as important aroma substances. Pyrazines have been considered the most important volatile flavor components in peanut butter [57].

As roasted peanuts are an intermediate product in the peanut butter production chain, the flavor of peanut butter is directly influenced by the volatile substances in roasted peanuts. 2,5-dimethylpyrazine, a characteristic aroma substance of roasted peanuts, was also found by Baker to be an important contributor to the roasted peanut aroma of peanut butter [49]. The same was true for 2-ethylpyrazine; 2,5-diethylpyrazine; 2-ethyl-5-methylpyrazine; and other ethyl-substituted pyrazines. Ethyl substitution of one or more methyl groups can significantly lower the threshold, and these compounds make an important contribution to the roasted flavor and nutty taste of peanut butter [49]. The substitution of pyrazine and the substitution of one or more methyl groups by ethyl groups significantly lowered the threshold, and these compounds contributed significantly to the roasted aroma and nutty taste of peanut butter. The key aroma active substances (OVA $\geq$ 1) in peanut butter were identified, including 2,5-dimethylpyrazine (grilled, grassy); 2,3,5-trimethylpyrazine (nutty, sweet); 3-ethyl-2,5-methylpyrazine (roasted aroma); and 4-

hydroxy-2,5-methylpyrazine (roasted aroma) [40]. Hathorn et al. focused on the phenolics in peanut butter, which generally make roasted peanut wood, hulls, and skins have a bitter odor. Peanut skins resulted in an overall reduction in flavor. The 10% peanut skins resulted in a reduction in overall flavor but were effective in enhancing the antioxidant properties of the peanut butter [58]. Similarly, the flavor of peanut butter is also influenced by the raw material and roasting process, as discussed in detail in Section 2.1.

3. Formation of Characteristic Flavor Substances of Heat-Processed Peanuts and the Mechanism of Aroma Presentation

Peanut contains about 50% fat and 25% protein and is rich in monounsaturated fatty acids, free amino acids, and other important flavor precursors [21,23]. These non-volatile substances can be converted into volatile compounds through processing methods, such as heat treatment, fermentation, and storage. In recent years, after research into the formation pathways of flavor substances in processed peanut products, there has been an updated description of the flavor presentation mechanisms of a wide range of processed peanut products. Unfortunately, the reactions involving the formation of typical peanut processed product flavor compounds and their corresponding precursors have not been fully elucidated. This paragraph reviews five reported pathways for the formation of flavor substances in processed peanut products, namely the MR, LOR, CR, and their interactions with each other.

3.1. Maillard Reaction

The MR is the most important reaction in the formation of flavor and color in processed peanut products [59]. Also known as the carbamide-ammonia reaction, MR is widespread in food processing, which has been studied worldwide for nearly a century. It was discovered and reported by Louis-Camile Maillard in 1922. The MR is the interaction of reducing sugars and amino compounds, such as amines, amino acids, peptides, and proteins in peanuts. The main products are heterocyclic nitrogen compounds such as furans, thiazoles, thiophenes, oxazoles, pyrroles, imidazoles, pyridines, and pyrazines [21,59]. Long baking and roasting at high temperatures is a necessary prerequisite for the occurrence of MR. Carbon module labeling (comla) technique with multi-reaction kinetics has been successfully used for MR kinetic evaluation and validated MR networks [60].

The generally accepted explanation of the Maillard reaction pathway is that the Maillard reaction is divided into three stages: initial, intermediate, and final [60]. The initial reactions include hydroxylamine condensation to produce cyclized Schiff's base and rearrangement of the Amodori molecule [61]. The mid-stage reactions mainly include the Strecker reaction, and fructosylamine converts to hydroxymethylfurfural (HMF) or reductone [62]. At the end of the reaction, the intermediates undergo further condensation and polymerization by hydroxyl aldehyde condensation, hydrogen sulfide polymerization, Strecker degradation, and other reactions to finally form N-, O-, and S-containing heterocyclic compounds and the colored pigment melanin (mrps) [63]. Strecker degradation is considered to be the main step in the production of characteristic flavor substances (alkyl pyrazines, aldehydes) in heat-treatment peanut products [32]. MR is very complex, with little known about the details of the reaction process, and its reaction pathway is briefly described in this paper, including the Amadori rearrangement, Heynes rearrangement, aldol condensation, Strecker thermal degradation, and the process of the heterocyclization reaction and the precursors and products in these reaction. The specific reaction pathways are shown in Figure 4 [61].

In order to further elucidate the complex mechanism of the MR in peanuts during heating, Newell tracked the concentrations of precursors and reaction products during different reaction stages of peanut roasting and demonstrated that aspartic acid, glutamic acid, glutamine, asparagine, histidine, and phenylalanine are the main precursors of typical peanut flavors involved in pyrazine formation [65]. Additionally, the role of monosaccharides is very important in the formation of pyrazines. The major soluble sugar component in peanuts is sucrose, followed by glucosamine, hydrosucrose, and raffinose [66]. Mason et al. found that sucrose is involved in the browning reaction by being hydrolyzed to fructose and glucose by converting enzymes during roasting [67].

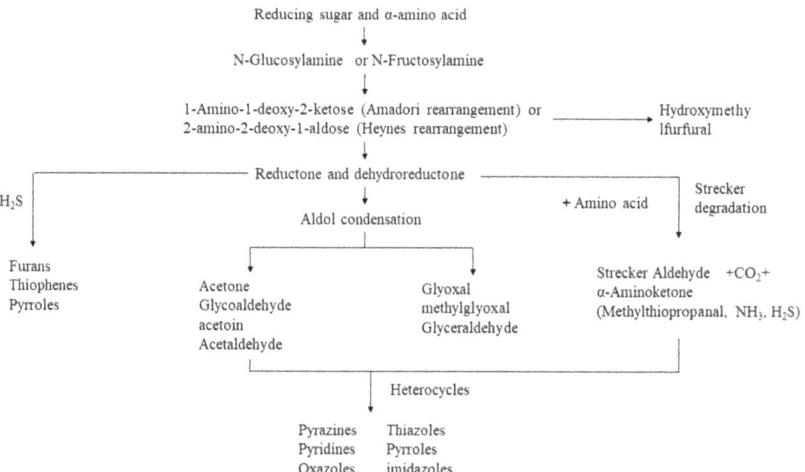

Figure 4. Maillard reaction scheme adapted from Hodge [64].

The reaction products generated by the MR have a characteristic odor profile that essentially affects the quality attributes of the processed peanut product, including flavor, color, and structure [47,60]. For example, pyrazine and thiols were found to be key flavor substances in most cases in roasted peanuts. Buckholz combined sensory evaluation with instrumental analysis of peanuts that had been roasted for various times and found that a decrease in carbonyl and an increase in pyrazine were signs of good peanut flavor, and the change in color was mainly due to the formation of melanin (mrp) [68]. The study demonstrated that one of the carcinogens-acrylamide is associated with the darker color of processed peanut products [59]. The products of the MR have antioxidant effects and also inhibit the oxidation of low-density lipoprotein (LDL) in the human body while reducing cardiovascular and cerebrovascular disease [62]. LDL is the important substrates in lipid-peroxidation reactions that are accelerated in the presence of iron ions and free radicals [69]. Vahid et al. found that oleic acid may reduce the risk of coronary heart disease up to 20–40% mainly via LDL cholesterol content-reducing activity [5].

Although the reaction results in a loss of nutritional quality (destruction of essential amino acids and reduced digestibility), it has beneficial effects on flavor, taste, color, and antioxidant activity.

3.2. Lipid Oxidation Reaction

LOR is usually the main cause of "fade" of peanut flavor, resulting in the loss of positive attributes associated with freshly roasted peanuts. The process produces a range of off-flavors, such as "putrid" and "sour". Pleasant smells, such as "baking aroma" and "sweet aroma", are masked by these off-flavors [70]. These off-flavors are mainly contributed by aldehydes [54]. Feussner reviewed the two main pathways for the production of aldehydes in peanut oil: thermal oxidation of LOR in crushed peanut seeds and auto-oxidation. The aliphatic aldehydes, alcohols, and ketones formed during heating are mainly produced by auto-oxidation of oils, while some acids, esters, and furans are formed by thermal oxidation of oils and fats [71].

Peanuts contain a large amount of unsaturated fatty acids (80%) and are highly susceptible to lipid oxidation. Prolonged high-temperature heating creates sufficient conditions for the oxidative degradation of fatty acids to form many iso-olfactory compounds, such as hexanal, heptanal, octanal, 1-octane-3-one, and unsaturated aldehydes, etc. [22]. The resulting mono-hydroperoxides act as the main precursors of iso-olfactory compounds. Of these, hexanal is one of the main oxidation products of linoleic acid esters and has long been used as an indicator of oxidative spoilage of food products. These compounds are directly related to the previously described "paint" and "cardboard" smell of processed peanut products [56]. Flavor "fade" of peanuts due to lipid thermal oxidation is associated with (1) carbonylamine reactions [72]; (2) flavor encapsulation between proteins and lipid hydroperoxides [73]; (3) degradation of heterocyclic nitrogen compounds by lipid radicals and hydroperoxides [74]; and (4) the possibility that the products of lipid thermal oxidation may interact with roasted peanut flavor compounds (pyrazine) and together cause changes in peanut flavor [75]. The reaction pathway for the thermal oxidation of peanut lipids is through auto-oxidation, photo-oxidation, and enzymatic oxidation during the heat process to produce hydroperoxides. Unsaturated fats in peanuts and precursors such as free radicals produce hydroperoxides in the presence of high temperatures and oxygen. The single hydroperoxides continue to oxidize to produce a variety of multiperoxides, which themselves continue to decompose to produce volatile and non-volatile alcohols, aldehydes, ketones, and other substances or undergo polymerization reactions to produce a series of polymers [76,77]. As in Figure 5, the reaction pathway of peanut autoxidation under heating conditions is described: the unsaturated fatty acid linolenic acid in peanuts reacts with oxygen and free radicals under heating conditions to produce hydroperoxides, which are decomposed under the catalytic action of enzymes to produce aldehydes, ketones, alcohols, and other flavor substances.

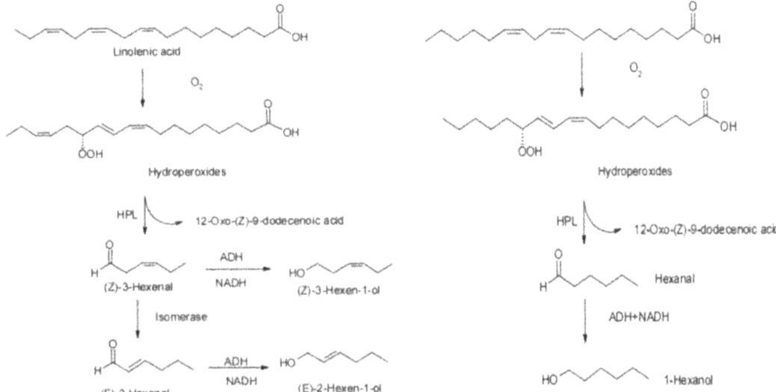

Figure 5. Flavor production process during peanut heating—lipid thermal oxidation reaction [78–80].

Furthermore, pyrazines in peanuts are wrapped in free radicals and hydroperoxides from lipid oxidation as thermal oxidation of lipids proceeds, resulting in changes in the flavor of peanut products [75]. The thermal stability of peanut flavor is influenced by peanut variety and water activity. Pattee found that high-oleic peanuts were shown to have better antioxidant properties and better flavor stability than regular peanuts, possibly due to the high production of polyphenolic antioxidants, namely coumaric acid and mrp, during oxidation [81]. Reed investigated the effect of water activity on the flavor stability of roasted peanuts and found that reducing water activity increased the formation of lipid oxidation products and the loss of pyrazine compounds [82]. This finding confirmed the conjecture of Mate et al. that the rate of lipid oxidation in peanuts with low relative humidity (~20%) was higher than that in peanuts with high relative humidity (~60%) [83]. Controlling the water activity in peanuts can effectively reduce peanut lipid thermal oxidation.

However, the effect of peanut lipid thermal oxidation on peanut flavor is not all unfavorable [75]. It is known that most of the non-hybrid compounds in processed peanut products are produced by lipid decomposition, and some aldehydes and ketones provide favorable aromatic odors [70]. For example, phenylacetaldehyde and p-vinyl guaiacol are important contributors to the "fruity aroma" of peanut flour [36]; benzaldehyde and 3-hydroxy-4,4-dimethyl-γ-butyrolactone are key aroma substances for the "floral aroma" flavor of peanut oil. 3-(methylthio) propanal; 3-methylthiopropionaldehyde; and 2,4-decadienal are important for the "frying aroma" of roasted peanuts [23,55]. Alcohols, alkanes, alkenes, and alkynes contribute less to the flavor of peanut heat processing products due to their higher odor thresholds [63]. LOR products make up a smaller proportion of the volatiles of heat-treatment foods and contribute less to flavor than the heterocyclic compounds produced by the MR but play a very important and integral role in the flavor of fats and oils of food [84].

3.3. Caramelization Reaction

CR is an important source of furans and their derivatives in peanut heat-treatment products. Furan derivatives are considered to be the second-largest group of compounds among heat-treatment volatiles, which usually impart caramel-like, sweet, fruity, and nutty flavors to foods [85]. The summary of typical proposed formation pathways of furan from heat-treatment peanuts is shown in Figure 6. Both the CR and the MR are browning reactions, which, at high temperatures (above 140–170 °C), produce a characteristic "baking aroma" flavor and a deepening of the color of the peanut product [86]. The difference with the MR is that the monosaccharides in peanuts are dehydrated and degraded under aerobic conditions by heat without the participation of amino acid compounds [87]. The reaction process includes dehydration, cleavage, and ring formation, resulting in low molecular weight open-chain oxygenates as well as heterocyclic oxygenates, such as furan, furanone, furfural, 5-hydroxymethylfurfural, maltol, and cyclopentenone [88]. The furfural and reduced ketones produced are further involved in the formation of mrp as precursors at the end of the MR [76].

Furfural not only forms the characteristic sweet aroma of caramel in heated foods but is also one of the important precursors for other furan derivatives and an important intermediate for other heterocyclic compounds [23]. Beksan found that hexose can be decomposed into 4-hydroxy-2,5-dimethyl-3(2h)-furanone during heat treatment, of which rhamnose and fructose 1, 6-diphosphate are the key precursors [89]. 4-hydroxy-2,5-dimethyl-3(2h)-furanone has long been recognized as the characteristic flavor substance of peanuts as the characteristic flavoring substance for the "caramel aroma" of peanuts [36]. Two other popcorn-like odor substances, 2-propionyl-1-pyrroline and 2-acetyl-tetrahydropyridine, were first isolated from peanuts and derived from proline and the sugar degradation products hydroxyacetone and 2-oxobutyraldehyde, respectively [54].

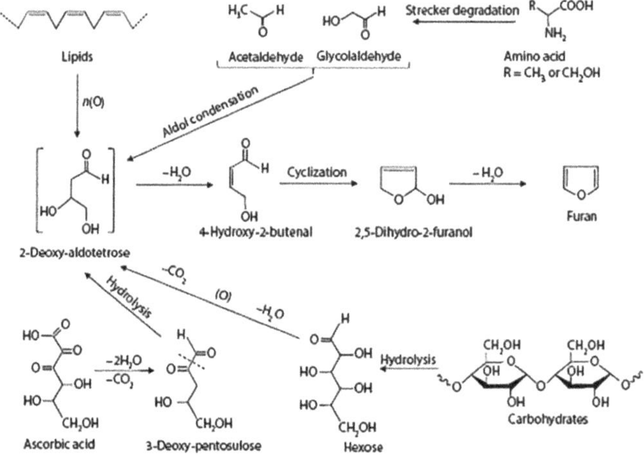

Figure 6. The typical pathway to form furan derivatives adapted from Kroh [86].

4. Effect of Heat Treatment on the Quality Characteristics of Peanut Protein

Heat treatment is the most effective way to purify peanut protein and enhance the sensory properties of the product. However, the functional properties of peanut protein in the heating process will change differently with the denaturation and structural changes of the protein, which will accordingly affect a series of indicators of peanut processed products. At the same time, the functional properties can also affect the quality of peanut products. These effects are both positive and negative mainly in the color, taste, flavor, dissolution, and dispersion ability and many other aspects of peanut products [36].

4.1. Thermal Degradation of Proteins

Protein makes an important contribution to peanut flavor. On the one hand, it can degrade to form flavor precursors peptides and amino acids [64]; on the other hand, it can interact with flavor compounds to change the head-space concentration of flavor substances and affect the overall flavor of food [54]. The interaction between proteins and flavors is specifically described as the denaturation of proteins by heat during heating; the breaking of hydrogen bonds; the destruction of secondary, tertiary, and quaternary structures of proteins; and the change in spatial structure caused by protein unfolding [90]. Peanut protein and peanut products in part of the flavor substances occurred in a series of reversible and irreversible binding; reversible binding includes ionic interaction, hydrogen bonding, and hydrophobic interactions, while irreversible reactions are mainly the role of covalent bonds, including lysine, carboxyl, disulfide bonds and aldehydes, amines, sulfur-containing compounds, and other flavor substances [47]. As mentioned in Section 2.2.1, thermal degradation of proteins may be an important source of undesirable flavor compounds in roasted peanuts [22]. In the following experiments, Basha conjectured that the protein fractions capable of producing odoriferous volatiles were mainly oleic-acid-rich lipoproteins in peanuts [52], which were thermally degraded and irreversibly covalently bonded to produce N-methylpyrrolopentane, acetone, dimethyl sulfide, 2-methylpropanol, pentanal, and hexanal and associated with musty, tongue-burning, and beany odors in roasted peanuts [35]. Hou identified volatile substances such as nonanal; hexanal; octanal; hexadecanal; octadecanal; 2,3-pentanedione; and 2-acetylfuran in the thermal reaction products of yak meat myogenic fibrin with sugar [91]. Unfortunately, the interaction between protein thermal degradation reactions and food flavor is not well explained. Currently, in the food processing industry, reversible binding of flavor substances through thermal degradation of proteins can be used to reduce flavor loss during processing and to re-release flavor components during consumption, and irreversible binding of covalent bonds for the removal of off-flavors from foods has been a very promising research component.

4.2. Effect of Heat Treatment on the Functional Properties of Peanut Proteins

The functional properties of proteins play a decisive role in the nutritional, organoleptic, physicochemical, and sensory properties of peanut products; meanwhile, their functional properties influence the behavior of proteins during processing and storage [92]. Peanut proteins have been shown to have ideal functional properties, which can be incorporated into food products to endue them better nutritional value and organoleptic properties and have been widely used in the food processing industry. However, the reduction of functional properties and nutritional value of the product due to protein denaturation caused by heat treatment seems to be inevitable.

The functional properties of proteins from various oilseed crops have been reviewed by a number of researchers. Koumanov reviewed the relationship between electrostatic interactions of proteins and their functional properties [93]; Schwenke illustrated the relationship between protein structure and functional properties [94]; In the study of the functional properties of peanut proteins, Yamada T. was the first to discover that peanut proteins contain about 10% whey protein and 90% globulins (66% peanut globulins and 23% peanut companion globulins) [95]. Peanut globulins and companion peanut globulins are the major protein fragments in peanut, both of which greatly determine the functional properties of peanut proteins [28]. According to Damodaran, the functional properties of peanut proteins are influenced by a variety of physicochemical properties, such as molecular size, molecular structure, amino acid composition and sequence, net charge, charge distribution, hydrophobicity, hydrophilicity, structure (secondary, tertiary, and quaternary), external environment (PH), etc. [96]. Functional properties of proteins were first explained at the molecular level by Morr et al. The presence and number of charged groups directly affect the solubility and foam stability of proteins due to electrostatic interactions between proteins [97]. Jiao used differential scanning calorimetry (DSC) to determine the thermal properties of the proteins after denaturation: 101.85 °C ± 0.47 °C and 89.68 °C ± 0.28 °C for peanut globulin and peanut companion globulin, respectively [98]. Yang determined the subunit structures of peanut globulin and peanut companion globulin using 2D-PAGE and found that peanut globulin is less soluble, less absorbent, and easily denatured by heat due to the absence of disulfide bonds, which is mainly attributed to the hydrophobic structure of the basic subunit and its heat resistance. Peanut companion globulin is composed of two subunits with different free isoelectric points but the same molecular weight, and the polypeptide composition of different varieties of concomitant arachidonic globulin is basically the same, and its heat resistance is better [99]. Liu found that peanut disulfide protein is mainly composed of polypeptides in the dissociated form but not in the subunit form. It has strong heat resistance that mainly affects the protein's water absorption, foaming, gelling, and emulsification [100]; in addition, the amphiphilicity of proteins is closely related to the distribution of polar and non-polar residues, interfacial interactions, and foaming and emulsifying activities. The size and composition of amino acids, on the other hand, can influence the sensory evaluation of product flavors by acting on odor receptors [101]. Moure reviewed the effects on the functional properties of oilseed crop proteins during protein processing and classified the functional properties of proteins according to the mechanism of action of the functional properties [28].

However, the functional properties of peanut protein in the process of heat treatment will change differently with the denaturation of protein, which accordingly will affect a series of indicators of peanut processed products. As a result of heat treatment caused by the denaturation of some peanut proteins, denatured proteins show reduced solubility, greatly limiting the use of peanut protein. The solubility of peanut proteins is mainly determined by molecular mass size, amino acid composition, size/composition of subunits, number of disulfide bonds, and hydrophilic/hydrophobic strength [94]. The methods available to assess denaturation of proteins were reviewed by Kilara and Sharkasi [101,102]. Changes in the organization and structure of peanut proteins occur, including structural changes, such as hydrolysis of peptide bonds, breakage of amino acid side chains, disulfide bonds being broken, and condensation of proteins with other molecules [33]. At the same time, the proteins become more compact due to polymerization between them, hydrophilic groups are encapsulated, water solubility is reduced, and they are less susceptible to enzymatic degradation. These variations depend on the intensity and duration of the heat treatment, water activity, pH, salt content, and other active substances [101]. Kinsella found that the solubility of BGPI proteins decreased as the hydrophobicity and sulfhydryl content of the protein surface increased after heat treatment [102]. Han studied the effect of different heat treatments on the solubility of peanut proteins and found that the treatment of roasting at 130 °C for 25 min made the most significant effect on the solubility of peanut proteins [103]. Li found that peanut flour was more difficult to dissolve in acidic beverages after baking than in alkaline conditions [104]. The isoelectric point of denatured peanut proteins ranged from pH 3.5 to 4.5, which is consistent with the experimental results of Chen.

The denaturation of peanut protein is characterized by protein inactivation, reduced water solubility, flocculation and precipitation in water, and reduced nutritional potency. Heat-induced denaturation of peanut proteins, with reduced nutritional value and solubility, limits the widespread use of peanut proteins in food systems. Interestingly, moderate heat treatment can loosen the protein structure and improve solubility. In addition, heat treatment can inactivate anti-nutritional factors, making the protein more easily digested and absorbed by the human body. Studies have shown that the many functional properties of proteins vary in a reciprocal manner; for example, good emulsification and foaming are predicated on higher solubility, while gelation is the opposite [92]. The processing of peanut protein involves physicochemical and thermal treatments that affect the nutritional value and functional properties of the product.

5. Research Progress on Solubilization Modification of Peanut Protein

In general, protein solubility is usually required to be above 80% nitrogen solubility index (NSI) for food applications [33]. Unfortunately, most peanut protein products do not exhibit the functional properties required by the food industry, especially solubility, which is attributed to the processing in order to enhance the sensory quality of the product; roasting, frying and other high-temperature treatment; as well as the common organic solvent leaching, high-temperature sterilization treatment, and other process operations. Heat treatment causes aggregation of protein molecules to varying degrees and changes in surface properties, especially the exposure of some hydrophobic groups, resulting in a subsequent decrease in functionality, especially solubility, which severely limits the application of peanut proteins. With the continuous maturation of protein modification techniques, it is possible to modify peanut proteins denatured after heating at the protein molecular level. The common means of solubilization modification mainly include physical modification, chemical modification, and biological modification [105]. As shown in Table 2, a summary of the studies related to the solubilization modification of proteins was conducted in recent years.

Table 2. Different forms of protein modification methods.

No.	Authors	Sample	Modification Method	Modification Mechanism	Modified Results	Reference
1	Dong	Pea protein	Cold Plasma Technology	New oxygen- or nitrogen-containing hydrophilic groups are formed on the surface of the protein.	Significant improvement of zein solubility in both neutral and acidic solutions could be observed after treatment with max solubility at 75 V.	[106]
2	Zhao	Peanut protein	Baking	Part of the globulin aggregates or decomposes, improving the solubility of the isolated protein under alkaline conditions.	The solubility at pH 7.0 increased gradually from 76% to 95%.	[107]
3	Zhang	Peanut protein	Microwave	Using microwave effect to change protein aggregation degree and spatial structure.	Under the conditions of microwave power 480 W, modification time 60 s, and pH value 9, the NSI of modified peanut protein concentrate was 53.26%.	[108]
4	Tu	Peanut protein	Dynamic high-pressure microfluidization	As the content of UV-absorbing groups in arachidon increased, the degree of molecular unfolding became larger; as the content of sulfhydryl groups decreased, the three-dimensional structure of arachisin changed locally.	The solubility of arachidrin increased significantly; foaming and foaming stability increased with the increase of homogenization pressure and reached the maximum when the treatment pressure was 120 V.	[109]
5	Li	Egg white protein	Microwave-assisted phosphorylation	The microwave technique can significantly shorten reaction times and accelerate phosphorylation process.	The 3 conditions for optimal phosphorylation modification of egg white are the concentration of sodium tripolyphosphate of 33.84 g/L, microwave power of 419.38 W, and microwave time 90 s for maximum functional properties (solubility, foaming ability, and foaming stability).	[110]
6	Miedzianka	Potato protein	Sodium trimetaphosphate (STMP)	By binding phosphate groups to the active groups of protein side chains, the electronegativity of protein molecules can be changed to increase the electrostatic repulsion between protein molecules and lower their isoelectric points.	The solubility of potato protein increases to 26% at pH 5.2.	[111]
7	Lu	Peanut protein	Sulfonated styrene cation exchange resin	The isoelectric point of the acylate peanut protein shifted, the main protein components are broken into subunits, and the amide group selectively deamidate the protein.	The solubility of modified peanut protein was improved, and the isoelectric point pH was reduced to 0.5–1; the emulsification, emulsion stability, and foaming properties were increased by 215%, 122%, and 538%.	[112]

Table 2. *Cont.*

No.	Authors	Sample	Modification Method	Modification Mechanism	Modified Results	Reference
8	Liu	Peanut protein	Dextran glycosylation	Glycosylation forms protein-polysaccharide complexes by covalent binding of proteins to polysaccharides and the introduction of sugar chains into protein polypeptide chains. Cross-linking of proteins with polysaccharides with hydrophilic hydroxyl groups increases the hydrophilicity of proteins.	Peanut protein nitrogen solubility index increased by 75%.	[113]
9	Qi	Soy protein	Pepsin and phytase complex enzymes	Enzymes modify the amino acid side chain groups of protein molecules by modifying the amino acid side-chain groups of protein molecules to partially degrade or cross-link the protein molecules to polymerize solubility and other functional properties of the protein	The nitrogen solubility index increased from 10.0% to 80.0% at pH 4.0 compared to the unmodified soybean isolate.	[114]
10	Ma	Peanut protein	Limited enzymatic hydrolysis and high-pressure homogenization; compound modification	High-pressure homogenization exposes internal groups of proteins and affects their secondary bonds, increases free sulfhydryl groups in solution, and destroys disulfide bonds, exposing more enzyme cleavage sites, making it easier for enzymes to act on peptide bonds and peptide bonds to break; accelerates protein breakdown.	The nitrogen solubility index of peanut protein concentrate increased to 96.57%.	[34]
11	Zang	Wheat protein	Ultrasonic and Glycosylation Compound Modification	Appropriate ultrasonic treatment is beneficial to the glycosylation modification of wheat gluten, and the surface hydrophobicity of the ultrasonically treated wheat gluten is reduced after grafting with glucose	The solubility of the modified wheat gluten protein is improved in the pH range of 4–7, and the solubility at the isoelectric point is 82.15%.	[115]

5.1. Physical Modification

Physical modification of proteins is the use of heating, microwave, ultrasound, high-pressure homogenization, and mechanical action to improve the functional properties of proteins by changing their secondary, tertiary, or quaternary structure and the way protein molecules are aggregated. Mechanical actions such as stirring and crushing have been shown to increase solubility by causing collisions and friction between materials to break up protein particles [116].

Heating, microwave, hygrothermal, and ultrasonic waves destroy the covalent bonds within and between peanut protein molecules through heat treatment, loosening the tightly bound internal structure, opening the hydrogen bonds, and increasing the hydration capacity. Han used hygrothermal treatment of peanut proteins and found that their solubility was increased, and the experimental results were consistent with those of Aminigo [6,117]. Zhao found by SDS-PAGE that with increasing roasting time, the soluble peanut isolate protein content continued to increase under alkaline conditions, and the foaming and emulsification properties first increased and then decreased [107]. Spray cooking has proven to be an effective means of heat-treatment solubilization. Combined with the conclusion of Section 4.2, it seems contradictory that heat treatment can both denature peanut proteins to reduce their solubility and loosen their structure to increase their solubility. A long time ago, Beuchat, Antonio, and many other researchers tried to simulate the industrial heat-treatment conditions of peanut, hoping to establish the relationship between heat-treatment conditions and changes in protein structure [111,118]. Cherry suggested that sustained high temperatures are a prerequisite for the aggregation of proteins into large insoluble aggregates [118]. Adrián showed that within a certain temperature range, the structure of proteins loosens with increasing temperature and solubility increases [119]. However, when this range is exceeded, the molecular motion is intense enough to break the bonds of the primary and secondary structures, and this denaturation leads to aggregation of the protein and thus a decrease in protein solubility. The mechanism of microwave treatment is similar to that of heating treatment. In the microwave field, the protein and water molecules can convert microwave energy into thermal energy, with the protein structure thermally denaturing, the structure loosening, and the solubility increasing. Zhang and Li's study proved that microwave modification has the advantages of uniform heating, high efficiency, and selectivity over ordinary heat treatment [108,110]; high-pressure homogenization is the most common method for protein solubilization modification. The "cavitation effect" generated by high pressure disrupts the protein peptide chain structure, resulting in a disorganized protein spatial structure and exposing a large number of charged groups leading to increased protein hydration and solubility. Tu used homogenization (0–40 MPa) dynamic microjets to increase solubility of peanut globulin from 3.97 g/L to 7.87 g/L, and the emulsification and foaming properties improved with increasing pressure [109].

Ma found that high-pressure homogenization not only improve the solubility, emulsification, and foaming of peanut protein but also reduce the allergenicity of peanut protein [34]. The mechanistically of homogenization is different from microwave treatment and heat treatment. Non-thermal plasma (NTP) is an emerging area in food research, which also shows that unfolding and exposing hydrophobic sites improve the functional properties of the proteins. Bubler and Dong used NTP to successfully improve the solubility of the soluble protein of corn and pea protein [106,120].

Physical modification has the advantages of simplicity and convenience, short time consumption, and low damage to proteins. However, physical modification mainly changes the advanced structure or intermolecular aggregation of proteins by mechanical action without involving the primary structure of proteins, which has the disadvantages of narrow modification range and limited modification effect. Physical modifications are often combined with chemical or bioenzymatic modifications as a pretreatment for complex modifications.

5.2. Chemically Modified

Chemical modification is a method of modifying the protein structure using chemical reagents. The site of action of the modification is the peptide bond, which breaks or reorganizes the internal chemical bonds of the protein or introduces various functional groups. Most studies on chemical modification of arachidonic proteins are based on the derivatization of lysine residues e-amino groups, which directly affect the charge distribution and net charge of protein molecules by introducing various functional groups and selectively derivatizing them with protein side-chain groups, thus affecting the solubility of proteins [121]. Chemical modification can be done by hydrocarbonation (Lys, Cys, Met, His, and Try), oxidation (Cys, Met, His, and Try), acylation (Lys and Tyr), esterification and amination (Glu and Asp), cross-linking, glycosylation, and filling. Acylation, phosphorylation, and glycosylation are the main modalities applied to the chemical solubilization modification of plant-based proteins. Acylation reactions are important for improving the functional properties of proteins, in particular succinylation and acetylation reactions [112]. Meanwhile, peanut and other plant proteins contain a large number of amide groups, and selective deamidation of proteins can improve the functional properties of peanut proteins. The introduction of succinyl groups can weaken the gravitational force between the amino and carbonyl groups in peanut protein molecules. The method change the molecular structure of peanut protein, making the protein molecules tend to stretch from folding, reducing the size of protein aggregates, increasing the net negative charge in peanut protein, moving the isoelectric point to low pH, and finally improving the solubility of peanut protein under acidic conditions significantly. Monteiro found that the main components of peanut protein, namely arachidonin, concomitant arachidonin I, and concomitant arachidonin II, could be modified by succinylation to lower their isoelectric points below 4 and improve their solubility and dispersibility [98]. Lu investigated the effects of succinylation modification factors on the modification effect and structural properties of peanut protein and found that the concentration of peanut protein and succinic anhydride addition were the significant influencing factors through the effect surface analysis [122]. Beuchat further reacted defatted peanut protein powder with different concentrations of succinic anhydride at pH 7.4–8.0. By using polyacrylamide gel electrophoresis, it was found that the main protein fraction broke into subunits [118]. Soluble nitrogen decreased at pH < 4 and increased between pH 6 and 7. Its water-holding capacity, water absorption, and emulsification increased, while viscosity increased significantly. The research work on the modification of peanut protein by succinylation is still in its infancy in this field. The modification of peanut proteins by protein phosphorylation has been shown to be safe and feasible. The mechanism is to use protein side-chain reactive groups to selectively bind phosphate groups to change the electronegativity and intermolecular electrostatic repulsion of protein molecules to make them more easily dispersed in food systems, thus improving the solubility and other functional properties of proteins [123]. Sung found that phosphorylation modification is the introduction of a large number of phosphate groups through the selective use of side chain groups such as hydroxyl. As the protein side chains continue to bind phosphate groups, the electrostatic repulsion and electronegativity between the peanut protein molecules change [124]. Pan, Xiong, and other experimental results proved that protein phosphorylation can effectively improve the solubility, emulsification, foam stability, and water absorption of peanut protein [125,126].

The MR can also be used to covalently bind proteins to polysaccharides (protein-glycan graft modification). By adding sugar compounds, protein polypeptide chains are covalently bound to sugar chains under heating conditions to form protein-polysaccharide complexes [90]. The cross-linking reaction between the protein and the polysaccharide with hydrophilic hydroxyl groups enhances the hydrophilicity of the protein and limits the interaction between the protein molecules, thus contributing to the improvement of the functional properties of the protein [127]. Liu graft-modified peanut isolate protein modified with dextran dry heat treatment for 7 d under heated conditions and showed a 75% increase in NSI of pH 4.0 compared to the peanut isolate protein control [128]. Protein-glycan graft modification resulted in a significant decrease in the α-helix and irregular curl of peanut proteins, a significant increase in the β-fold, and denaturation of proteins, resulting in reduced sensitivity to pH and increased solubility under acidic conditions [129]. In the process of MR, polysaccharides and proteins can also be combined by electrostatic interactions to form soluble complexes, which is conducive to improving the solubility of peanut proteins. Chitosan, because of its positive charge, can be combined with the negatively charged groups of protein relying on electrostatic interactions so that peanut protein has a certain amount of positive charge at the isoelectric point [98]. Yuan and Guo shifted the isoelectric point of peanut isolate by adjusting the protein–chitosan complex ratio, which improved its stability and solubility under acidic conditions [130,131].

Chemical modification can effectively alter the functional properties of proteins, but chemical solvent residues and toxicity of derivatized proteins are difficult to avoid. The protein–glycan graft modification associated with MR is currently the most promising chemical modification modality, but clearly predicting the cross-linking conditions of glycan components with peanut proteins is difficult [113].

5.3. Enzymatic Modification

Enzymatic modification is considered to be the most researched and promising way of protein modification [132]. The enzymatic modification has the advantages of high specificity, mild reaction conditions, controllable influencing factors, and good functionality of hydrolysis products [133]. Modification of the side-chain groups of amino acid residues of proteins by selected proteases results in peptides with desirable functional properties. Restriction enzymatic digestion has been widely reported to effectively improve the solubility of peanut proteins. Beucha reported the changes in solubility, water absorption, and emulsification of defatted peanut protein after hydrolysis by pancreatic protease, pepsin, and pineapple protease [118]; Sekul et al. found that papain-treated peanut protein had higher solubility and hydration capacity than the corresponding untreated peanut protein by blank control [134]. Li successfully improved the solubility of concentrated peanut protein in water by using neutral protease [135]. The application of enzymatic modification can solve the limitation of the modification caused by the uniqueness of the enzyme and some studies have shown that the use of alkaline protease and flavorzyme for deep hydrolysis of peanut protein can significantly improve the degree of hydrolysis of peanut protein, and the hydrolyzed protein has better solubility and better properties. It should be noted that excessive enzymatic hydrolysis tends to produce bitter peptides, which adversely affects the sensory quality of peanut protein products [136].

6. Conclusions

As people's consumption levels continue to rise, food with a single nutritional composition can no longer meet consumer demand for health. Theoretically, adding plant protein to traditional food products not only enhances the flavor of the product but also improves the nutritional value of the food, which is currently a hot spot in food development research. As described in Section 1, peanut protein has advantages that other vegetable proteins cannot match, and the use of peanut meal, a by-product of oil extraction, as a raw material for the production of IFPP can not only save costs and increase the added value of peanut meal but also give the food a pleasant, flavorful taste and nutritional value.

The importance of peanut proteins in the human diet is based on the nutritional quality and the functional properties. However, there is not a unique methodology or protocol for leading to optimal products. In fact, food development based on consumer preferences is very important for the production of goods. Even if there are adequate nutritional and physiological functions, consumers will not accept them if their preferred functions do not meet their expectations. Flavor, as an indicator that must be taken into account in the processing of peanut flour, is key to controlling the factors that influence flavor during the milling process, especially as the heat-treatment process involves the MR, CR, and LOR that are primary in the production of flavor substances. The review summarizes the recent investigations on how compounds formed during these reaction influence the sensorial properties (color, flavor, and texture) of heat-treatment peanut products and also summarizes that processing of peanuts affects the functional properties of the proteinic products and their improvement through protein modification, which should be addressed, including physics, chemical, and enzymatic technologies applied to obtain products with desirable properties for food applications.

In summary, heat treatment is a necessary way to develop and produce IFPP. The functional properties of IFPP can be modulated by carefully selecting both original seed, adjustment of heat-treatment time and temperature, and the operational variables during protein modification (pH, temperature, solvent, presence of salts, ionic strength, etc.) to establish the interaction between peanut flavoring mechanism and protein functional properties to achieve the balance between sensory properties and nutritional value of peanut thermally processed products. In conclusion, an effective protein modification approach can balance the content of protein denaturation due to heat treatment to enhance the functional properties of peanut products. Moreover, the study of the interaction between protein thermal degradation and flavor components is the technical key to establish the link between flavor substances and protein functional properties to rationally control the changes of functional properties during heat treatment and to develop IFPP.

Author Contributions: Conceptualization, Y.L. and H.H.; methodology, Y.L.; software, H.H. and H.L.; validation, Y.L.; formal analysis, Y.L. and H.H.; investigation, Y.L. and H.H.; resources, Q.W.; data curation, Y.L.; writing—original draft preparation, Y.L.; writing—review and editing, H.H., H.L. and Q.W.; visualization, Y.L.; supervision, Q.W.; project administration, Q.W.; funding acquisition, Q.W. All authors have read and agreed to the published version of the manuscript.

Funding: This research was funded by National Key R&D Program, grant number 2021YFD2100400; Hebei Oil Crop Innovation Team of Modern Agro-industry technology Research System of China (HBCT2019090203); National peanut industry technology system of China (CARS-13-08B).

Institutional Review Board Statement: Not applicable.

Informed Consent Statement: Not applicable.

Data Availability Statement: The data presented in this study are available on request from the corresponding author.

Conflicts of Interest: The authors declare no conflict of interest.

References

1. USDA. Available online: https://www.fas.usda.gov/data/oilseeds-world-markets-and-trade/ (accessed on 9 November 2021).
2. Wang, Q.; Liu, L.; Wang, L.; Guo, Y.; Wang, J. Introduction. In *Peanuts: Processing Technology and Product Development*, 1st ed.; Wang, Q., Ed.; Academic Press: Cambridge, MA, USA, 2016; pp. 1–22.
3. Liu, X.Q.; Yi, H.; Yao, L.; Ma, H.W.; Zhang, J.Y.; Wang, Z.M. Advances in plants of Polygonatum and discussion of its development prospects. *J. Chin. Pharm.* **2017**, *52*, 530–534.
4. Matthäus, B.; Musazcan Özcan, M. Oil Content, Fatty Acid Composition and Distributions of Vitamin-E-Active Compounds of Some Fruit Seed Oils. *Antioxidants* **2015**, *4*, 124–133. [CrossRef] [PubMed]
5. Suchoszek-Łukaniuk, K.; Jaromin, A.; Korycińska, M.; Kozubek, A. Chapter 103: Health Benefits of Peanut (*Arachis hypogaea* L.) Seeds and Peanut Oil Consumption. *Nuts Seeds Health Dis. Prev.* **2011**, 873–880. [CrossRef]
6. Hu, H.; Liu, H.; Shi, A.; Liu, L.; Fauconnier, M.L.; Wang, Q. The Effect of Microwave Pretreatment on Micronutrient Contents, Oxidative Stability and Flavor Quality of Peanut Oil. *Molecules* **2018**, *24*, 62. [CrossRef]
7. Colombo, A.; Ribotta, P.D.; León, A.E. Differential scanning calorimetry (DSC)studies on the thermal properties of peanut proteins. *J. Agric. Food Chem.* **2010**, *58*, 4434–4439. [CrossRef]
8. Molino Artiz, S.E.; Wagner, J.R. Hydrolysates of native and modified soy protein isolates: Structural characteristics, solubility and foaming properties. *Food Res. Int.* **2002**, *35*, 511–518. [CrossRef]
9. Dong, X.H.; Zhao, M.M.; Jiang, Y.M. Research progress of peanut protein modification. *Chin. J. Cereals Oils* **2011**, *26*, 109–117.
10. Cherry, J.P. Peanut protein and product functionality. *J. Am. Oil Chem. Soc.* **1990**, *67*, 5473–5481. [CrossRef]
11. Dong, X.H.; Zhao, M.M.; Yang, B.; Jiang, Y.M. Effect of high-pressure homogenization on the functional property of peanut protein. *J. Food Process Eng.* **2010**, *34*, 2191–2204. [CrossRef]
12. Feng, X.L.; Liu, H.Z.; Shi, A.M.; Liu, L.; Wang, Q.; Adhikari, B. Effects of transglutaminase catalyzed crosslinking on physico-chemical characteristics of arachin and conarachin-rich peanut protein fractions. *Food Res. Int.* **2014**, *62*, 84–90. [CrossRef]
13. Gao, Y.; Liu, C.; Yao, F.; Chen, F. Aqueous enzymatic extraction of peanut oil body and protein and evaluation of its physicochemical and functional properties. *Int. J. Food Eng.* **2021**, *17*, 897–908. [CrossRef]
14. Boukid, F. Peanut protein—An underutilised by-product with great potential: A review. *Int. J. Food Sci. Technol.* **2021**, 111438. [CrossRef]
15. Gong, K.J.; Shi, A.M.; Liu, H.Z.; Liu, L.; Hu, H.; Adhikari, B.; Wang, Q. Emulsifying properties and structure changes of spray and freeze-dried peanut protein isolate. *J. Food Eng.* **2016**, *170*, 33–40. [CrossRef]
16. Zhao, Z.; Shi, A.; Wang, Q.; Zhou, J. High oleic acid peanut oil and extra virgin olive oil supplementation attenuate metabolic syndrome in rats by modulating the gut microbiota. *Nutrients* **2019**, *11*, 3005. [CrossRef]
17. Leunissen, M.; Davidson, V.J.; Kakuda, Y. Analysis of volatileflavor components in roasted peanuts using supercriticalfluid extraction and gas chromatography mass spectrometry. *Agric. Food Chem.* **1996**, *44*, 2694–2699. [CrossRef]
18. Mu, J.; Yu, X. Determination of the composition of peanut protein powder and its properties. *Molecules* **2018**, *24*, 163–168.
19. Ros, E. Health Benefits of Nut Consumption. *Nutrients* **2010**, *2*, 652–682. [CrossRef]
20. Wang, Q. *Introduction to Peanut Bioactive Substances*; China Agricultural University Press: Beijing, China, 2012.
21. Chetschik, I.; Granvogl, M.; Schieberle, P. Comparison of the Key Aroma Compounds in Organically Grown, Raw West-African Peanuts (*Arachis hypogaea*) and in Ground, Pan-Roasted Meal Produced Thereof. *J. Agric. Food Chem.* **2008**, *56*, 10237. [CrossRef]
22. Qian, D.; Yao, L.; Deng, Z.; Li, H.; Li, J.; Fan, Y.; Zhang, B. Effects of hot and cold-pressed processes on volatile compounds of peanut oil and corresponding analysis of characteristic flavor components. *LWT* **2019**, *112*, 107648.
23. Walradt, J.P.; Pittet, A.O.; Kinlin, T.E.; Muralidhara, R.; Sanderson, A. Volatile components of roasted peanuts. *J. Agric. Food Chem.* **1971**, *19*, 972–979. [CrossRef]
24. Brown, D.F.; Stanley, J.B.; Senn, V.J.; Dollear, F.G. Comparison of carbonyl-compounds in raw and roasted runner peanuts. I. Major qualitative and some quantitative differences. *J. Agric. Food Chem.* **1972**, *20*, 700–706. [CrossRef]
25. Balagiannis, D.P.; Parker, J.K.; Pyle, D.L.; Desforges, N.; Mottram, D.S. Modelling the generation of flavour in a real food system. In *Expression of Multidisciplinary Flavour Science: Proceedings of the 12th Weurman Symposium*; Blank, I., Wüst, M., Yeretzian, C., Eds.; Zürcher Hochschule für Angewandte Wissenschaften: Wädenswil, Switzerland, 2010; pp. 284–287.
26. Ajandouz, E.H.; Tchiakpe, L.S.; Ore, F.D.; Benajiba, A.; Puigserver, A. Effects of pH on caramelization and Maillard reaction kinetics in fructose—Lysine model systems. *J. Food Sci.* **2001**, *66*, 926–931. [CrossRef]
27. Shi, A.; Wang, Q.; Liu, H.; Wang, L.; Zhang, J.; Du, Y.; Chen, X. Peanut Processing Quality Evaluation Technology. In *Peanuts: Processing Technology and Product Development*; Academic Press: Cambridge, MA, USA, 2016; pp. 23–61. [CrossRef]
28. Li, J.; Shi, A.; Liu, H. Research progress on the modification method and application of vegetable protein for solubility enhancement under acidic conditions. *China Fats Oils* **2019**, *44*, 59–65.
29. Moure, A.; Sineiro, J.; Dominguez, H.; Parajó, J.C. Functionality of oilseed protein products: A review. *Food Res. Int.* **2006**, *39*, 945–963. [CrossRef]
30. Zhang, S. Study on the Production Technology of Low Denatured Peanut Protein Powder. *J. Wuhan Univ. Light Ind.* **2003**, *22*, 10–11.

31. Qin, M. *Research on the Production of Low Denatured Peanut Protein Powder by Low Temperature Pressing and Low Temperature Leaching Process*; Qingdao Liangquan Group Co., Ltd.: Qingdao, China, 2006.
32. Sun, Q. *Preparation and Modification of Low Denatured Peanut Protein*; Qingdao Dongsheng Group Co., Ltd.: Qingdao, China, 2010.
33. Wang, Q. *Research on Key Technologies for the Preparation of Functional Short Peptides from Peanut*; Institute of Agricultural Products Processing, Chinese Academy of Agricultural Sciences: Beijing, China, 2007.
34. Ma, T.; Zhu, H.; Wang, J.; Wang, Q.; Yu, L.; Sun, B. Influence of extraction and solubilizing treatments on the molecular structure and functional properties of peanut protein. *LWT Food Sci. Technol.* **2017**, *79*, 197–204. [CrossRef]
35. Dkv, A.; As, B.; Akn, B.; Mt, C.; Ns, D.; Ss, E. Recent trends in microbial flavour compounds: A review on chemistry, synthesis mechanism and their application in food. *Saudi J. Biol. Sci.* **2022**, *29*, 1565–1576.
36. Chetschik, I.; Granvogl, M.; Schieberle, P. Quantitation of key peanut aroma compounds in raw peanuts and pan-Roasted peanut meal. Aroma Reconstitution and Comparison with Commercial Peanut Products. *J. Agric. Food Chem.* **2010**, *58*, 11018–11026. [CrossRef]
37. Yun, L.; Hu, H.; Liu, H.; Wang, Q. Flavor analysis of peanut cake meal and peanut shell baking. *J. Food Sci.* **2017**, *38*, 146–153.
38. Liu, X.; Jin, Q.; Liu, Y.; Liu, Y.; Huang, J.; Wang, X.; Mao, W.; Wang, S. Changes in volatile compounds of peanut oil during the roasting process for pduction of aromatic roasted peanut oil. *Food Sci.* **2011**, *76*, C404–C412. [CrossRef] [PubMed]
39. Ho, C.T.; Lee, M.H.; Chang, S.S. Isolation and identification of volatile compounds from roasted peanuts. *Food Sci.* **1981**, *47*, 127–133. [CrossRef]
40. Yu, H.; Liu, H.; Erasmus, S.W.; Zhao, S.; Wang, Q.; van Ruth, S.M. An explorative study on the relationships between the quality traits of peanut varieties and their peanut butters. *LWT* **2021**, *151*, 112068. [CrossRef]
41. Zhang, J.; Sun, D.; Feng, Y.; Su, G.; Zhao, M.; Lin, L. The umami intensity enhancement of peanut protein isolate hydrolysate, and its derived factions and peptides by Maillard reaction and the analysis of peptide (EP) Maillard products. *Food Res. Int.* **2019**, *120*, 895–903. [CrossRef]
42. Shen, J.; Liu, L.; Qian, J. Effects of Polygonatum polysaccharide on immunological activity of immunosuppressive mouse model. *Drug Eval. Res.* **2012**, *35*, 328–331.
43. Nepote, V.; Olmedo, R.H.; Mest Ra Llet, M.G.; Grosso, N.R. A Study of the Relationships among Consumer Acceptance, Oxidation Chemical Indicators, and Sensory Attributes in High-Oleic and Normal Peanuts. *Food Sci.* **2009**, *74*, 218–225. [CrossRef] [PubMed]
44. Ku, K.L.; Lee, R.S.; Young, C.T.; Chiou, R.Y.Y. Roasted peanutflavor and related compositional characteristics of peanut kernels of spring and fall crops grown in Taiwan. *J. Agric. Food Chem.* **1998**, *46*, 3220–3224. [CrossRef]
45. Settaluri, V.S.; Kandala, C.; Puppala, N.; Sundaram, J. Peanuts and their nutritional aspects—A review. *Food Nutr. Sci.* **2015**, *3*, 1644–1650. [CrossRef]
46. Maga, J.A. Phytate: Its chemistry, occurrence, food interactions, nutritional significance, and methods of analysis. *J. Agric. Food Chem.* **1982**, *30*, 1–9. [CrossRef]
47. Hu, H.; Shi, A.; Liu, H.; Liu, L.; Fauconnier, M.L.; Wang, Q. Study on Key Aroma Compounds and Its Precursors of Peanut Oil Prepared with Normal- and High-Oleic Peanuts. *Foods* **2021**, *10*, 3036. [CrossRef]
48. Mason, M.E.; Johnson, B.; Hamming, M. Flavor components of roasted peanuts. Some low molecular weight pyrazines and a pyrrole. *Agric. Food Chem.* **1966**, *14*, 454–460. [CrossRef]
49. Baker, G.L.; Cornell, J.A.; Gorbet, D.W.; O'Keefe, S.F.; Sims, C.A.; Talcott, S.T. Determination of pyrazine and flavor variations in peanut genotypes during roasting. *Food Sci.* **2003**, *68*, 394–400. [CrossRef]
50. Johnson, B.R.; Waller, G.R.; Foltz, R.L. Volatile components of roasted peanuts: Neutral fraction. *Agric. Food Chem.* **1971**, *19*, 1025–1027.
51. Buckholz, L.L.; Withcombe, D.A.; Daun, H. Application and characteristics of a polymer adsorption method used to analyzeflavor volatiles from peanuts. *J. Agric. Food Chem.* **1980**, *28*, 760–765. [CrossRef]
52. Schirack, A.V.; Drake, M.A.; Sanders, T.H.; Sandeep, K.P. Characterization of aroma-active compounds in microwave blanched peanuts. *Food Sci.* **2006**, *71*, 513–520. [CrossRef]
53. Basha, S.M.; Young, C.T. Protein fraction producing off-flavor headspace volatiles inpeanut seed. *J. Agric. Food Chem.* **1996**, *44*, 3070–3074. [CrossRef]
54. Young, C.T.; Hovis, A.R. A method for the rapid analysis of headspace volatiles of raw and roasted peanuts. *Food Sci.* **1990**, *55*, 279–280. [CrossRef]
55. Coelho, G.; Mendes, M.F.; Pessoa, F. *Handbook of Fruit and Vegetable Flavors*; John Wiley & Sons, Inc.: Hoboken, NJ, USA, 2010.
56. Shu, Y.; Liu, Y.; Jiang, Y. Effect of roasting temperature of fresh peanut kernels on flavor and overall quality of peanut butter. *Food Sci.* **2020**, *41*, 8.
57. Lou, F.; Liu, Y.; Sun, X. Pan, Y.; Zhao, J.; Zhao, Y. Identification of volatile flavor components in peanut butter. *J. Food Sci.* **2009**, *30*, 393–396.
58. Hathorn, C.S.; Sanders, T.H. Flavor and Antioxidant Capacity of Peanut Paste and Peanut Butter Supplemented with Peanut Skins. *Food Sci.* **2012**, *77*, S407–S411. [CrossRef]
59. Starowicz, M.; Zieliński, H. How Maillard Reaction Influences Sensorial Properties (Color, Flavor and Texture) of Food Products? *Food Rev. Int.* **2019**, *35*, 1–19. [CrossRef]
60. Schieberle, P. The carbon Module labeling (CAMOLA) technique: A useful tool for identifying transient intermediates in the formation of Maillard-type target molecules. *Ann. N. Y. Acad. Sci.* **2005**, *1043*, 236–248. [CrossRef] [PubMed]

61. Yu, H.; Zhang, R.; Yang, F.; Xie, Y.; Guo, Y.; Yao, W.; Zhou, W. Control strategies of pyrazines generation from Maillard reaction. *Trends Food Sci. Technol.* **2021**, *112*, 795–807. [CrossRef]
62. Yu, H.; Zhong, Q.; Xie, Y.; Guo, Y.; Cheng, Y.; Yao, W. Kinetic study on the generation of furosine and pyrraline in a Maillard reaction model system of dglucose and l-lysine. *Food Chem.* **2020**, *317*, 126458. [CrossRef]
63. Yu, H.; Seow, Y.-X.; Ong, P.K.C.; Zhou, W. Effects of high-intensity ultrasound on Maillard reaction in a model system of d-xylose and l-lysine. *Ultrason. Sonochem.* **2017**, *34*, 154–163. [CrossRef] [PubMed]
64. Martins, S.; Jongen, W.; Boekel, M. A review of Maillard reaction in food and implications to kinetic modelling. *Trends Food Sci. Technol.* **2000**, *11*, 364–373. [CrossRef]
65. Newell, J.A.; Mason, M.E.; Matlock, R.S. Precursors of typical atypical roasted peanut flavor. *Agric. Food Chem.* **1967**, *15*, 767–772. [CrossRef]
66. Suri, K.; Singh, B.; Kaur, A.; Singh, N. Impact of roasting and extraction methods on chemical properties, oxidative stability and Maillard reaction products of peanut oils. *J. Food Sci. Technol.* **2019**, *56*, 2436–2445. [CrossRef]
67. Tian, Y.; Liu, C.; Zhang, K.; Tao, S.; Xue, W. Glycosylation between recombinant peanut protein Ara h 1 and glucosamine could decrease the allergenicity due to the protein aggregation. *LWT* **2020**, *127*, 109374. [CrossRef]
68. Mason, M.E.; Newell, J.A.; Johnson, B.R.; Koehler, P.E.; Waller, G.R. Nonvolatile flavor components of peanut. *Agric. Food Chem.* **1969**, *17*, 728–732. [CrossRef]
69. Buckholz, L.L. Microwave Browning and Baking to Give Surface Crust. US4943697-A, 24 July 1990.
70. Momin, A.H.; Acharya, S.S.; Gajjar, A.V. Corriandrum Sativum review of advances in phytopharmacology. *Int. J. Pharm. Sci. Res.* **2012**, *13*, 1233–1239.
71. Andreou, A.; Feussner, I. Lipoxygenases—Structure and reaction mechanism. *Phytochemistry* **2009**, *70*, 1504–1510. [CrossRef] [PubMed]
72. Alzagtat, A.A.; Alli, I. Protein-lipid interactions in food systems: A review. *Int. Food Nutr.* **2002**, *53*, 249–260. [CrossRef]
73. Funes, J.A.; Weiss, U.; Karel, M. Effects of reaction conditions and reactant concentrations on polymerization of lysozyme reacted with peroxidizing lipids. *Agric. Food Chem.* **1982**, *30*, 1204–1208. [CrossRef]
74. Vercellotti, J.R.; Mills, O.E.; Bett, K.L.; Sullen, D.L. Gas chromatographic analyses of lipid oxidation volatiles in foods. In *Lipid Oxidation in Food*; St. Angelo, A., Ed.; American Chemical Society: Washington, DC, USA, 1992; pp. 232–263.
75. Williams, J.P.; Duncan, S.E.; Williams, R.C.; Mallikarjunan, K.; Eigel, W.N.; O'Keefe, S.F. Flavor fade in peanuts during short-term storage. *Food Sci.* **2006**, *71*, S265–S269. [CrossRef]
76. Rey, F.; Melo, T.; Lopes, D.; Couto, D.; Marques, F.; Domingues, M.R. Applications of lipidomics in marine organisms: Progress, challenges and future perspectives. *Mol. Omics* **2022**. [CrossRef]
77. Girotti, A.W.; Korytowski, W. Intermembrane Translocation of Photodynamically Generated Lipid Hydroperoxides: Broadcasting of Redox Damage. *Photochem. Photobiol.* **2022**. [CrossRef]
78. Zhang, W.; Cao, X.; Liu, S.Q. Aroma modulation of vegetable oils—A review. *Crit. Rev. Food Sci. Nutr.* **2022**, *62*, 1740–1751. [CrossRef]
79. Min, D.B.; Boff, J.M. *Lipid Oxidation of Edible Oil*; Department of Food Science and Technology, The Ohio State University: Columbus, OH, USA, 2002.
80. Shahidi, F. *Bailey's Industrial Oil and Fat Products, Volume 3, Edible Oil and Fat Products: Specialty Oils and Oil Products, Part 2*, 6th ed.; John Wiley & Sons: Hoboken, NJ, USA, 2005.
81. Pattee, H.E.; Isleib, T.G.; Giesbrecht, F.G.; Cui, Z. Prediction of parental genetic compatibility to enhance flavor attributes of peanuts. In *Crop Biotechnology*; Rajasekaran, K., Jacks, T.J., Finley, J.W., Eds.; American Chemical Society: Washington, DC, USA, 2002; pp. 217–230.
82. Reed, K.A.; Sims, C.A.; Gorbet, D.W.; O'Keefe, S.F. Storage water activity affects flavor fade in high and normal oleic peanuts. *Food Res. Int.* **2002**, *35*, 769–774. [CrossRef]
83. Mate, J.I.; Salveit, M.E.; Krochta, J.M. Peanut and walnut rancidity: Effects of oxygen, concentration and relative humidity. *Food Sci.* **1996**, *61*, 465–472. [CrossRef]
84. Kamal-Eldin, A. *Lipid Oxidation Pathways*; AOCS: Champaign, IL, USA, 2003.
85. Zhang, W.; Rui, W.; Yuan, Y.; Yang, T.; Liu, S. Changes in volatiles of palm kernel oil before and after kernel roasting. *LWT Food Sci. Technol.* **2016**, *73*, 432–441. [CrossRef]
86. Kroh, L.W. Caramelisation in Food and Beverages. *Food Chem.* **1994**, *51*, 373–379. [CrossRef]
87. Hwang, H.I.; Hartman, T.G.; Rosen, R.T.; Lech, J.; Ho, C.T. Formation of pyrazines from the Maillard reaction of glucose and lysine-alpha-amine-N-15. *Agric. Food Chem.* **1994**, *42*, 1000–1004. [CrossRef]
88. Friedman, M. Food browning and its prevention: An overview. *Agric. Food Chem.* **1996**, *44*, 631–653. [CrossRef]
89. Beksan, E.; Schieberle, P.; Robert, F.; Blank, I.; Fay, L.B.; Schlichtherle-Cerny, H.; Hofmann, T. Synthesis and sensory characterization of novel umami-tasting glutamate glycoconjugates. *Agric. Food Chem.* **2003**, *51*, 5428–5436. [CrossRef] [PubMed]
90. Chumngoen, W.; Chen, C.F.; Tan, F. Effects of moist- and dry-heat cooking on the meat quality, microstructure and sensory characteristics of native chicken meat. *Anim. Sci. J.* **2018**, *89*, 193–201. [CrossRef]
91. Hou, C.; Li, X.; Wang, Z.; Huang, C.; Zhang, Q.; Luo, Z.; Zhang, D. Analysis of amino acid and fatty acid contents and nutritional value evaluation of different parts of yak meat. *Meat Res.* **2019**, *33*, 52–57.
92. Kinsella, J.E. Functional properties of food proteins: A review. *Crit. Rev. Food Sci. Nutr.* **1976**, *7*, 219–280. [CrossRef]

93. Koumanov, A.; Ladenstein, R.; Karshikoff, A. Electrostatic interactions in proteins: Contribution to structure–function relationships and stability. *Recent Res. Dev. Protein Eng.* **2001**, *1*, 123–148.
94. Schwenke, K.D. Reflections about the functional potential of legume proteins. *Nahrung* **2001**, *45*, 377–381. [CrossRef]
95. Shichijo, K.; Yamada, T.; Shimizu, T. Independence of the goiter development to the concentration of circulating protein bound iodine. *Endocrinol. Jpn.* **1957**, *4*, 120. [CrossRef] [PubMed]
96. Damodaran, S. Food proteins: An overview. In *Food Proteins and Their Applications*; Damodaran, S., Paraf, A., Eds.; Marcel Dekker: New York, NY, USA, 1997; pp. 1–21.
97. Morr, C.V. Current status of soy protein functionality in food systems. *J. Am. Oil Chem. Soc.* **1990**, *67*, 265–271. [CrossRef]
98. Jiao, B. *Preparation and Stabilization Mechanism of Peanut Protein-Polysaccharide Pickering Emulsion*; Chinese Academy of Agricultural Sciences: Beijing, China, 2018.
99. Monteiro, P.V.; Prakash, V. Alteration of functional properties of peanut (*Arachis hypogaea* L.) protein fractions by chemical and enzymatic modifications. *J. Food Sci. Technol.* **1996**, *33*, 19–26.
100. Liu, L.; Shi, A.-M.; Liu, H.-Z.; Hu, H.; Wang, Q. Research progress on the structure and properties of peanut protein subunits. *Chin. J. Cereals Oils* **2016**, *31*, 151–156.
101. Birch, G.G.; Kemp, S.E. Apparent specific volumes and tastes of amino acids. *Chem. Sens.* **1989**, *14*, 249–258. [CrossRef]
102. Kilara, A.; Sharkasi, T.Y. Effects of temperature on food proteins and its implications on functional properties. *Crit. Rev. Food Sci. Nutr.* **1986**, *23*, 323–395. [CrossRef]
103. Han, Z.; Guo, F.; Chen, J. Effects of different heat treatments on the main functional properties of peanut protein. *Food Ind. Sci. Technol.* **2000**, *21*, 37–38.
104. Li, H.; Li, X.; Zhang, C.-H.; Wang, J.-Z.; Tang, C.-H.; Chen, L.-L. Flavor compounds and sensory profiles of a novel Chinese marinated chicken. *J. Sci. Food Agric.* **2016**, *96*, 1618–1626. [CrossRef]
105. Wang, Q. *Peanut Processing Characteristics and Quality Evaluation*; Springer: Singapore, 2018.
106. Bußler, S.; Steins, V.; Ehlbeck, J.; Schlüter, O. Impact of thermal treatment versus cold atmospheric plasma processing on the technofunctional protein properties from Pisum sativum 'Salamanca'. *J. Food Eng.* **2015**, *167*, 166–174. [CrossRef]
107. Zhao, G.; Zhao, M.; Liu, Y.; Cui, C. Effects of roasting on the structure and functional properties of peanut protein isolate. *Food Ferment. Ind.* **2009**, *35*, 20–23.
108. Zhang, X.; Liu, Y.; Wang, X. Study on microwave modification of peanut protein concentrate. *Grain Oil Process.* **2010**, *6*, 19–22.
109. Tu, Z.; Zhang, X.; Liu, C. Effects of ultra-high pressure microfluidics on the structure of peanut protein. *Chin. J. Agric. Eng.* **2008**, *24*, 306–308.
110. Li, P.; Sheng, L.; Jin, Y. Using microwave-assisted phosphorylation to improve foaming and solubility of egg white by response surface methodology. *Poult. Sci.* **2019**, *98*, 5. [CrossRef] [PubMed]
111. Beuchat, L.R. Functional and electrophoretic characteristics of succinylated peanut flour protein. *J. Agric. Food Chem.* **1977**, *25*, 258–261. [CrossRef]
112. Schnabelrauch, M.; Schiller, J.; Möller, S.; Scharnweber, D.; Hintze, V. Chemically modified glycosaminoglycan derivatives as building blocks for biomaterial coatings and hydrogels. *Biol. Chem.* **2021**, *402*, 1385–1395. [CrossRef]
113. Liu, Y.; Zhao, G.; Zhao, M.; Ren, J.; Bo, Y. Improvement of functional properties of peanut protein isolate by conjugation with dextran through Maillard reaction. *Food Chem.* **2012**, *131*, 901–906. [CrossRef]
114. Qi, J.; Weng, J.; Kang, Y. Preparation and characterization of soybean acid-soluble protein/soybean polysaccharide nanoemulsion. *J. Mod. Food Technol.* **2015**, *6*, 136–141.
115. Zang, Y.; Zhao, Y.; Luo, S. Effects of ultrasonic and glycosylation compound modification on the properties and structure of wheat gluten. *J. Food Sci.* **2017**, *38*, 122–128.
116. Liu, C.; Pei, R.; Heinonen, M. Faba bean protein: A promising plant-based emulsifier for improving physical and oxidative stabilities of oil-in-water emulsions. *Food Chem.* **2022**, *369*, 130879. [CrossRef]
117. Aminigo, E.R.; Ogundipe, H.O. Effect of heat treatment on functional characteristics of peanut (*Arachis hypogeae*) meal. *J. Food Sci. Technol.* **2003**, *40*, 205–208.
118. Beuchat, L.R.; Cherry, J.P.; Quinn, M.R. Physicochemical properties of peanut flour as affected by proteolysis. *J. Agric. Food Chem.* **2015**, *23*, 616–620. [CrossRef] [PubMed]
119. Lomelí-Martín, A.; Martínez, L.M.; Welti-Chanes, J.; Escobedo-Avellaneda, Z. Induced Changes in Aroma Compounds of Foods Treated with High Hydrostatic Pressure: A Review. *Foods* **2021**, *10*, 878. [CrossRef] [PubMed]
120. Dong, S.; Gao, A.; Zhao, Y.; Li, Y.; Chen, Y. Characterization of physicochemical and structural properties of atmospheric cold plasma (ACP) modified zein. *Food Bioprod. Process.* **2017**, *106*, 65–74. [CrossRef]
121. Wang, Z. Crosslinked Recombinant-Arah1 Catalyzed by Microbial Transglutaminase: Preparation, Structural Characterization and Allergic Assessment. *Foods* **2020**, *9*, 1508.
122. Lu, Y.X.H.; Zheng, Z. Study on functional improvement of peanut protein processing. *Food Sci.* **1995**, *16*, 6.
123. Matheis, G. Phosphorylation of food proteins with phosphorus oxychloride-Improvement of functional and nutritional properties: A review. *J. Food Chem.* **1991**, *39*, 13–26. [CrossRef]
124. Nam, S.C.; Sung, H.R.; Kang, S.H.; Joo, J.Y.; Lee, S.J.; Chung, Y.B.; Lee, C.K.; Song, S.G. Phosphorylation-dependent septin interaction of Bni5 is important for cytokinesis. *J. Microbiol.* **2007**, *45*, 227–233.
125. Xiong, L.; Sun, G.; Wang, J. Study on phosphorylation modification of peanut protein isolate. *J. Food Sci.* **2010**, *10*, 35–41.

126. Pan, Q.; Shen, B.; Cheng, S. Phosphorylation modification of peanut protein. *J. China Oils Fats* **2017**, *22*, 25–27.
127. Mouécoucou, J.; Fremont, S.; Sanchez, C.; Villaume, C.; Méjean, L. In vitro allergenicity of peanut after hydrolysis in the presence of polysaccharides. *Clin. Exp. Allergy* **2004**, *34*, 1429–1437. [CrossRef]
128. Liu, C.C.; Tellez Garay, A.M.; Castell Perez, M.E. Physical and mechanical properties of peanut protein films. *LWT-Food Sci. Technol.* **2004**, *37*, 731–738. [CrossRef]
129. Marcela, J.P.; Zeng, T.; Temelli, F.; Chen, L. Understanding the stability mechanisms of lentil legumin like protein and polysaccharide foams. *Food Hydrocolloid* **2016**, *61*, 903–913.
130. Yuan, Y. *Study on the Interaction between Food Protein and Chitosan and Its Application in Food System*; South China University of Technology: Guangzhou, China, 2014.
131. Guo, R. *Preparation and Application of Functional Soybean Protein*; South China University of Technology: Guangzhou, China, 2011.
132. Zhao, G.; Yan, L.; Zhao, M.; Ren, J.; Bo, Y. Enzymatic hydrolysis and their effects on conformational and functional properties of peanut protein isolate. *Food Chem.* **2011**, *127*, 1438–1443. [CrossRef]
133. Ambrosi, V.; Polenta, G.; Gonzalez, C.; Ferrari, G.; Maresca, P. High hydrostatic pressure assisted enzymatic hydrolysis of whey proteins. *Innov. Food Sci. Emerg. Technol.* **2016**, *38*, 294–301. [CrossRef]
134. Sekul, A.A.; Vinnett, C.H.; Ory, R.L. Some functional properties of peanut proteins partially hydrolyzed with papain. *J. Agric. Food Chem.* **1978**, *26*, 855–858. [CrossRef]
135. Li, N.; Liu, H.; Liu, L.; Wang, Q. Study on the preparation of peanut peptides by step-by-step enzymatic hydrolysis of peanut protein isolate by neutral protease. *J. China Agric. Sci.* **2013**, *46*, 5237–5247.
136. Zhao, G.; Liu, Y.; Ren, J.; Zhao, M.; Yang, B. Effect of protease pretreatment on the functional properties of protein concentrate from defatted peanut flour. *J. Food Process Eng.* **2013**, *36*, 9–17. [CrossRef]

Review

Changes of Soybean Protein during Tofu Processing

Xiangfei Guan [1,2] , Xuequn Zhong [1], Yuhao Lu [1], Xin Du [1], Rui Jia [1], Hansheng Li [2] and Minlian Zhang [1,*]

[1] Department of Chemical Engineering, Institute of Biochemical Engineering, Tsinghua University, Beijing 100084, China; gxf0413@163.com (X.G.); xq.behappy@gmail.com (X.Z.); luyh17@tsinghua.org.cn (Y.L.); du-x18@mails.tinghua.edu.cn (X.D.); jiar19@mails.tsinghua.edu.cn (R.J.)

[2] School of Chemistry and Chemical Engineering, Beijing Institute of Technology, Beijing 102488, China; hanshengli@bit.edu.cn

* Correspondence: zhangminlian@mail.tsinghua.edu.cn; Tel./Fax: +86-10-6279-5473

Abstract: Tofu has a long history of use and is rich in high-quality plant protein; however, its production process is relatively complicated. The tofu production process includes soybean pretreatment, soaking, grinding, boiling, pulping, pressing, and packing. Every step in this process has an impact on the soy protein and, ultimately, affects the quality of the tofu. Furthermore, soy protein gel is the basis for the formation of soy curd. This review summarizes the series of changes in the composition and structure of soy protein that occur during the processing of tofu (specifically, during the pressing, preservation, and packaging steps) and the effects of soybean varieties, storage conditions, soybean milk pretreatment, and coagulant types on the structure of soybean protein and the quality of tofu. Finally, we highlight the advantages and limitations of current research and provide directions for future research in tofu production. This review is aimed at providing a reference for research into and improvement of the production of tofu.

Keywords: tofu; protein; structure; mechanism

Citation: Guan, X.; Zhong, X.; Lu, Y.; Du, X.; Jia, R.; Li, H.; Zhang, M. Changes of Soybean Protein during Tofu Processing. *Foods* **2021**, *10*, 1594. https://doi.org/10.3390/foods10071594

Academic Editors: Qiang Wang and Aimin Shi

Received: 31 May 2021
Accepted: 1 July 2021
Published: 9 July 2021

Publisher's Note: MDPI stays neutral with regard to jurisdictional claims in published maps and institutional affiliations.

Copyright: © 2021 by the authors. Licensee MDPI, Basel, Switzerland. This article is an open access article distributed under the terms and conditions of the Creative Commons Attribution (CC BY) license (https://creativecommons.org/licenses/by/4.0/).

1. Introduction

Tofu originated in the time of the Western Han dynasty and has been consumed for more than two thousand years. Tofu is rich in soy protein and has high nutritional value [1]. Processing of soy products can remove most of the anti-nutritional factors in soy and significantly improve the digestibility of soy protein. Studies have shown that the digestibility of whole ripe soybeans is only 65.3%; after processing it into soy milk and tofu, the digestibility becomes 85% and 92–98%, respectively [2]. Furthermore, the FDA authorized a "Soy Protein Health Claim" on 26 October 1999 stating that 25 g of soy protein a day may reduce the risk of heart disease. In addition to protein, tofu contains lipids, carbohydrates, crude fiber, isoflavones, minerals, and saponins, which can lower cholesterol, alleviate the symptoms of cardiovascular and kidney diseases, and reduce the incidence of cancer and tumors [3].

Several types of tofu are available in the market to meet the different needs of consumers, each produced via a different complex process. Here, we review the changes that occur in soybean protein in each step of tofu production (specifically, during soaking and refining). The types, composition, and structure of soy protein and the effects of soy varieties and storage conditions on soy protein are summarized. In addition, the effects of pretreatment steps and coagulants on soy protein structure and tofu quality are introduced. We also summarize the advantages and disadvantages of current research and provide directions for future research into soy protein structural changes during tofu processing. Overall, this review is expected to serve as a foundation for research into the curdling mechanism of tofu and as a theoretical guide for the actual production process.

2. Soy Protein

Soy is composed of approximately 40% proteins, 20% lipids, 25% carbohydrates, and 5% crude fibers [2,4]. They are also rich sources of isoflavones, minerals, and other components. The nutrient contents of soybean are shown in Figure 1.

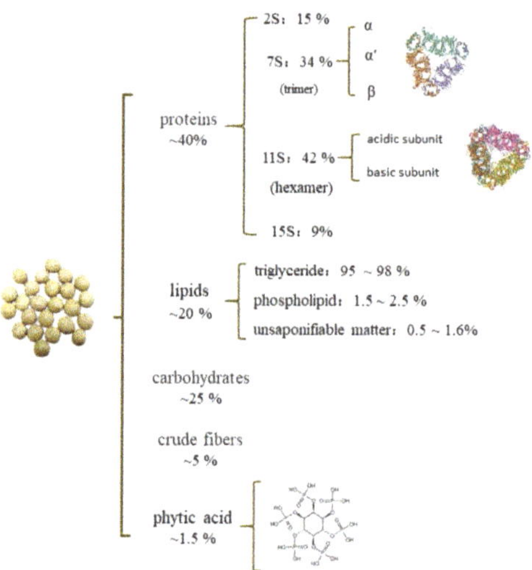

Figure 1. Nutritional composition of soybean [5,6]; Protein Data Bank.

Soy protein can be classified according to its solubility, physiological function, and centrifugal sedimentation speed. Based on solubility, soy protein is divided into globulin and albumin. Based on physiological functions, it is divided into storage protein and biologically active protein. Based on centrifugal sedimentation speed, it is divided into 2S, 7S, 11S, and 15S components (S refers to the sedimentation coefficient, where $1S = 10^{-13}$ s = 1 Svedberg), and each component is composed of protein molecules with similar molecular weights [4]. The graded composition of soy protein is shown in Table 1.

The 7S conglycinin and 11S glycinin are the key components of tofu curd, accounting for more than 70% of the total soy protein content [11]. 7S conglycinin is a trimer composed of an α subunit, α' subunit, and β subunit and accounts for approximately 30% of the soy protein content [12]. 11S glycinin is a hexamer composed of acidic polypeptides (A) and basic polypeptides (B) linked by disulfide bonds and accounts for approximately 40% of the soy protein content. The composition and functions of 7S conglycinin and 11S glycinin are shown in Table 2. Ren et al. [13] analyzed the interaction between protein subunits in soy milk and proposed a model for protein subunit structure. The β subunit and the B subunit form the hydrophobic core of the granule protein through electrostatic interaction, in which the B subunit is covalently connected by disulfide bonds. Other hydrophilic subunits, such as α, α', and A subunits, are distributed around this hydrophobic core through hydrophobic interactions and hydrogen bonds.

The functional properties of soy protein mainly include gelling, emulsification, and foaming; of these, gelling is the basis for the formation of tofu. During tofu production, soy protein undergoes dissociation or association reactions during the acid-base treatment and heat treatment, which changes the ionic strength of the solution. Through these association–dissociation reactions, 11S glycinin polymerizes to form dimers, oligomers, or multimers or dissociates to form 7S and 3S components [14].

Table 1. Composition, structure, and physiological functions of soy protein [5–10]; Protein Data Bank.

Component	Ingredient	Structure	pH	Relative Molecular Mass	
				A	B
2S (15~22%)	Trasylol		4.5	8000~21,500	15,000~30,000
	Cytochrome C		10.2~10.8	12,000	
7S (34~37%)	Hemagglutinin		-	102,000	100,000~200,000
	Lipoxygenase		5.7~6.4	102,000	
	β-amylase		5.0~6.5	61,700	
	β-Conglycinin		5.07~5.88	180,000~210,000	
11S (31~42%)	Glycinin		5.28~5.78	350,000	350,000
15S (9~10%)	-	-	-	600,000	600,000

-: Not Available.

Table 2. The composition and functions of 7S and 11S [15,16].

Protein	Subunit	Molecular Weight (kDa)	Isoelectric Point
7S β-Conglycinin	α	57~72	5.23
	α'	57~68	5.07
	β	45~52	5.88
11S Glycinin	A1aB1b	53.6	-
	A2B1a	52.4	-
	A1bB2	52.2	-
	A5A4B3	61.2	-
	A3B4	55.4	-
	A1a	-	5.78
	A1b	-	5.28
	B1a	-	5.46
	B1b	-	5.73
	A2	-	5.46
	A3	-	5.60
	A4	-	5.29

-: Not Available.

3. Changes in Soy Protein during Tofu Processing

To obtain raw soy milk for tofu production, soybeans are soaked, pulped, and filtered. For tofu production using the raw soy milk, the milk is first heated and then a coagulant is added to form tofu curd. The curd is then pressed to obtain a tofu product. As shown in Figure 2, the structure and content of soy protein undergo numerous changes throughout the tofu production process. The quality of the final tofu product is affected by soybean varieties, storage conditions, soaking, grinding, soymilk pretreatment, types of coagulants, operating conditions, pressing, and packaging.

3.1. Influence of Soybean Varieties and Growing Conditions

The protein, the 11S/7S ratio, and the methionine and cysteine content in soybeans have a significant impact on the hardness and the water-holding capacity of tofu [17]. Therefore, these indices are important indicators for screening soybeans [18].

Different soybean varieties differ in their soy protein subunit compositions, leading to changes in their denaturation temperature and gel network structure [19–21]. Stanojevic [22] and Cai et al. [23] assessed the effects of soybean varieties on the quality of tofu. They found that soybeans with a low 11S/7S ratio formed a uniform spherical aggregated gel, whereas the gel formed by beans with a higher 11S/7S ratio had higher macroscopic phase separation, a coarser network structure, and larger pores [24].

The content of the 11SA4 subunit in soybean can affect the overall protein content and seed size. Soybeans with lower 11SA4 content have higher protein content and smaller seeds, and the gel structure of the obtained tofu is denser. Therefore, the protein content in soybeans can be increased by removing the 11SA4 subunit through genetic breeding, thereby improving the gel properties, hardness, and water-holding capacity of the tofu produced [18].

The growth environment and growing period of soybean also affect the 7S and 11S content and the composition of their subunits. The total protein content in soybeans is negatively correlated with latitude (12–32 N°) and rainfall in the growing season (61–956 mm) and positively correlated with the daily average temperature in the growing period (19.0–26.7 °C) [25]. Yang et al. [26] showed that soybeans from different growth environments presented obvious differences in protein subunits, which in turn influenced the yield, color, hardness and water loss of tofu. Poysa et al. [27] observed that the effects of soybean genetic profile and growth year significantly affected soymilk and tofu yield, solids levels, and pH and effected tofu color, hardness, and firmness more than the growth environment. Moreover, the effects of the interaction of genotype with location and year were minor relative to the effects of genotype and year individually.

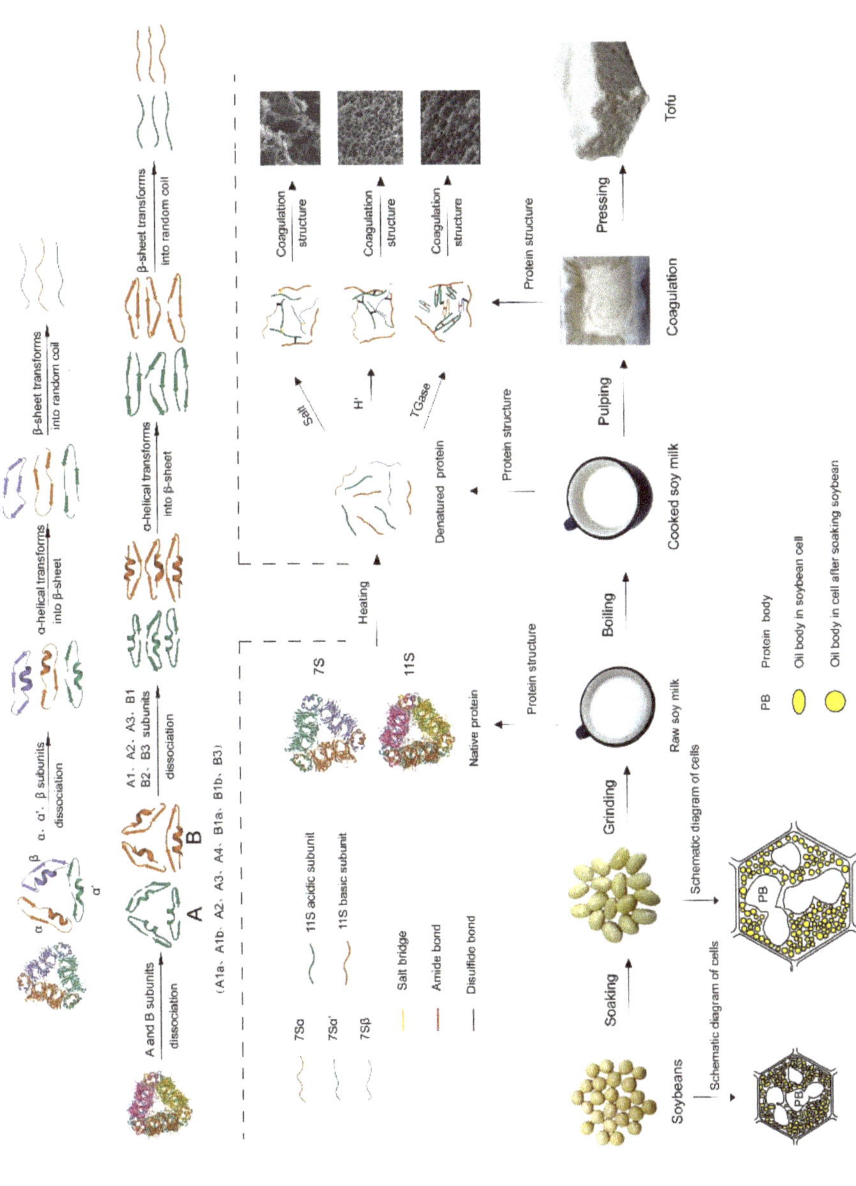

Figure 2. Changes of soybean protein during tofu processing [5,6]; Protein Data Bank.

3.2. Influence of Storage Conditions

Generally, the newly harvested soybeans are unripe and have lower oil and protein content than fully mature seeds, making processing difficult. After storage, soybeans mature, resulting in improved tofu yield, color, hardness, and water loss. However, extended storage can also lead to a decline in tofu quality.

The main factors affecting soybean storage include the relative humidity of the storage environment, seed moisture content, storage temperature, and duration. Kong et al. [28] found that long-term storage leads to a decrease in the water-holding capacity of soy protein, thus increasing the yield and protein content of tofu. Saio et al. [29] showed that the relative humidity of the soybean protein storage environment has a greater impact on soybean protein components than the storage temperature.

In summary, different soybean varieties have different genes, protein composition, and 11S/7S protein ratios. The growth environment affects the gene expression of soybeans, which influences the composition and structure of their protein. Storage conditions, on the other hand, cause physical, chemical, and biological changes in soy protein. Therefore, both storage conditions and storage time influence the protein content of tofu.

3.3. Influence of Soaking and Refining

Soaking and refining are important steps in tofu processing. Soaking changes the structural characteristics and crushing performance of soybeans, accelerating the extraction of soybean protein and thereby increasing the protein content of tofu [30]. Conversely, the limited swelling of soy protein doubles the absorption of water and increases soybean volume. Refining can dissolve the protein in soybeans and disperse them evenly in water. In the refining process, the extraction rate of soy protein is approximately 85% [2].

The soybean soaking process is affected by several factors, such as soybean particle size and variety, soaking water quality, water temperature, pressure, and soaking method and time. Therefore, choosing the right soaking conditions is crucial in the processing of tofu. High-quality water and suitable temperature and soaking time lead to a higher protein extraction rate and content in soymilk, increasing the gel strength and water-holding capacity of tofu [31–33]. Yang et al. [30] and Pan et al. [34] showed that the optimal soaking time decreases as the soaking temperature increases. However, as the temperature rises from 30 °C to 40 °C, protein and carbohydrates leak significantly and the solid content of the soaked soybean seeds decreases. Guo et al. [35] showed that the order of factors affecting the yield of soy protein during the soaking process was soaking time > soaking temperature > pH of the soaking solution. Furthermore, the optimal conditions of soaking may also be affected by soybean varieties.

In conclusion, soaking and refining dissolve the protein, oil, and other substances in soybeans from a solid to liquid phase. The treatment conditions of soaking and refining, such as soaking water quality, soaking water temperature and time, refining temperature, and the material-to-water ratio of refining, affect the dissolution of soy protein, oil, and other components and change the content of each component in soymilk, affecting the quality of tofu.

3.4. Pretreatments

3.4.1. Soybean Pretreatment

Okara is rich in dietary fiber, protein, fat, and isoflavones. However, the tofu production process usually removes the okara in soymilk, causing nutrient loss. Therefore, researchers have developed a preparation method that produces whole soybean curd while retaining the okara. However, the dietary fiber, gel, and other impurities contained in soybeans have an adverse effect on the texture and flavor of tofu.

Studies have shown that the protein and oil content of dehulled soybeans is higher than whole soybeans [36]. The use of dehulled soybeans to produce tofu not only improves product quality but also facilitates the extraction of soy protein. Moreover, the network structure of whole bean curd is mostly irregular, discontinuous, large, and uneven. This is because some insoluble dietary fiber particles are embedded in the network structure, which destroys the continuity of the soy protein gel network [37].

Tofu prepared from frozen soybeans showed a more orderly and denser network structure compared to unfrozen soybeans, inducing an increase in some textural parameters, such as hardness, springiness, gumminess, chewiness, and syneresis. Freezing also enhanced tofu quality with a lower yield, lower fat, and higher protein content [38].

3.4.2. Soymilk Pretreatment

Denaturation of soy protein is a prerequisite for curd formation of tofu and is generally applied in the form of heat in tofu processing technology [39]. Studies have found that the hydrophobicity, emulsification, and gel strength of soy protein can be enhanced by ultrasonic treatment of soy milk.

(1)   Heat Treatment

Raw soymilk is relatively stable because the natural soybean protein molecule has a hydrophobic group inside and a hydrophilic group on the surface of the molecule. As the raw soymilk is heated, the energy in the system increases, the thermal motion of protein molecules intensifies, and the vibration frequency of certain groups in the molecule increases. This change leads to the breaking of the secondary bonds that hold the protein molecular structure, while the spatial structure also changes [40]. During the thermal denaturation of soy protein, hydrophobic groups such as sulfhydryl groups, disulfide bonds, and hydrophobic amino acid side chains are exposed, and the hydrophobicity of the surface of soy protein increases, which intensifies the protein molecule aggregation.

Soybean protein aggregates during soybean milk heat treatment. At 80 °C, the solubility of 11S globulin decreases, and the α-helical structure of the protein gradually transforms into β-sheet and random coil structures. After heat treatment at 90 °C and 100 °C, the solubility of protein increases slightly, and the α-helix and β-sheet structures change to β-turn and random coil structures, which plays an important role in the formation of aggregates [41,42]. In the formation of thermal aggregates and network structures, β-sheets have a greater effect than α-helixes [43]. The decrease in β-sheet content exposes the hydrophobic area of the protein [44]. Guo et al. [40] showed that 7S produced soluble limited aggregations, while 11S formed insoluble aggregations. After heating, 7S terminated the assembly of 11S and restored the solubility of 11S. The three-dimensional network structure of agglomerates prepared by heated soybean protein shows low sedimentation and a high curdling rate, suitable water-holding capacity, low hardness, and high elasticity.

Currently, tofu is prepared with soymilk at approximately 100 °C. Nevertheless, as the heat denaturation temperature of 11S protein (85~95 °C) is estimated to be 20 °C higher than the heat denaturation temperature of 7S protein [45], such a heating method denatures both proteins almost simultaneously. Studies have shown that the two-step heating method (i.e., first denaturing 7S protein, then 11S protein) is conducive to the effective denaturation of soy protein and the formation of curd, maximizing the use of the two-storage protein gel characteristics to obtain the best quality curd [46].

During the heating process, in addition to the structural changes of the protein itself, the oil in the soymilk also has a certain impact. Peng et al. [47] described this process in detail. Between 65–75 °C, oil molecules, 7S, and 11S are released into soluble components from the storage protein–oil body complex. At 75–95 °C, the oil enters the floating components on the surface of the soymilk. The β and B subunits then aggregate to form protein particles, while the α, α', and A subunits remain in the soluble fraction.

Heating induces protein aggregation and protein–polysaccharide interaction, leading to the modification of protein particle size distribution, viscosity, surface hydrophobicity, and solubility [48] and altering the structure of soybean protein. Moreover, soy protein has hydrophobic interactions with flavor substances, such as aldehydes and ketones. In addition to the aggregation of subunits, the conformation of the polypeptide chain also varies during thermal denaturation. The exposed hydrophobic areas on the surface of the heat-denatured protein particles provide active binding sites for flavor molecules and affect the sensory quality of soymilk [49].

(2)   Ultrasound Pretreatment

Ultrasound is a phenomenon of cavitation that exceeds the threshold of human hearing. The cavitation effect of ultrasound can change the structure of protein molecules. After ultrasonic treatment, the polypeptide chain inside the protein molecule is partially expanded, the protein structure becomes more stretched, the hydrophobic group is exposed, the surface activity enhances, and the emulsification increases [50].

Liu et al. [51] found that the 11S globulin aggregates were broken into small uniform particles after ultrasonic treatment, which narrowed the distribution and increased the surface charge density. Proteins can be completely distributed in the oil–water interface by appropriate ultrasound and heat treatment, thus reducing surface tension. As the sonication time extends, the emulsification of soy protein tends to be stable after its peak [52]. In this regard, Chen et al. [53] and Karki et al. [54] suggested that with a sonication time extension the protein structure becomes loose, the polar part shifts to water, and the non-polar part shifts to lipids. The emulsion becomes evenly dispersed and the emulsification performance is improved. However, if it is processed for a long time, the insoluble protein content increases and the emulsification decreases.

Different ultrasound powers have different effects on proteins. Liu et al. [55] found that low-power ultrasound treatment weakens the ability of soy protein gel formation, and high-power ultrasound delays the formation of the gel. However, under the treatment of medium-power ultrasound (200–600 W), the hydrophobic interaction and hydrogen bonding positions during the formation of the thermal gel increase, forming a stronger gel with a denser three-dimensional network structure. This improves the gel properties of soy protein. Li et al. [56] found that with the increase of ultrasound power, the particle size of soy protein gradually increased, reaching the highest value at 300 W. However, if the ultrasonic power continued to be enhanced, the particle size of soybean protein decreased, hydrophobic groups were exposed, and the emulsification and stability of soybean protein were improved.

Hu et al. [57] found that with ultrasonic treatment under 400 W, the calcium ion-induced soy protein formed a compact and uniform three-dimensional gel structure, which improved its water-holding capacity and gel strength. Zhang et al. [58] showed that high-intensity ultrasound can open the spatial structure of soy protein isolate, exposing the site of action of transglutaminase, thus enhancing the strength of the gel formation induced. Therefore, the modification of soy protein by ultrasound can be applied to tofu processing.

3.5. Coagulants

A key step in tofu production is the addition of a coagulant to make the soybean protein form a gel network structure that is macroscopically reflected in the formation of tofu coagulation. This process is mainly influenced by the type of coagulant and processing conditions. During the coagulation process, the protein–protein and protein–water interactions cause soy protein to aggregate and form a honeycomb-like gel [59]. Currently, typical coagulants include salt, acid, and enzyme and composite coagulants. They have different coagulation mechanisms and interfere with the quality of tofu.

3.5.1. Salt Coagulants

Salt coagulants are the most traditional and widely used in tofu production; mainly include magnesium chloride, calcium chloride, magnesium sulfate, calcium sulfate, calcium acetate, and so on. Regarding the curdling mechanism of salt-based coagulants to make tofu, researchers believe that the gelatinization process of tofu can be divided into two steps [60]: (1) the heat denaturation process of protein and (2) a hydrophobic condensation process promoted by metal ions.

There are three main theories for the mechanism of salt coagulants: (1) ion bridge theory [61], (2) salting-out theory [62], and (3) isoelectric point theory [63]. In recent years, researchers have also proposed a new explanation based on the ion bridge theory, which states that the formation of protein particles varies in the presence of specific metal ions [64]. These four explanations, however, have their limitations.

In order to explore the solidification mechanism of salt coagulants, researchers have studied the contributions of various interaction forces in the solidification process. Lee et al. [61] used optical and scanning electron microscopy to observe the microstructure of soy protein aggregates during heat treatment and spotting. They found that the isoelectric point precipitation and calcium ion aggregation did not change the globular structure of soy protein, but heating could change the protein structure. Zhou et al. [65] studied the changes in the texture characteristics of the gel by adding different types of additives, including NaCl, thiothreitol, sodium dodecyl sulfate, and urea, during the preparation of the marinated tofu gel. Their results showed that electrostatic interactions, disulfide bonds, hydrophobic interactions, and hydrogen bonds have important effects on the formation of tofu gel. Liu et al. [66] analyzed freeze-dried tofu samples using chemical methods and studied the influence of intermolecular forces in the curding process of different coagulants on this basis. They highlighted that hydrophobic interaction and disulfide bonds play a dominant role in the formation of soybean gel. Yang et al. [67] studied the changes in the secondary structure and moisture state of the protein during the formation of tofu gel and concluded that electrostatic interaction and hydrophobic interaction mainly affect the aggregation of protein molecular chains; moreover, hydrogen bonds and disulfide bonds mainly affect the connection of molecular chains. Jin et al. [68] showed that as the solidification temperature increases, the proportions of ionic and hydrogen bonds decrease significantly, while the proportions of hydrophobic interactions and disulfide bonds increase. In summary, hydrogen and disulfide bonds and electrostatic and hydrophobic interactions play a certain role in the curdling process of tofu, but there are still some controversies about their specific mode of action.

The type and concentration of salt coagulants play a decisive role in the properties of tofu curd. Lu et al. [63] found that a variety of calcium salts (calcium chloride, calcium lactate, calcium acetate, calcium gluconate, etc.) can induce gelation of soy protein; thus, it precipitated when the pH of soymilk was 6. Liu et al. [69] found that as the concentration of coagulant increased, the gel strength of tofu increased, while the water retention decreased.

The type of ions affects the coagulation characteristics of tofu, and the influence of anions is greater than that of cations. Other than soy protein, lipids are also the main components of soymilk. With the addition of the coagulant, the lipid particles incorporate into the protein gel network and disperse in the gel network. Based on previous research, Peng et al. [47] proposed a specific model that combines soy protein, lipids, and small molecules.

3.5.2. Acid Coagulant

Acid coagulants are another important type of tofu coagulants. Acid coagulants include glucolactone (GDL), physalis, lactic acid, acetic acid, succinic acid, citric acid, malic acid, and tartaric acid; of these, GDL is the most widely used. Acid coagulants can provide hydrogen ions that reduce the pH value of soymilk to the isoelectric point of soy protein, thus promoting the isoelectric precipitation of protein.

At a certain temperature, GDL slowly hydrolyzes the gluconic acid and releases protons, which is a suitably gradual process for forming a continuous soy protein network structure through hydrophobic and electrostatic interactions [70]. Excessive gel rate can result in gel with an uneven structure and low strength [71]. Kaoru et al. [60] showed that the gelation process of GDL is similar to that of salt-based coagulants, and its gelation curve conforms to first-order reaction kinetics. However, the coagulation rate induced by GDL is lower than that of salt-based coagulants [60]. Due to the difference in isoelectric point, there are more protons recombined in 7S than 11S. The addition of GDL promotes aggregation through hydrophobic interactions, thereby inducing gelation, whereas the interaction between charges may be secondary.

During the tofu production process, yellow water containing certain nutrients is produced. Under suitable conditions, the yellow water can be fermented to obtain physalis. Using physalis as a coagulant can save costs and reduce pollution. Liu et al. [72] showed that the physalis coagulant produces a large amount of hydrogen ions. The ions reduce the negative charge of protein molecules, increase the content of free sulfhydryl groups, and gradually reduce the surface hydrophobicity, thus inducing the formation of protein aggregates.

3.5.3. Enzyme Coagulant

Coagulant enzymes, widely present in animal and plant tissues and microorganisms, have great development potential as bean curd coagulants. At present, the most studied enzyme coagulants include transglutaminase (TGase), pepsin, acalase, papain, and bromelain.

From the 1980s to the beginning of this century, researchers tried to use natural proteases derived from plants and animals as a tofu coagulant. Fuke et al. [73] confirmed the role of bromelain in the aggregation and gelation of heated soymilk by measuring sulfhydryl content and hydrophobicity. Luan et al. [74,75] studied 13 different proteases and found that alcalase, papain, and bromelain have a strong soymilk solidifying ability.

TGase is a type of enzyme that catalyzes the acyl transfer reaction between the γ-hydroxylamine group (acyl donor) of peptide glutamine residues and a variety of primary amines (acyl acceptor). This process is mainly realized in three ways: introduction of amines, intermolecular and intramolecular cross-linking, and deamination. Yang et al. [76] believed that the effect of TGase in improving the strength of tofu is mainly related to 7S and 11S protein. In this process, the α' and α subunits in 7S and the A3 peptide chain in 11S have the greatest influence on the action of TGase, followed by the β subunit in 7S and the A peptide chain in 11S. They analyzed the amino acid content of these subunits and peptide chains and concluded that the activity of TGase is closely related to the lysine in soy protein.

3.5.4. New Coagulants

(1) Emulsion Coagulant

Magnesium chloride has high solubility and can be used in a rapid and intense tofu gelation process. The coagulation rate of tofu curd influences the characteristics of the curd. Overly quick coagulation results in coagulation with poor water retention and coarse particles. As a slow-release platform, emulsion coagulant can achieve controlled release of coagulant, solving the above problems [77,78]. The main components of emulsion coagulant include a brine-based water phase, a natural fat-based oil phase, emulsifier, and protein. The current emulsion coagulants can be divided into two types: water-in-oil (W/O) and water-in-oil-in-water (W/O/W).

Li et al. [79,80] studied the preparation methods of W/O and W/O/W emulsion coagulants and their influence on the characteristics of tofu curd and found that the emulsion coagulant can improve the spatial structure of tofu gel. Compared with the traditional marinated tofu, the tofu gel network prepared by the emulsion coagulant is finer, and its water content and water-holding capacity are improved. Zhu et al. [81] used a differential calorimetry scanner and low-field nuclear magnetic resonance to study the change of water in the W/O tofu gel process and found that tofu prepared using this coagulant had higher free water content than did traditional marinated tofu.

(2) Complex Coagulants

Common coagulants such as gypsum, brine, and GDL present disadvantages compared to other coagulants. The coagulation speed using a small amount of gypsum is slow, whereas tofu products made with a large amount of gypsum contain residues and a bitter taste. Tofu made with bittern has poor water-holding capacity, resulting in short shelf life. Lactone tofu is soft, and not suitable for frying [82–84]. Obtaining a complex coagulant that can overcome the shortcomings of a single coagulant and ensure the quality and taste of tofu has become a major research focus in the field.

Different coagulants have different abilities to induce the coagulation of soy protein, and the ratio of complex coagulants also influences the aggregation state and gel properties of the protein. Wang et al. [85] used magnesium sulfate and calcium sulfate as a complex coagulant to induce gelation of the soybean protein emulsion. The results showed that different concentrations of magnesium ions changed the hardness and strength of the gel due to different aggregation forces. At low concentration, magnesium ions were conducive to the formation of dense protein aggregates and promoted the uniformity and deformation resistance of the gel; when the concentration of magnesium ions increased, the synergistic effect of calcium and magnesium ions promoted the coarsening of the protein gel structure, and the emulsification performance was significantly improved.

Ramy et al. [86] added nano fish bones to the citric acid-induced soymilk curdling system, which significantly enhanced the compactness and uniformity of the three-dimensional gel network. The combination of citric acid and salt coagulants increases the hardness of tofu, probably due to the increase of ionic bonds caused by the addition of salt ions. Salt can change the structures of water and polar groups and provide electric charge, affecting the electrostatic and hydrophobic interactions [87]. The complex coagulant formed by TGase and lactic acid bacteria can induce the formation of a denser gel network [88]. Shi et al. [89] used a composite coagulant of TGase and GDL to prepare tofu, resulting in enhanced water content, water-holding capacity, and microstructure density under certain operating conditions.

3.5.5. Additives

(1)   Carbohydrates

Adding carbohydrates and other additives to the coagulant can significantly improve the performance of tofu curd. The interaction between polysaccharides and protein polymers has been shown to effectively improve the properties of the curd [90]. Common carbohydrate additives include chitosan, guar gum, carrageenan, acacia gum, and konjac gum.

Li et al. [90] compared the tofu made with a compound coagulant (magnesium chloride and guar gum) with traditional tofu (gypsum and marinated tofu) and found that the addition of guar gum affected the gel structure and texture characteristics of tofu. Researchers speculated that the higher viscosity of guar gum increases soymilk viscosity, resulting in a slower coagulation rate. In addition, they compared the performance of carrageenan, guar gum, and acacia gum mixed with magnesium chloride to make tofu [91]. The texture data show that the addition of guar gum significantly reduced the hardness and protein content of tofu, and carrageenan increased the hardness of the curd, while the protein content remained unchanged.

Cao [92] showed that salt coagulants and polysaccharides have a synergistic promotive effect on soy protein curd formation, which significantly improves the texture properties of tofu. Zhao et al. [93] added konjac gum, gellan gum, and Kotlan gum to the calcium sulfate-induced soy protein isolate gel system and found that the addition of polysaccharides enhanced the gel structure, accelerated gelation, improved the microstructure of the gel, and lowered the starting temperature of gelation. Chitosan as a coagulant lowers tofu's ash content and improves its protein content [94], and in the production process of pressurized lactone tofu, improves its water holding capacity [95]. Jun et al. [96] used acetic acid-treated crab shell extract as a coagulant, and the texture of tofu produced was comparable to that of commercially available tofu.

(2)   Phytic Acid

Phytic acid or phytate added to soymilk has a significant effect on the texture characteristics of tofu curd. Schaefer et al. [97] studied the relationship between the content of various components in soybeans and the properties of tofu. They concluded that phytic acid can preferentially combine with calcium coagulants, thereby changing the yield, composition, texture, and microstructure of tofu curd. Saio et al. [98] found that when the amount of calcium salt added was in a certain range, as the phytic acid content in soymilk increased, the effect of calcium ions on protein coagulation decreased and tofu gel formation became increasingly difficult. Therefore, tofu made from soymilk with high phytic acid content has a high yield but low gel hardness. Ishiguro et al. [99] performed curdling experiments and measured the phytic acid content of 27 types of soybeans. The results showed that tofu with a higher phytic acid content and prepared at a general coagulant range had a softer texture. In conclusion, the higher the phytic acid content during the curding process, the lower the hardness of the tofu and the higher the viscosity and fracture stress.

3.6. Compression, Preservation, and Packaging

Compression is the process of stabilizing the tofu gel network structure. The pressing operation applies pressure to the formed curd, expelling the excess yellow water and reducing the syneresis of the tofu during the storage process. During the pressing process, the ratio of the β-sheet structure of soy protein increases, the disordered structure decreases, and the gel system gradually stabilizes [72]. The amount of pressure and the pressing time affect the structure of tofu. The moisture around the soy protein gel network cannot be completely released under low pressure, which leads to uneven shaping of tofu and easy syneresis. Under high pressure, the gel structure of tofu is greatly damaged, resulting in excessive loss of yellow water. Studies have shown that the water retention of tofu is negatively correlated with the loss of yellow water [100], and a certain amount of soluble protein and other nutrients are also dissolved in yellow water. Therefore, once the amount of yellow water decreases during the pressing process, the water retention and nutritional value of tofu also decrease [101]. Tofu is preserved by adding preservatives to increase product shelf life [102]. Tofu preservation requires a comprehensive preservation technology. Freezing can achieve the effect of freshness, and the freezing temperature and time greatly influence the texture of tofu. Freezing treatment converts the water in the tofu into ice crystals, expands the mesh inside the original tofu tissue, and improves the texture characteristics of the tofu. However, a very low temperature makes the ice crystals in the protein gel network too dense, and the pores become smaller after thawing, which can decrease the texture quality of tofu [103]. Kobayashi et al. [104] found that the ice recrystallization and dehydration of frozen tofu with a shelf life of 0–7 days causes changes in the balance of hydrophilic and hydrophobic zones. This change induces the formation of new protein interactions, resulting in a firmer tofu texture.

Tofu packaging is the last step in tofu processing. The choice of packaging materials affects the water retention and shelf life of tofu. Tofu is rich in water and protein, which can deteriorate rapidly. Appropriate packaging can prevent the growth of microorganisms and slow down protein deterioration and water loss, thereby extending the shelf life of tofu [105].

4. Conclusions

The production of tofu includes a series of processes, such as soybean screening, soaking, grinding, filtering, boiling, coagulating, pressing, preserving, and packaging. The composition, structure, and content of soy protein are constantly changing during the production process. The gelation properties of soy protein are the basis for the preparation of tofu.

Different soybean varieties have different genotypes, protein composition, 11S/7S protein ratio, etc. The growth environment and storage conditions of soybeans also have a significant impact on the composition and structure of the protein. Soybeans become raw soymilk after soaking and grinding, and soy protein is dissolved from the solid phase to the liquid phase to form an emulsion. In the boiling process, the soymilk protein is denatured and the hydrophobic groups are exposed. By adding a coagulant to soymilk, a curd network structure with protein as the backbone is formed, and the pressing process stabilizes the curd network structure. The subsequent preservation and packaging operations further affect the structure of the curd.

The soymilk system is complex and changeable. Tofu products are mature and diverse. The research in this field covers a wide area, but it is not easy to go deep. The composition structure and spatial configuration of soy protein have not yet been fully resolved. At present, analyses of the interaction between the components and the curdling process are based on model predictions rather than actual observations. Most of the established models require adjustments because there are still contradictions between different models.

We suggest that further clarifying the curd formation mechanism using different coagulants and analyzing the changes at a molecular level can contribute greatly toward improving tofu quality. In addition, the specific changes that occur during the interaction and assembly of soy protein, oil, phytic acid, and other components in the curd are also an important research direction. The application of molecular simulation technology in the analysis of the composition and structural changes of soybean protein and other important components should be the focus of future research.

Author Contributions: Conceptualization, M.Z.; writing—original draft preparation, X.G. and X.Z.; writing—review and editing, X.G., X.Z., Y.L., X.D., R.J., H.L. and M.Z.; visualization, X.G., Y.L. and M.Z.; funding acquisition, M.Z., supervision, H.L. and M.Z. All authors have read and agreed to the published version of the manuscript.

Funding: This research was funded by the National Key R&D Program of China (2016YFD0400203).

Acknowledgments: We thank our editor Cecilia for providing language help and writing assistance.

Conflicts of Interest: The authors declare no conflict of interest.

References

1. Shi, Y.G.; Liu, L.L. Research progress on correlation between soybean protein and tofu quality. *J. Food Technol.* **2018**, *36*, 1–8.
2. Du, L.Q. *New Technology of Tofu Production*, 1st ed.; Chemical Industry Publishing House: Beijing, China, 2018; pp. 1–64. ISBN 978-7-122-31433-8.
3. Hui, E.; Henning, S.M.; Park, N.; Heber, D.; Liang, V.; Go, W. Genistein and daidzein/glycitein content in tofu. *J. Food Compos. Anal.* **2001**, *14*, 199–206. [CrossRef]
4. Shi, Y.G. *Soybean Products Technology*, 2nd ed.; China Light Industry Press: Beijing, China, 2011; pp. 7–90. ISBN 978-7-5019-4807-9.
5. Maruyama, Y.; Maruyama, N.; Mikami, B.; Utsumi, S. Structure of the core region of the soybean β-conglycinin α′ subunit. *Acta Crystallogr. Sect. D* **2004**, *60*, 1–6. [CrossRef]
6. Tandang-Silvas, M.R.; Fukuda, T.; Fukuda, C.; Prak, K.; Cabanos, C.; Kimura, A.; Itoh, T.; Mikami, B.; Utsumi, S.; Maruyama, N. Conservation and divergence on plant seed 11S globulins based on crystal structures. *Bba-Proteins Proteom.* **2010**, *1804*, 1432–1442. [CrossRef]
7. Minor, W.; Steczko, J.; Stec, B.; Otwinowski, Z.; Axelrod, B. Crystal structure of soybean lipoxygenase L-1 at 1.4 A resolution. *Biochemistry* **1996**, *35*, 10687–10701. [CrossRef] [PubMed]
8. Song, H.K.; Suh, S.W. Kunitz-type soybean trypsin inhibitor revisited: Refined structure of its complex with porcine trypsin reveals an insight into the interaction between a homologous inhibitor from *Erythrina caffra* and tissue-type plasminogen activator. *J. Mol. Biol.* **1998**, *275*, 347–363. [CrossRef] [PubMed]
9. Hirata, A.; Adachi, M.; Sekine, A.; Kang, Y.N.; Mikami, B. Structural and enzymatic analysis of soybean-amylase mutants with increased pH optimum. *J. Biol. Chem.* **2004**, *279*, 7287–7295. [CrossRef]
10. Olsen, L.R.; Dessen, A.; Gupta, D.; Sabesan, S.; Brewer, C.F. X-ray crystallographic studies of unique cross-linked lattices between four isomeric biantennary oligosaccharides and soybean agglutinin. *Biochemistry* **1997**, *36*, 15073–15080. [CrossRef]
11. Taski-Ajdukovic, K.; Djordjevic, V.; Vidic, M.; Vujakovic, M. Subunit composition of seed storage proteins in high-protein soybean genotypes. *Pesqui. Agropecuária Bras.* **2010**, *45*, 721–729. [CrossRef]
12. Utsumi, S.; Matsumara, Y.; Mori, T. Structure-function relationships of soy proteins by using recombinant systems. *Enzym. Microb. Techol.* **2002**, *30*, 284–288. [CrossRef]
13. Ren, C.; Tang, L.; Zhang, M.; Guo, S. Structural characterization of heat-induced protein particles in soy milk. *J. Agric. Food Chem.* **2009**, *57*, 1921–1926. [CrossRef]
14. Zeng, J.H.; Liu, L.L.; Yang, Y.; Zhang, N.; Shi, Y.G.; Zhu, X.Q. Research progress on thermal modification and its dissociation association action of soy proteins. *Soybean Sci.* **2019**, *38*, 142–147+158. [CrossRef]
15. Thanh, V.H.; Shibasaki, K. Major proteins of soybean seeds. Subunit structure of β-conglycinin. *J. Agric. Food Chem.* **1978**, *26*, 692–695. [CrossRef]
16. Fukushima, D. Recent progress in research and technology on soybeans. *Food Sci. Technol. Res.* **2001**, *7*, 8–16. [CrossRef]

17. Pazdernik, D.L.; Plehn, S.J.; Halgerson, J.L.; Orf, J.H. Effect of temperature and genotype on the crude glycinin fraction (11S) of soybean and its analysis by near-infrared reflectance spectroscopy (Near-IRS). *J. Agric. Food Chem.* **1996**, *44*, 2278–2281. [CrossRef]
18. James, A.T.; Yang, A. Interactions of protein content and globulin subunit composition of soybean proteins in relation to tofu gel properties. *Food Chem.* **2016**, *194*, 284–289. [CrossRef]
19. Li, M.; Dong, H.; Wu, D.; Chen, H.; Zhang, Q. Nutritional evaluation of whole soybean curd made from different soybean materials based on amino acid profiles. *Food Qual. Saf.* **2020**, *4*, 1. [CrossRef]
20. Asrat, U.; Horo, J.T.; Gebre, B.A. Physicochemical and sensory properties of tofu prepared from eight popular soybean [glycine max (L.) merrill] varieties in ethiopia. *Sci. Afr.* **2019**, *6*, 1–13. [CrossRef]
21. Bainy, E.M.; Tosh, S.M.; Corredig, M.; Woodrow, L.; Poysa, V. Protein subunit composition effects on the thermal denaturation at different stages during the soy protein isolate processing and gelation profiles of soy protein isolates. *J. Am. Oil Chem. Soc.* **2008**, *85*, 581–590. [CrossRef]
22. Stanojevic, S.P.; Barac, M.B.; Pesic, M.B.; Vucelic-Radovic, B.V. Assessment of soy genotype and processing method on quality of soybean tofu. *J. Agric. Food Chem.* **2011**, *59*, 7368–7376. [CrossRef]
23. Cai, T.; Chang, K. Processing effect on soybean storage proteins and their relationship with tofu quality. *J. Agric. Food Chem.* **1999**, *47*, 720–727. [CrossRef]
24. Wu, C.; Hua, Y.; Chen, Y.; Kong, X.; Zhang, C. Effect of 7S/11S ratio on the network structure of heat-induced soy protein gels: A study of probe release. *RSC Adv.* **2016**, *6*, 11981–11987. [CrossRef]
25. Kumar, V.; Rani, A.; Solanki, S.; Hussain, S.M. Influence of growing environment on the biochemical composition and physical characteristics of soybean seed. *J. Food Compos. Anal.* **2006**, *19*, 188–195. [CrossRef]
26. Yang, A.; James, A.T. Influence of globulin subunit composition of soybean proteins on silken tofu quality. 1. Effect of growing location and 11SA4 and 7Sα' deficiency. *Crop Pasture Sci.* **2014**, *65*, 259. [CrossRef]
27. Poysa, V.; Woodrow, L. Stability of soybean seed composition and its effect on soymilk and tofu yield and quality. *Food Res. Int.* **2002**, *35*, 337–345. [CrossRef]
28. Kong, F.; Chang, S.K.C.; Liu, Z.; Wilson, L.A. Changes of soybean quality during storage as related to soymilk and tofu making. *J. Food Sci.* **2008**, *73*, S134–S144. [CrossRef] [PubMed]
29. Saio, K.; Kobayakawa, K.; Kito, M. Protein denaturation during model storage studies of soybeans and meals. *Cereal. Chem.* **1982**, *59*, 408–412. [CrossRef]
30. Yang, A.; James, A.T. Comparison of two small-scale processing methods for testing silken tofu quality. *Food Anal. Method.* **2016**, *9*, 385–392. [CrossRef]
31. Li, L.T.; Cao, W. Influence of different soaking methods on tofu processing. *Sci. Technol. Food Ind.* **1998**, *3*, 19–21.
32. Li, L.T.; Cao, W. Effect of soybean soaking temperature on tofu processing. *Food Sci.* **1998**, *6*, 29–32. [CrossRef]
33. Shi, Y.G.; Li, G.; Hu, C.L.; Zhao, J.Y. Effect of soaking time on the quality of tofu. *Food Sci.* **2006**, *12*, 167–169. [CrossRef]
34. Pan, Z.; Tangratanavalee, W. Characteristics of soybeans as affected by soaking conditions. *Food Sci. Technol.* **2003**, *36*, 143–151. [CrossRef]
35. Guo, X.F.; Guo, Q.Q.; Lin, X.Z.; Liang, Z.C.; He, Z.G. Isothermal water absorption model of peeled soybean and optimization of process parameters of rehydration. *Sci. Technol. Food Ind.* **2020**, *41*, 207–211. [CrossRef]
36. Zhang, H.; Jiang, Y.Z.; Xu, G.H.; Sun, D.S.; Liu, L.J.; Dong, S.K. Study on the difference of protein and fat content under soybean peeling condition. *New Agric.* **2019**, *15*, 29–32.
37. Cui, J.; Ye, F.Y.; Zhao, G.H. Preparation of high fiber content soymilk and tofu using dehusked soybean. *Mod. Food Sci. Technol.* **2016**, *32*, 164–169. [CrossRef]
38. Noh, E.J.; Park, S.Y.; Pak, J.I.; Hong, S.T.; Yun, S.E. Coagulation of soymilk and quality of tofu as affected by freeze treatment of soybeans. *Food Chem.* **2005**, *91*, 715–721. [CrossRef]
39. Cao, Y.; Mezzenga, R. Design principles of food gels. *Nat. Food* **2020**, *1*, 106–118. [CrossRef]
40. Guo, J.; Yang, X.; He, X.; Wu, N.; Wang, J.; Gu, W.; Zhang, Y. Limited aggregation behavior of β-Conglycinin and its terminating effect on glycinin aggregation during heating at pH 7.0. *J. Agric. Food Chem.* **2012**, *60*, 3782–3791. [CrossRef]
41. Qi, B.K.; Zhao, C.B.; Li, Y.; Xu, L.; Ding, J.; Wang, H.; Jiang, L.Z. Effect of heat treatment on solubility and secondary structure of soybean 11S glycinin. *Food Sci.* **2018**, *39*, 39–44. [CrossRef]
42. Shilpashree, B.G.; Arora, S.; Chawla, P.; Tomar, S.K. Effect of succinylation on physicochemical and functional properties of milk protein concentrate. *Food Res. Int.* **2015**, *72*, 223–230. [CrossRef]
43. Przybycien, T.M.; Bailey, J.E. Secondary structure perturbations in salt-induced protein precipitates. *Biochim. Biophys. Acta* **1991**, *1076*, 103. [CrossRef]
44. Mills, E.N.; Huang, L.; Noel, T.R.; Gunning, A.P.; Morris, V.J. Formation of thermally induced aggregates of the soya globulin beta-conglycinin. *Biochim. Biophys. Acta* **2001**, *1547*, 339–350. [CrossRef]
45. German, B.; Damodaran, S.; Kinsella, J.E. Thermal dissociation and association behavior of soy proteins. *J. Agric. Food Chem.* **1982**, *30*, 807–811. [CrossRef]
46. Liu, Z.; Chang, S.K.C.; Li, L.; Tatsumi, E. Effect of selective thermal denaturation of soybean proteins on soymilk viscosity and tofu's physical properties. *Food Res. Int.* **2004**, *37*, 815–822. [CrossRef]
47. Peng, X.; Ren, C.; Guo, S. Particle formation and gelation of soymilk: Effect of heat. *Trends Food Sci. Techol.* **2016**, *54*, 138–147. [CrossRef]
48. Yang, Y.; Ji, Z.; Wu, C.; Ding, Y.; Gu, Z. Effect of the heating process on the physicochemical characteristics and nutritional properties of whole cotyledon soymilk and tofu. *RSC Adv.* **2020**, *1*, 4625–4636. [CrossRef]

49. Damodaran, S.; Kinsella, J.E. Interaction of carbonyls with soy protein: Thermodynamic effects. *J. Agric. Food Chem.* **1981**, *29*, 1249–1253. [CrossRef]
50. Wang, X.B.; Wang, L.; Zhou, G.W.; Qiao, J.W.; Zhang, A.Q.; Wang, Y.Y. Effect of different ultrasound time on the structure and emulsifying property of soybean-whey mixed protein. *Trans. Chin. Soc. Agric. Mach.* **2020**, *51*, 358–364.
51. Liu, L.; Zeng, J.; Sun, B.; Zhang, N.; He, Y.; Shi, Y.; Zhu, X. Ultrasound-assisted mild heating treatment improves the emulsifying properties of 11S globulins. *Molecules* **2020**, *25*, 875. [CrossRef]
52. Sun, B.Y.; Shi, Y.G. Effect of ultrasonic on emulsification of soybean protein concentrate by alcohol method. *J. Chin. Cereals Oils Assoc.* **2006**, *4*, 60–63. [CrossRef]
53. Chen, L.; Chen, J.; Ren, J.; Zhao, M. Effects of ultrasound pretreatment on the enzymatic hydrolysis of soy protein isolates and on the emulsifying properties of hydrolysates. *J. Agric. Food Chem.* **2011**, *59*, 2600–2609. [CrossRef] [PubMed]
54. Karki, B.; Lamsal, B.P.; Grewell, D.; Pometto, A.L.; van Leeuwen, J.; Khanal, S.K.; Jung, S. Functional properties of soy protein isolates produced from ultrasonicated defatted soy flakes. *J. Am. Oil Chem. Soc.* **2009**, *86*, 1021–1028. [CrossRef]
55. Liu, R.; Zeng, Q.H.; Wang, Z.Y.; Cheng, S.; Mu, H.J.; Liang, R. Effects of ultrasonic treatment on gel rheological properties and gel formation of soybean protein isolate. *Sci. Technol. Food Ind.* **2020**, *41*, 87–92. [CrossRef]
56. Li, Y.; Tian, T.; Liu, J.; Lou, B.B.; Li, S.X.; Wang, Z.J. Effect of ultrasound on structure and emulsification of soy protein isolate. *Food Ind.* **2019**, *40*, 184–188.
57. Hu, H.; Li-Chan, E.C.Y.; Wan, L.; Tian, M.; Pan, S. The effect of high intensity ultrasonic pre-treatment on the properties of soybean protein isolate gel induced by calcium sulfate. *Food Hydrocoll.* **2013**, *32*, 303–311. [CrossRef]
58. Zhang, P.; Hu, T.; Feng, S.; Xu, Q.; Zheng, T.; Zhou, M.; Chu, X.; Huang, X.; Lu, X.; Pan, S.; et al. Effect of high intensity ultrasound on transglutaminase-catalyzed soy protein isolate cold set gel. *Ultrason. Sonochem.* **2016**, *29*, 380–387. [CrossRef] [PubMed]
59. Cao, F.; Li, X.; Luo, S.; Mu, D.; Zhong, X.; Jiang, S.; Zheng, Z.; Zhao, Y. Effects of organic acid coagulants on the physical properties of and chemical interactions in tofu. *LWT* **2017**, *85*, 58–65. [CrossRef]
60. Kohyama, K.; Sano, Y.; Doi, E. Rheological characteristics and gelation mechanism of tofu (soybean curd). *J. Agric. Food Chem.* **1995**, *43*, 1808–1812. [CrossRef]
61. Lee, C.H.; Rha, C. Microstructure of soybean protein aggregates and its relation to the physical and textural properties of the curd. *J. Food Sci.* **1978**, *43*, 79–84. [CrossRef]
62. Cheng, R.D. Changes of soy protein in the process of making tofu. *China Brew.* **1993**, *4*, 8–12.
63. Lu, J.Y.; Carter, E.; Chung, R.A. Use of calcium salts for soybean curd preparation. *J. Food Sci.* **1980**, *45*, 32–34. [CrossRef]
64. Zhang, Q.; Wang, C.Z.; Li, B.K.; Lin, D.R.; Chen, H. Research progress in tofu processing: From raw materials to processing conditions. *Crit. Rev. Food Sci. Nutr.* **2018**, *58*, 1–85. [CrossRef]
65. Zhou, S.H.; Chen, Y.; Zhang, M.; Liu, J.; Wang, J.Y.; Guo, S. Study of molecular forces in formation of brine concentrated tofu gelatin. *Food Res. Dev.* **2013**, *34*, 15–19. [CrossRef]
66. Liu, L.S.; Jin, Y.; Zhang, X.F.; Zhang, Q.; Bai, J.; Guo, H.; Peng, Y.J. Comparative Study on Structure and Chemical Interaction Between Brine Tofu and GDL Tofu. *Food Sci. Technol.* **2020**, *45*, 60–64. [CrossRef]
67. Yang, F.; Pan, S.Y.; Zhang, C.L. Structural Change of Protein during Tofu Gelation Process. *Food Sci.* **2009**, *30*, 120–124. [CrossRef]
68. Jin, Y.; Liu, L.S.; Zhang, X.F.; Zhang, Q.; Bai, J.; Guo, H.; Peng, Y.J. Effects of Coagulation Temperature on Gelling Properties and Chemical Forces of Lactone Tofu. *Food Sci.* **2020**, *636*, 58–64.
69. Liu, Z.S.; Li, L.T.; Eizo, T. Study on properties of tofu salt-coagulant and mechanism of tofu coagulation. *Cereals Oils Assoc.* **2000**, *3*, 39–43. [CrossRef]
70. Liu, H.H.; Kuo, M.I. Effect of microwave heating on the viscoelastic property and microstructure of soy protein isolate gel. *J. Texture Stud.* **2011**, *42*, 1–9. [CrossRef]
71. Li, Z.; Regenstein, J.M.; Fei, T.; Yang, L. Tofu products: A review of their raw materials, processing conditions, and packaging. *Compr. Rev. Food Sci. Food Saf.* **2020**, *19*, 1–8. [CrossRef]
72. Liu, L.L.; Zhu, X.Q.; Sun, B.Y.; Zeng, J.H.; Yang, Y.; Shi, Y.G. Change of protein in the processing of acid wheytofu. In Proceedings of the Abstracts of Food Summit in China and 16th Annual Meeting of CIFST, Wuhan, China, 13–14 November 2019.
73. Fuke, Y.; Sekiguchi, M.; Matsuoka, H. Nature of stem bromelain treatments on the aggregation and gelation of soybean proteins. *J. Food Sci.* **1985**, *50*, 1283–1288. [CrossRef]
74. Luan, G.Z.; Li, L.T. Study on coagulation of soybean milk by protease. *Food Ind. Sci. Technol.* **2006**, *1*, 71–74. [CrossRef]
75. Luan, G.Z.; Cheng, Y.Q.; Lu, Z.H.; Li, L.T. Development of soymilk clotting enzyme researching. *Acad. Period. Farm Prod. Process.* **2006**, *10*, 41–43. [CrossRef]
76. Yang, H.P.; Hua, Y.F.; Chen, Y.M.; Zhang, C.M.; Kong, X.Z. Effect of transglutaminase on rupture strength of GDL Tofu and its mechanism. *Food Ferment. Ind.* **2018**, *44*, 8–12. [CrossRef]
77. Li, J.L.; Cheng, Y.Q.; Jiao, X.; Zhu, Q.M.; Yin, L.J. Effect of W/O and W/O/W controlled-release emulsion coagulants on characteristic of bittern-solidified tofu. *Trans. Chin. Soc. Agric. Mach.* **2013**, *44*, 162–168. [CrossRef]
78. Li, J.L. *Preparation of W/O and W/O/W Emulsion Coagulation and Their Improvement to the Quality of Traditional Bittern-Solidified Tofu;* China Agricultural University: Beijing, China, 2014.
79. Li, J.; Qiao, Z.; Tatsumi, E.; Saito, M.; Cheng, Y.; Yin, L. A novel approach to improving the quality of bittern-solidified tofu by w/o controlled-release coagulant. 1: Preparation of W/O bittern coagulant and its controlled-release property. *Food Bioprocess Tech.* **2013**, *6*, 1790–1800. [CrossRef]

80. Li, J.; Qiao, Z.; Tatsumi, E.; Saito, M.; Cheng, Y.; Yin, L. A novel approach to improving the quality of bittern-solidified tofu by w/o controlled-release coagulant. 2: Using the improved coagulant in tofu processing and product evaluation. *Food Bioprocess Techol.* **2013**, *6*, 1801–1808. [CrossRef]

81. Zhu, Q.M.; Li, J.L.; Liu, Y.; Yin, L.Y. Effect of new W/O halogen coagulant on moisture change in soybean protein gel. *J. Chin. Cereals Oils Assoc.* **2014**, *29*, 100–105. [CrossRef]

82. Yu, X.; Huang, X.D. *Traditional Soybean Products Processing Technology*, 1st ed.; Chemical Industry Press: Beijing, China, 2011; pp. 11–90. ISBN 978-7-122-10594-3.

83. Yang, M.; Zhang, Q.C. Discussion on improving the quality of lactone tofu. *Food Sci.* **1997**, *2*, 72–73.

84. He, Y.D. *Production Technology and Deep Processing Technology of Soybean Products*, 1st ed.; Agricultural Press: Beijing, China, 1990; pp. 1–50. ISBN 7-109-01670-6.

85. Wang, X.; Luo, K.; Liu, S.; Adhikari, B.; Chen, J. Improvement of gelation properties of soy protein isolate emulsion induced by calcium cooperated with magnesium. *J. Food Eng.* **2019**, *244*, 32–39. [CrossRef]

86. Khoder, R.M.; Yin, T.; Liu, R.; Xiong, S.; Huang, Q. Effects of nano fish bone on gelling properties of tofu gel coagulated by citric acid. *Food Chem.* **2020**, *332*. [CrossRef]

87. Zhao, Y.Y.; Cao, F.H.; Li, X.J.; Mu, D.D.; Zhong, X.Y.; Jiang, S.T.; Zheng, Z.; Luo, S.Z. Effects of different salts on the gelation behaviour and mechanical properties of citric acid-induced tofu. *Int. J. Food Sci. Technol.* **2019**, *55*, 785–794. [CrossRef]

88. Xing, G.; Giosafatto, C.V.L.; Carpentieri, A.; Pasquino, R.; Dong, M.; Mariniello, L. Gelling behavior of bio-tofu coagulated by microbial transglutaminase combined with lactic acid bacteria. *Food Res. Int.* **2020**, *134*. [CrossRef] [PubMed]

89. Shi, N.; Xu, H.W.; Chen, H.Q.; Zhang, Y.Y.; Guo, K.Y.; Liu, S.; Dong, B.; Ma, Y.L.; Tan, J.X. Optimization of preparation process of colored tender tofu with compound coagulant. *J. Food Sci. Technol.* **2019**, *37*, 93–99. [CrossRef]

90. Li, M.; Chen, F.S.; Yang, H.S.; Wang, M.L.; Lai, S.J. Effect of magnesium chloride guar gum mixture on the coagulation process of tofu. *Grain Fats* **2014**, *27*, 30–33. [CrossRef]

91. Li, M.; Chen, F.; Yang, B.; Lai, S.; Yang, H.; Liu, K.; Bu, G.; Fu, C.; Deng, Y. Preparation of organic tofu using organic compatible magnesium chloride incorporated with polysaccharide coagulants. *Food Chem.* **2015**, *167*, 168–174. [CrossRef]

92. Cao, F.H. *Study on Formation and Gel Mechanism of Bean Curd Induced by Organic Acids*; Hefei University of Technology: Hefei, China, 2018.

93. Zhao, H.; Chen, J.; Hemar, Y.; Cui, B. Improvement of the rheological and textural properties of calcium sulfate- induced soy protein isolate gels by the incorporation of different polysaccharides. *Food Chem.* **2020**, *310*. [CrossRef]

94. No, H.K.; Meyers, S.P. Preparation of tofu using chitosan as a coagulant for improved shelf-life. *Int. J. Food Sci. Technol.* **2004**, *39*, 133–141. [CrossRef]

95. Zhao, X.Y.; Wang, S.H.; Deng, C.K. Applied research of chitosan in processing of pressure lactone bean curd. *Sci. Technol. Food Ind.* **2012**, *4*, 177–180+186.

96. Jun, J.Y.; Jung, M.J.; Jeong, I.H.; Kim, G.W.; Sim, J.M.; Nam, S.Y.; Kim, B.M. Effects of crab shell extract as a coagulant on the textural and sensorial properties of tofu (soybean curd). *Food Sci. Nutr.* **2019**, *7*, 547–553. [CrossRef]

97. Schaefer, M.J.; Love, J. Relationships between soybean components and tofu texture. *J. Food Qual.* **2010**, *15*, 53–66. [CrossRef]

98. Saio, K.; Watanabe, T.; Koyama, E.; Yamazaki, S. Protein-Calcium-Phytic acid relationships in soybean: Part III. Effect of phytic acid on coagulative reaction in tofu-making. *Agric. Biol. Chem.* **1969**, *33*, 36–42. [CrossRef]

99. Ishiguro, T.; Ono, T.; Wada, T.; Tsukamoto, C.; Kono, Y. Changes in soybean phytate content as a result of field growing conditions and influence on tofu texture. *Biosci. Biotechnol. Biochem.* **2006**, *70*, 874–880. [CrossRef]

100. Zhao, L.; Zhu, J.; Su, E.Y.; Yang, H.W.; Hu, Z.Y.; Li, L. Effect of processing conditions on quality and protein secondary structure in southern tofu. *Food Sci.* **2019**, *40*, 62–69.

101. Shimizu, S.; Stenner, R.; Matubayasi, N. Statistical thermodynamics of biomolecular denaturation and gelation from the Kirkwood-Buff theory towards the understanding of tofu. *Food Hydrocolloid* **2017**, *62*, 128–139. [CrossRef]

102. Wu, H.; Wu, T.; Chen, Z.J. Research on the new preservative technology of tofu. *Mod. Food Sci. Technol.* **2005**, *21*, 1–4.

103. Obatolu, V.A. Effect of different coagulants on yield and quality of tofu from soymilk. *Eur. Food Res. Technol.* **2008**, *226*. [CrossRef]

104. Kobayashi, R.; Ishiguro, T.; Ozeki, A.; Kawai, K.; Suzuki, T. Property changes of frozen soybean curd during frozen storage in "Kori-tofu" manufacturing process. *Food Hydrocolloid* **2020**, *104*. [CrossRef]

105. Ward, G. Modified atmosphere packaging for extending storage life of fresh fruits and vegetables. *Ref. Modul. Food Sci.* **2016**, *2*, 1–8.

MDPI

St. Alban-Anlage 66

4052 Basel

Switzerland

Tel. +41 61 683 77 34

Fax +41 61 302 89 18

www.mdpi.com

Foods Editorial Office

E-mail: foods@mdpi.com

www.mdpi.com/journal/foods